Mushrooms

Imran Ul Haq • Siddra Ijaz
Hayssam Mohamed Ali
Editors

Mushrooms

Nature's Miracle for Sustainability

Springer

Editors
Imran Ul Haq
Department of Plant Pathology
University of Agriculture
Faisalabad, Pakistan

Siddra Ijaz
Centre of Agricultural Biochemistry and Biotechnology (CABB)
University of Agriculture
Faisalabad, Pakistan

Hayssam Mohamed Ali
Department of Botany and Microbiology,
College of Science
King Saud University
Riyadh, Saudi Arabia

ISBN 978-981-95-2538-6   ISBN 978-981-95-2539-3 (eBook)
https://doi.org/10.1007/978-981-95-2539-3

This Springer imprint is published by the registered company Springer Nature Singapore Pte Ltd.
The registered company address is: 152 Beach Road, #21-01/04 Gateway East, Singapore 189721, Singapore

Preface

Mushrooms (edible fungi) also have been classified as vegetables; however, technically, they are not considered plants, cultivated on several substrates. They are carbohydrates, minerals, vitamins, and protein-enriched and low-fat food. They have tremendous medicinal and nutritional properties and vast applications in the food and pharmaceutical industry. The growth of mushrooms has spread almost all over the world. More than 200 species have been cultivated as resourceful food, but, in contrast, 35 species have been commercially cultivated. Mushrooms have antineoplastic, antibacterial, antiviral, hypoglycemic, and hypocholesterolemic properties.

Cultivation of mushrooms is a profitable business, ranging from cottage industry to multimillion business. The global mushroom market size was 14.35 million tons, exhibited a higher growth of 6.3% in 2020, which is expected to grow from 15.25 million tons in 2021 to 24.05 million tons in 2028. Rising health consciousness, awareness of the consumption of high nutritious food, and advancement and funding opportunities in the R&D of mushroom cultivation are the driving factors for the growth of the mushroom industry.

Cultivation of mushrooms using crop and industrial residues is the best way of recycling agricultural and industrial wastes into edible protein and safe decomposition of these wastes. It helps greatly to reduce the pollution and keep the environment healthy. Mushrooms, on the other hand, can generate more export opportunities and primarily contribute to the economy of developing countries. In recent years, the interest of the world has been attracted to the technique of bioconversion in which industrial wastes and agriculture remnants are converted into food items and animal feed.

Considering the abovementioned facts, the authors have planned to compile valuable, strong scientific, logical, and updated information to make it available in the form of a state-of-the-art book on mushrooms. This book will comprise themes including edible fungi and their marvelous and intimate roles in human life; world over edible mushrooms: types, identification, and distribution; domesticated mushrooms: techniques and technologies; insects pests and diseases of cultivated mushrooms; mushrooms that cannot be farmed; mushroom cultivation: an integral part of circular agriculture; use of mushrooms: future lines of modern research; biotechnological interventions revitalizing the mushroom industry; mushrooms: the miraculous medicine; mushrooms to avoid.

Faisalabad, Pakistan — Imran Ul Haq
Faisalabad, Pakistan — Siddra Ijaz
Riyadh, Saudi Arabia — Hayssam Mohamed Ali

Contents

About the Editors

Dr. Imran Ul Haq with a bright career in agriculture, plant pathology, and fungal molecular biology, has a postdoc degree from University of California Davis, USA. He is a Professor in the Department of Plant Pathology, University of Agriculture Faisalabad, Pakistan. He has supervised more than 40 graduate students and established the Fungal Molecular Biology Laboratory Culture Collection (FMB-CC-UAF); he is an affiliated member of the World Federation for Culture Collections (WFCC). He has published more than 60 articles, 9 books, 4 laboratory manuals, and several book chapters. He has made colossal contributions to fungal taxonomy by reporting novel species of fungal pathogens in plants. His research interests are fungal molecular taxonomy and nanotechnology integration with other control strategies for sustainable plant disease management.

Dr. Siddra Ijaz with a vibrant career in agriculture and biotechnology, has a PhD from the University of Agriculture Faisalabad, Pakistan, and a postdoc degree from the Plant Reproductive Biology Laboratory, University of California Davis, USA. She is a Professor at the Center of Agricultural Biochemistry and Biotechnology (CABB), University of Agriculture, Faisalabad, Pakistan. She is also serving as Managing Editor of an international high-impact-factor journal, the *Pakistan Journal of Agricultural Sciences*. She has supervised more than 60 MPhil and PhD students. She was a National Curriculum Revision Committee member from 2013 to 2014. She has published over 50 articles, 10 books, 1 laboratory manual, and several book chapters. She is in charge of the Molecular Biology of Plant Disease Resistance Laboratory, CABB, UAF. She is the Curator of the Fungal Molecular Biology Laboratory Culture Collection (FMB-CC-UAF) and an affiliated member of the World Federation for Culture Collections (WFCC). Her research focuses on plant biotechnology, including genomics, genome probing for identifying disease resistance genes, genome engineering for fungal disease resistance in plants, and disease management through nanobiotechnology.

Hayssam Mohamed Ali is working at the Botany and Microbiology Department, College of Science, King Saud University. Dr. Hayssam M. Ali has a WORLD RANKING OF TOP 2% SCIENTISTS according to the Stanford Scientists' Ranking released by Stanford University in 2021, 2022, 2023, 2024 and 2025. He has more than 350 published papers, 20 published books, and he has reviewed more than 1700 papers in highly impacted journals. He is interested in studying plant physiology, biotic and abiotic stress, sustainability, and environmental stresses. Plant Biotechnology, Crop Improvement, Plant Breeding, Forestry, Plant Biology.

Chapter 1
World Over Edible Mushrooms: Types, Identification, and Distribution

Saifullah, Imran Ul Haq, Siddra Ijaz, Jamila Fatima, and Barket Hussain

Abstract Nutrition is critical to the community's survival, so food security is an economic and ecological priority. A nutritionally balanced diet abundant in essential macronutrients like vitamins, minerals, proteins, and polyunsaturated fatty acids is critical for good health, especially in developing countries where malnutrition is common. Edible mushrooms are high in essential bioactive components, including proteins, dietary fibers, vitamins, and phenolic compounds, which contribute to their antioxidant, immune-stimulatory, cholesterol-lowering, and antimicrobial properties. The ancient Egyptians, Greeks, and Romans once considered mushrooms divine food, as well as a health elixir in Chinese culture, due to their nutritional and medicinal properties. Modern nutraceutical therapy increasingly acknowledges their role in combating health issues. At the same time, their culinary appeal and bioactive potential continue to pique interest in commercial cultivation and their contribution to global food security. This chapter thoroughly examines edible mushrooms worldwide, their types, key identification steps, and distribution

Saifullah
Department of Botany, University of Agriculture, Faisalabad, Pakistan

I. U. Haq (✉)
Department of Plant Pathology, University of Agriculture, Faisalabad, Pakistan
e-mail: Imran_1614@uaf.edu.pk

S. Ijaz
Centre of Agricultural Biochemistry and Biotechnology (CABB), University of Agriculture, Faisalabad, Pakistan

J. Fatima
Department of Environmental Sciences and Safety Engineering, Jiangsu University, Zhenjiang, China

B. Hussain
Department of Botany, Aligarh Muslim University, Aligarh, India

I. U. Haq et al. (eds.), *Mushrooms*, https://doi.org/10.1007/978-981-95-2539-3_1

patterns, highlighting their significance in commercial cultivation to enhance attention to the global food supply.

Keywords Polyunsaturated fatty acids · Edible mushrooms · Phenolic compounds Nutraceutical therapy

1.1 Introduction

Fungi are considered one of the most diverse and ecologically important groups of organisms on Earth. They play a significant role in nutrient cycling among ecosystems by acting as primary decomposers, facilitating the breakdown of organic matter and the release of essential mineral nutrients (Dighton 2018; Sudharsan et al. 2023). Different members of the phyla Basidiomycota and Ascomycota produce visible fruiting bodies known as "mushrooms" under specific abiotic conditions and interactions with the surrounding flora. These macrofungi, visible to the human eye, can grow above ground (epigeous) or below ground (hypogeous) (Virágh et al. 2021; Bunyard 2024). They may have evolved from an ancient ancestry around 400 million years ago and thrived as saprobes and parasites together with land plants (Nuñez Otaño et al. 2021).

Given the broad spectrum of fungal diversity, the overall predicted diversity of mushrooms available to science is quite limited. Only 50% of documented mushrooms (7000 species) have various ranges of edibility, with less than 3000 species belonging to 31 genera and around 1–10% being poisonous (Antonelli et al. 2020; Niazi and Ghafoor 2021; Mwilenga 2023; Acharya and Khatua 2024). Over 14,000 mushroom species have been identified, with roughly 10% classified as macrofungi. Edible mushrooms, visible to the naked eye, are made up of fleshy fruiting bodies from several mushroom species. These mushrooms are recognized for their particular flavor and their wide range of bioactive characteristics, making them useful in culinary and medical applications (Qing et al. 2021). The fruiting bodies of around 700 edible mushroom species are safe to consume and may provide human health advantages (Li et al. 2021b).

1.1.1 Importance of Edible Mushrooms

Mushrooms are known for their diverse bioactive compounds, which have significant medicinal value against pathogens and various health disorders (Bell et al. 2022). Mycophagy, or the consumption of mushrooms, has been an important part of human diets since ancient times (Table 1.1). The terms "mycophilia" and "mycophobia," coined by British mycologist Hay (1887), distinguish societies that enjoy mushrooms from those that dislike them (Peintner et al. 2013). Edible mushrooms

Table 1.1 Application of different edible mushrooms in daily functional foods and their possible roles

Sr. no.	Name of mushroom	Name of product	Mushroom part/ ingredient	Biological activity	Main results	Reference
	A. auricula	Bread	Mushroom powder	Source of Nutrients	Improve Better quality (5%) and dough functionality as well as acceptable	Yuan et al. (2017)
1.	A. bohusii Bon	Cream cheese	Mushroom extract	Antifungal activity	Improve antioxidant, phenolic content, and preserving properties, inhibit the P. verucosum var. cyclopium growth	Reis et al. (2012b)
2.	L. edodes	Beef burgers	Mushroom homogenate	Replacement of salt	Improve taste and sensory acceptance Reduced salt content	Mattar et al. (2018)
3.	B. edulis	Frankfurter	Mushroom decoction	Antimicrobial and Antioxidant	improves color, and moisture and decreases hardness Antimicrobial properties	Novakovic et al. (2020)
4.	L. edodes	Pasta	Dried mushroom	Source of Nutrients	Increase the dietary fiber, protein and antioxidant contents and decrease starch degradation modulate the glycaemic responses	Lu et al. (2018)
5.	Winter mushroom	Cantonese sausages	Mushroom powder	Source of Nutrients	Increase overall acceptance and protein content Improve color and reduce hardness issues Inhibit the oxidation of protein and lipid	Wang et al. (2019c)

(continued)

Table 1.1 (continued)

Sr. no.	Name of mushroom	Name of product	Mushroom part/ ingredient	Biological activity	Main results	Reference
6.	P. eryngii	Pork sausage	deep-fried, boiled, fresh mushroom	Replacement of fat (pork back)	Increase protein and moisture content, improve color decrease cooking and hardness losses	Wang et al. (2019b)
7.	Flammulina velutipes	Canned ground ham	Mushroom powder	Replacement of Synthetic nitrite	The same result as in the control	Jo et al. (2020)
8.	L. edodes	Rce noodles (Gluten-free)	Beta-glucan	improver noodles quality	improve cooking qualities and thermo-mechanical properties and reduce the cooking loss and swelling index	Heo et al. (2014)
9.	P. sajor-caju	Papad (Indian snack food)	Mushroom powder	Source of Nutrients	Improve contents of protein, minerals and crude fiber (15.8%, 38.87%, 218%), respectively.	Parab et al. (2012)
10.	L. edodes	Emulsion-type sausage	hydrated mushroom	Replacement of meat (Pork)	Increase antioxidant and moisture content, decrease cooking losses	Wang et al. (2019a)

are very important due to their high nutritional profile, low caloric content, and cholesterol-free status, making them ideal for health-conscious diets (El-Ramady et al. 2022; Wang and Zhao 2023).

Mushrooms contain high-quality proteins (20–30% dry weight), similar to animal sources such as milk, eggs, and meat, and all essential amino acids. Their dietary fiber content primarily comprises carbohydrates, specifically glycogen, chitin, and glucans. They have a low lipid content (2–8%), are cholesterol-free, and are high in B-complex vitamins, vitamin D, and essential minerals like copper, selenium, and magnesium. These properties make mushrooms useful for controlling blood pressure, regulating glucose levels, lowering cholesterol, and providing antioxidant and anti-inflammatory properties, among other health benefits (Mwangi et al. 2022; Navarro-Simarro et al. 2024).

1.1.2 World Production Overview of Edible Mushrooms

The global mushroom industry can be categorized into three main groups: wild, medicinal, and edible mushrooms. Over the last 20 years, the production of edible mushrooms and truffles has increased by over threefold, representing a huge growth in global cultivation and consumption (Royse et al. 2017). Of this total, Asia dominates global mushroom production, followed by Europe. China, the United States, Italy, Poland, and the Netherlands are among the largest producers. Over 100 countries engage in commercial mushroom cultivation, employing various production systems and scales. Global mushroom production is growing at an annual rate of about 7%. Global consumption currently stands at 12.74 million tons, with projections indicating a rise to 20.84 mt in 2026 (Shirur et al. 2021).

1.2 Historical Background of Edible Mushroom

According to historical records, the earliest intentionally cultivated mushroom species was *Auricularia auricula*, which was grown in China around 600 A.D. (Chang et al. 1993). Edible mushroom cultivation dates back around 700 years in China, with *Lentinula edodes* grown on wood logs in forested areas. *Agaricus bisporus* (button mushroom) was widely cultivated throughout Europe in the eighteenth century, using nutrient-rich substrates such as horse manure placed in caves to optimize growth conditions (van Griensven 1988). Chang et al. (1993) reported that, of over 2000 mushroom species with edible potential, approximately 80 had been effectively cultivated, with 40 of these being cultivated commercially. Among them, five to six species are produced industrially in multiple countries. The majority of cultivated mushrooms are saprophytes. Mushroom consumption is thought to have originated in the early stages of human food gathering, and it served as a staple food source for ancient societies (Ahmed et al. 2023). For millennia, many communities have valued mushrooms as both medicine and food sources (Wasson 1971; Power et al. 2015).

1.2.1 Radiocarbon Study

A radiocarbon study proved that certain mushrooms are approximately 18,700 years old. Dietary evidence preserved in dental calculus from the "Red Lady" reveals the existence of fungi from the order Boletales and other unidentified agaric species, verifying their consumption during that time (Kotowski 2019). According to research, bioactive compounds in *Fomitopsis betulina*, such as agaric acid and toxic oils, can potentially treat whipworm infections. Historically, *Fomitopsis betulina* held cultural importance, especially in Mayan traditions. Hallucinogenic fungi, such as *Amanita muscaria* and Psilocybe species, were revered in these cultures and referred to as "Teonanácatl," which means "the flesh and nourishment of the gods," emphasizing their ceremonial and spiritual significance (Wasson 1971; Niksic et al. 2016). *Amanita muscaria* is part of a traditional preparation that improves cognitive and physical abilities, allowing people to perform extraordinary feats (Hajicek-Dobberstein 1995). The Greeks believed that mushrooms provided power to soldiers during battle (Daba et al. 2008). Mushrooms were historically used in certain Mayan regions for religious and ceremonial practices (Matsushima et al. 2009). In ancient Egypt, mushrooms were considered divine gifts, thought to be bestowed by the God Osiris (Maihara et al. 2012). Mushrooms were revered in Ancient Egypt as "plants of immortality" and thought to be a divine gift from Osiris. They were regarded as a delicacy only for the nobility due to their distinct flavor. Their consumption and even physical contact were strictly forbidden to the general public, emphasizing their high status in Egyptian culture (Niksic et al. 2016).

1.2.2 Back to the Period of the Roman Empire

The Romans revered edible mushrooms, referring to them as "Food of the Gods" and reserving them for special events (Rahi and Malik 2016; Ahmed et al. 2023). This veneration raised mushrooms to a luxury, consumed primarily by the elite (Reginald Buller 1914). Puffballs, truffles, porcini, and Caesar's mushrooms were especially popular with the elite (Reginald Buller 1914). Historical records also indicate a link between mushrooms and toxicity. Ancient Roman literature depicts their role in political intrigue, with rulers exploiting their popularity as a means of assassination. Emperor Claudius, who ruled from A.D. 41 to 54, was particularly fond of boletes and Caesar's mushrooms, and according to Pliny the Elder's stories, he consumed them on the day of his death, allegedly poisoning himself (Grimm-Samuel 1991).

1.2.3 Beginning of Mushroom Domestication

The first artificially grown mushroom was *Auricularia auricula-judae*, followed by *F. velutipes* and *L. edodes* (Atila et al. 2017). *A. bisporus* was first cultivated in France in the seventeenth century, and Pleurotus species were introduced to the United States in the early twentieth century (Chang 2008). Approximately 35 mushroom species have been domesticated for cultivation, with only about 20 being grown on an industrial scale (Table 1.2) (Atila et al. 2017). The most commonly cultivated edible mushroom species in different parts of Egypt are *Pleurotus ostreatus* and *A. bisporus*. The dried *P. ostreatus* contains 33.85% crude protein, 2.41%

Table 1.2 List of reported edible mushroom genera and its associated edible mushroom species

Sr. No	Genra of edible mushrooms	Edible mushroom species	
1.	*Agaricus*	*Bisporus, brasiliensis, silvaticus, silvicola, arvensis, semotus*	Heim (1977), Singer (1986), Aida et al. (2009), Hall et al. (2010), Ao et al. (2016a), Kalac (2016), Diego and Pardo-Giménez (2017), Zied and Pardo-Giménez (2017), Adeniyi et al. (2018), Ye et al. (2019), Kumar et al. (2021b), Li et al. (2021a), Khastini et al. (2023)
2.	*Agrocybe*	*Aegerita, cylindracea*	
3.	*Antrodia*		
4.	camphorate		
5.	*Auricularia*	*Auricular, cornea, polytricha, mesenterica, fuscosuccinea*	
6.	*Armillariella*	*Mellea,*	
7.	*Amanita*	*Rubescens, ovoidea*	
8.	*Scleroderma*	*Flavidum,*	
9.	*Suillus*	*Bellini, luteus, granulatus*	
10.	*Russula*	*Fellea, cyanoxantha, delica, brevipes,vinosa*	
11.	*Pleurotus*	*Subtilis, floridanus, fockei, strigosus, Levis, tuber- regium, salignus, calyx, squarrulosus, ostreatus, dryinus, florida, pulmonarius, cystidiosus, djamor, sajor-caju*	
12.	*Ganoderma*	*Lucidum, atrum, applanatum, tsugae*	
13.	*Boletus*	*sensibilis, subvelutipes, edulis, badius*	
14.	*Clavariadelphus*	*Pistillaris,*	
15.	*Cantharellus*	*Lutescens, cibarius, clavatus*	
16.	*Cortinarius*	*Caperatus,*	
17.	*Cerrena*	*Unicolor,*	
18.	*Calvatia*	*Gigantea,*	
19.	*Cordyceps*	*Sinensis,*	
20.	*Cyanoboletus*	*Pulverulentus,*	
21.	*Dictophora*	*Indusiata,*	
22.	*Flammulina*	*velutipes(yellow),*	
23.	*Inonotus*	*Obliquus,*	
24.	*Sparassis*	*Crispa,*	
25.	*Ramaria*	*Botrytis,*	
26.	*Trametes*	*Hirsute, orientalis, versicolor*	
27.	*Phaeolus*	*schweinitzii,*	
28.	*Pluteus*	*Salicinus, petasatus*	
29.	*Pulveroboletus*	*ravenelii*	
30	*Helvella*	*Crispa, marmoreus, acetabulu*	
31.	*Hericium*	*Coralloides, erinaceus,*	
32.	*Hydnum*	*repandum*	
33.	*Hypholoma*	*fasciculare*	
34.	*Hygrophorus*	*agathosmus*	
35.	*Hypsizigus*	*marmoreus*	

(continued)

Table 1.2 (continued)

Sr. No	Genra of edible mushrooms	Edible mushroom species	
36.	*Gyroporus*	*castaneus*	
37.	*Geastrum*	*saccatum*	
38.	arenarius		
39.	*Grifola*	*frondosa*	
40.	*Lepista*	*Nuda, irina*	
41.	*Lactarius*	*Piperatus, deliciosus, sanguifluus, pubescens, repraesentaneus, rufus, uvidus, zonarius,*	
42.	*Lycoperdon*	*molle*	
43.	perlatum		
44.	*Laetiporus*	*sulphureus*	
45.	*Leucopaxillus*	*giganteus*	
46.	*Lentius*	*sajor- caju*	
47.	*Lentinus*	*Edodes, squarrosulus, conatus*	
48.	*Phellinus*	*Linteus, merrillii, rimosus*	
49.	*Tirmania*	*nivea*	
50.	*Sarcodon*	*imbricatus*	
51.	*Volvopluteus*	*gloiocephalus*	
52.	*Strobilomyces*	*strobilaceus*	
53.	*Leucoagaricus*	*leucothites*	
54.	*Schizophyllum*	*commune*	
55.	*Morchella*	*Conica, Esculenta, angusticeps*	
56.	*Macrolepiota*		
57.	procera		
58.	*Schizophyllum*	*commune*	
59.	*Tricholoma*	*Acerbum, equestre, giganteum*	
60.	*Tricholomopsis*	*rutilans*	
61.	*Termitomyces*	*Microcarpus, schimperi, mummiformis, tylerance, heimii, albuminosus, robustus*	
62.	*Terfezia*	*claveryi*	
63.	*Tremella*	*fuciformis*	
64.	*Verpa*	*conica*	
65.	*Volvariella*	*volvacea*	
66.	*Neoboletus*	*luridiformis*	
67.	*Neolentinus*	*lepideus*	
68.	*Piptoporus betulinus*	*Piptoporus betulinus*	
69.	*Paecilomyces japonica*	*Paecilomyces japonica*	
70.	*Picoa juniperi*	*Picoa juniperi*	
71.	*Polyporus squamosus*	*Polyporus squamosus*	

crude fat, and 57.05% carbohydrates, whereas *A. bisporus* contains 26.05% crude protein, 2.79% crude fat, and 42.56% carbohydrates (Abou Raya et al. 2014).

1.3 Classification/Types/Diversity of Edible Mushroom

Mushrooms are widely categorized into three functional groups: saprophytes, parasites, and mycorrhizae (Fig. 1.1) (Rosemary Kinge et al. 2017). According to Li et al. (2021a), 2180 species of edible mushrooms have been reported worldwide, with 1181 being ectomycorrhizal, 940 saprophytic, and 59 parasitic. Saprophytic fungi, which include many gourmet mushrooms, are important decomposers that break down complex organic materials. This breakdown recycles carbon, nitrogen, hydrogen, and minerals, contributing significantly to ecosystem nutrient cycling.

Fig. 1.1 Classification of edible mushroom into three groups

Notably, some saprophytic fungi show weak parasitic activity under certain conditions. Parasitic mushrooms, though uncommon, contain edible species such as *Armillaria mellea*. Mycorrhizal fungi create mutualistic relationships with plant roots, giving them water and critical minerals while obtaining carbohydrates and vitamins produced during photosynthesis. Many terrestrial plants rely on mycorrhizal connections for optimal growth and survival (Singh et al. 2021; El-Gharabawy 2023; Stojek et al. 2024).

1.3.1 Saprophytic Mushrooms

Agaricus subrufescens is a member of basidiomycete and belongs to the family Agaricaceae and order Agaricales. This species, along with other members of its genus, is commonly found in forest ecosystems and also in grasslands, pastures, and along roadsides. It is a saprotrophic organism that obtains nutrients from decomposing organic matter, specifically leaf litter, in ecotonal zones between forests and open parklands (Joshi et al. 2021). This mushroom, also known as the "almond mushroom" due to its distinct almond-like flavor, was widely consumed and cultivated in various regions of the United States during the late nineteenth and early twentieth centuries (Kour et al. 2021). When the fungus was discovered in 1960 in Piedade, a town in the Brazilian province of São Paulo, it was named the "Piedade mushroom" in honor of the location of its discovery (Kerrigan 2005). Champignon mushrooms (*Agaricus bisporus*) were commercially grown in Brazil in the early 1990s and have since spread to international markets. Industrial-scale production is also prevalent in China, Japan, South Korea, and Taiwan. Sugarcane bagasse, forage grasses such as Brachiaria spp., *Cynodon dactylon*, and *Panicum maximum*, and cereal straws from *Triticum aestivum*, *Avena sativa*, and *Oryza sativa* are among the substrates used for the production of fruiting bodies. These substrates are frequently added with nitrogen sources such as wheat, soybean, urea, cottonseed meal, and ammonium sulfate, as well as calcium and phosphorus sources, to improve growth conditions (Llarena-Hernández et al. 2014). The fruiting bodies of A. subrufescens contain between 89% and 91% water and are high in essential nutrients like phosphorus, potassium, magnesium, zinc, and calcium. Very small amounts of cadmium have also been identified. Glucose (57.7%) is the most abundant polysaccharide, followed by galactose (27.7%), mannose/xylose (7.3%), and fucose (4%) (Györfi et al. 2010).

Agrocybe aegerita is a white-rot basidiomycete widely grown in Italy and considered a gourmet delicacy, commonly called the Pioppino mushroom. It has high commercial value due to its distinct flavor and texture, contributing to its popularity in culinary applications (Katoch et al. 2023). The life cycle of *Agrocybe aegerita* is managed by a tetrapolar homogenous incompatibility system consisting of two unlinked genetic loci, A and B. Under favorable climatic conditions, the dikaryotic mycelium develops into fruiting bodies (basidiocarps) that undergo karyogamy before meiosis (Herzog et al. 2016). The chestnut mushroom is scientifically recognized for its saprophytic growth on decaying wood, particularly poplar logs. It usually grows in clustered fruiting bodies on stumps in temperate regions of Southern Europe, parts of the Far East, and the Southeastern United States. In cultivation, this species has thrived on supplemented alder sawdust or wood chips and oak, with robust fruiting also supported by willow, poplar, and maple. It is widely cultivated in Asia and valued for its functional properties, which include antioxidant, antifungal, antitumor, hypolipidemic, and hypocholesterolemic effects (Jasińska et al. 2014; Jakopovic et al. 2021).

Auricularia auricula-judae belongs to the class Heterobasidiomycetes, order Auriculariales, and family Auriculariaceae, and is also called jelly ear mushroom. It is widely grown throughout East and Southeast Asia, including China, Thailand, Taiwan, Malaysia, and Indonesia (Tang et al. 2012). The waxy fruiting body is cartilaginous in form, turning purplish-brown to black when dried. Its septate basidia are distinguished morphologically and taste-wise from other cultivated mushrooms, such as Pleurotus species and *Agaricus bisporus*. Auricularia mushrooms are among the four most extensively grown mushrooms worldwide, with Southeast Asia and China producing about 420,000 tons annually. This species thrives on various industrial and agricultural organic wastes (Yun and Hall 2004). *A. auricula-judae* is prized for its culinary and medicinal properties, which include high levels of manganese (1.66 g/kg) and potassium (172.03 mg/kg). Significant amounts of cysteine (0.34 mg/kg), methionine (0.80 mg/kg), and glutamic acid (10.09 mg/kg) have also been detected. Methyl ester (9, 12-octadecadienoic acid) is the most common lipid component (Ohiri and Bassey 2017). Its aroma is musty, resembling raw compost (Ohiri and Bassey 2017; Bandara 2019).

Coprinus comatus is a member of the family Agaricaceae and order Agaricales, and is commonly called the shaggy ink cap mushroom. Agaricales members, including *C. comatus*, are widely distributed and frequently found in gravel roads, lawns, and disturbed areas worldwide, especially in temperate regions during late fall and summer (Nowakowski et al. 2020). This species has a farinaceous, mildly sweet aroma (Winkler 2023). *C. comatus* has long been valued in China for its superior flavor and nutritional value. The fruiting body is white, fragile, and thin, with a fibrous stipe that hardens with time. As it matures, it turns pink to black, releasing a distinctive black ink-like liquid. *C. comatus* is widely grown in China, with an annual production of over 380,000 tons. Cotton waste, rice straw, corncobs, urea, lime, and steer manure are some examples of cultivation substrates (Cheek et al. 2020; Sun et al. 2022a; Mallick et al. 2023). Nutritional analyses show that wild and cultivated mushrooms have similar energy, carbohydrate, and ash content (368 kcal, 76%, 10%). According to Stojković et al. (2013), cultivated mushrooms contain slightly more protein (11.84% versus 10.9% in wild types) and fat (1.98% compared to 1.8% for wild kinds).

Flammulina velutipes is a basidiomycete from the family Physalacriaceae and is also known as "enokitake," commonly referred to as the winter or golden needle mushroom in Japan. It is a saprophytic fungus that primarily inhabits hardwoods but can also be found on conifers. This species thrives across a wide altitude range, from sea level to the tree line. It is especially popular in Asia due to its ability to grow in low-temperature environments (Parkash et al. 2023). Light and temperature affect morphological traits such as stalk length and cap size (Sakamoto et al. 2007). *F. velutipes* is among the world's six most widely commercially cultivated mushrooms, with an annual production of over 300,000 tons. Its white-rot abilities let it degrade cellulose and lignin, allowing it to thrive on decaying wood. Sawdust substrates are most commonly used in commercial cultivation and frequently combined with agricultural residues such as cottonseed husks, sugarcane bagasse, or corn cobs. Reduced growth cycles, cold treatment, and waste material composition are all important cultivation materials. Various strains need cold shock or low temperatures (4.4–13 °C) for optimal growth, with varying light and CO_2 sensitivity influencing development (Park et al. 2014; Xie et al. 2017).

Grifola frondosa is commonly known as the hen of the woods or sheep's head; it derives its name (Japanese) from Maitake, meaning "dancing mushroom," from folklore, indicating that its discovery brings great joy due to its high market significance. This mushroom is a fleshy polypore distinguished by its undulating pileus of smoky brown color, which forms large rosettes from a single branched stipe. The fruiting body is initially dark gray-brown, then becomes light gray with time (Miles and Chang 2004). It is a saprotrophic species that lives on the trunks of decayed, aging, or dead hardwoods like maple, elm, chestnut, and oak. Its first large-scale cultivation began in Japan (1981), followed by China and the United States in later years. This mushroom is a valuable dietary resource due to its nutritional content, which includes 13.1% pure protein, 19.7% crude protein, 61.06% carbohydrates, 3.2% lipids, and 9.7% fiber (Miles and Chang 2004).

Hericium erinaceus is frequently called bear's head and belongs to the family Hericiaceae, order Russulales, and class Agaricomycetes (Thongbai et al. 2015; Index Fungorum 2017). This fungus lives primarily as a weak parasite or saprophyte, feeding on knots of dead wood or cracks in living hardwood trees. The fruiting body is fleshy, semispherical, and white initially, darkening to yellowish or brownish hues with age (He et al. 2017). Historically, from 1959 to 1960, *H. erinaceus* was only found in the wild, primarily in deep forests on decomposing wood. The macroscopic features of *H. erinaceus* are usually sufficient for identification, but distinguishing it from *Hericium coralloides* can be difficult due to the similar basidiospore sizes. However, *H. coralloides* has more branched fruiting bodies and is more commonly associated with conifers, whereas *H. erinaceus* grows on deciduous trees (Stamets 2005). The Shanghai Agricultural Academy of Science's domestication efforts facilitated cultivation on various substrates, including artificial media and agricultural waste like bagasse (sugarcane), cottonseed hulls, sawdust, and corncobs. It can also be grown on unconventional substrates such as tofu, whey, and cheese. According to Rodrigues et al. (2015), its fruiting body contains 61.3% carbohydrates, 18.8% protein, 6.91% ash, and 2.9% lipids, making it a nutritious food source.

Pleurotus spp.: The genus Pleurotus includes several species famous for their edibility and nutritional importance. *P. djamor* var. *roseus*, *Pleurotus ostreatus*, *P. pulmonarius*, *P. sajor-caju*, and *P. eryngii* are well-known species. These mushrooms grow naturally in temperate regions and thrive on decaying hardwoods like elm, poplar, and oak, but they can also adapt to subtropical climates during colder seasons (Index Fungorum 2017). The global production of Pleurotus species increased significantly, with China as the leading country, followed by the United States (de Carvalho et al. 2010). Their cultivation has spread worldwide, including Germany, Italy, India, China, Tanzania, Belgium, Mexico, Singapore, Nigeria, Japan, Brazil, Colombia, Canada, Taiwan, Thailand, Malawi, Zambia, the United States, and Indonesia. This increase is attributed to their nutritional content, good flavor, and capacity to degrade agricultural and forestry biomass (hemicellulose, lignin, and cellulose) via enzyme secretion, allowing for cost-effective production (Miles and Chang 2004). Popular cultivation substrates include agricultural residues (wheat, corn, rice husks), sugarcane bagasse, and plant by-products (sawdust, coffee, pulp), allowing for sustainable recycling of organic waste and reducing environmental pollution (Ritota and Manzi 2019; Doroški et al. 2022).

Volvariella volvacea is commonly referred to as the straw mushroom and is a basidiomycete fungus from the family Pluteaceae (Index Fungorum 2017). This saprophytic, thermophilic species thrives in tropical and subtropical Asia but grows seasonally in temperate climates. It is classified as a homothallic basidiomycete, with homokaryotic mycelium capable of converting to a dikaryotic state to complete its lifecycle. Notably, dikaryotic hyphae lack clamp connections, which are morphological markers that distinguish the dikaryon from the homokaryon (Bao et al. 2013; Ali et al. 2024). According to historical records, straw cultivation of *V. volvacea* was started in the eighteenth century by Buddhist monks at the Nanhua Temple in Guangdong Province. The mushroom was grown on fermented rice straw to supplement their diet and was regarded as a tribute to Chinese royalty (Chang 1977). Different substrates, such as bananas, cotton waste, water lilies, rice straw, coconuts, pineapples, sugarcane bagasse, and palm oil residues, are among the agricultural residues used in its cultivation over time. Cotton waste compost has proven particularly effective, allowing semi-industrialized production in Thailand, China, Indonesia, Hong Kong, and Taiwan (Miles and Chang 2004). By 2010, China's annual production had reached around 330,000 tons, accounting for above 80% of the global output (Bao et al. 2013).

1.3.2 Mycorrhizal Mushrooms

Tuber spp. It is an edible mushroom of genus Tuber that grows hypogeously, producing fruiting bodies underground. They form ectomycorrhizal symbioses with a variety of angiosperms and gymnosperms and in a wide range of environments, including temperate, boreal, and subtropical forests (Bulman et al. 2010; Zambonelli et al. 2017). The formation of tuber fruiting bodies is influenced by biotic factors such as fungi, mesofauna, yeast, bacteria, and plant hosts, alongside abiotic factors like humidity, soil composition, pH, and temperature. *Tuber melanosporum* is primarily harvested in Southern and Western Europe, particularly in France, Spain, and Italy, whereas *Tuber magnatum* grows in Slovenia, Croatia, Hungary, and Italy.

These truffles are usually collected between September and January with trained pigs or dogs (Mello et al. 2006). *Tuber melanosporum* has a high economic value, with prices ranging from €150 to €800 per kilogram on farms and up to €2000–4000 in urban markets like Paris and London. The price reached €950 per kilogram in Australia in 2012 (Duell 2012). Cultivation efforts started in France in the nineteenth century, and while progress has been made, domestication of these species is still incomplete (Zambonelli and Bonito 2013). Advances in genomic research, such as sequencing the *Tuber melanosporum* genome, show promise for improving cultivation methods (Martin et al. 2010).

Tricholoma matsutake or "matsutake" It is an ectomycorrhizal fungus and a member of the family Tricholomataceae and order Agaricales. It is commonly associated with *Pinus densiflora* (Japanese red pine) in Japanese forest ecosystems. This species forms subterranean mycelial networks known as "shiros," which produce fruiting bodies in circular patterns called "fairy rings" (Dai et al. 2021). *T. matsutake's* life cycle includes a monokaryotic-to-dikaryotic phase transition, characteristic of most basidiomycetes. Its fruiting bodies are widely used as edible mushrooms in Japan. Still, domestic production is insufficient to meet demand, resulting in approximately 3000 tons of imports each year, primarily from North America, China, North Africa, and Korea. Natural production is affected by different factors of the environment, such as temperature and precipitation, which limit yields and hinder forecasting (Amend et al. 2009). However, this has not yet proven successful; therefore, optimization of *T. matsutake* cultivation processes is intended to determine the factors that influence the formation of fruiting bodies, with a particular emphasis on the relationship between ecological variables such as soil properties, host plants, climate conditions, as well as the role of associated microbial communities.

Amanita caesarea It is commonly called Caesar's mushroom and is a highly valued edible species with great social and economic importance that has been prized for its culinary properties since Roman times (Sitta and Davoli 2012; Mittu et al. 2023). This ectomycorrhizal fungus can be found alone or in groups in deciduous woods, creating symbiotic associations with several tree species such as Pinus spp., Castanopsis spp., Quercus spp., and *Castanea sativa* (Sánchez-Ramírez et al. 2015). It is found in tropical and temperate parts of Asia, Europe, Africa, North America, and Australia. *A. caesarea*, known for its stem, yellow gills, and orange head, is a thermophilic mushroom and thrives in soils with a pH of 6–6.5 (Sánchez-Ramírez et al. 2015). Efforts to increase fruiting body production have included research into the role of bacteria in mycorrhizal establishment. The research by Cano et al. (2017) demonstrated that bacteria such as Bacillus cereus and Pseudomonas fluorescens can influence mycorrhization in vitro. The study found that P. fluorescens was spatially and temporally compatible with A. caesarea while not inhibiting fungal growth, whereas B. cereus was incompatible, slightly reducing fungal growth.

Boletus edulis *Boletus* is cosmopolitan and is predominantly found in temperate regions across both hemispheres, particularly within tropical and mid-latitude forests. *Boletus edulis* sensu lato, or cep or penny bun, is an edible ectomycorrhizal fungus highly valued for its culinary qualities. This *Boletus edulis* sensu lato (species complex), includes four key European species: *B. aestivalis*, *B. pinophilus, B. aereus*, and *B. edulis* (Leonardi et al. 2005; Águeda et al. 2008a). Unlike many mushrooms, *Boletus edulis* retains its distinct flavor and aroma when cooked and dried exhibiting a rich, nutty, earthy, and meaty taste (Tsai et al. 2007). Global demand for edible mushrooms has gone up over the last three decades, leading to increased imports, particularly from Eastern Europe and Asia. The market for *Boletus edulis* is large, with estimated annual consumption ranging from 20,000 to 100,000 metric tons and wholesale prices for fresh fruiting bodies in the United States ranging from $60 to $200 per kilogram in 2009 (Dentinger et al. 2010). Efforts to improve fruiting body production have been going on with studies indicating that an increase in 40%soil moisture and precipitation are critical for fruiting bodies. Moreover, temperature fluctuations play a role, with soil temperatures and a 5 °C cold shock required for Boletus edulis fruiting, which may occur over a period of 6 to 25 days depending on the environment (Savoie and Largeteau 2011).

1.3.3 Parasitic/Pathogenic Mushrooms

Armillaria mellea It is a member of basidiomycete commonly referred to as the honey mushroom, and belongs to the Physalacriaceae family. It is a parasitic and saprophytic mushroom, infecting various woody hosts, including broadleaf plants and conifers, commonly during the spring to autumn season. The fruiting body of A. mellea is rich in protein complexes, contributing to its nutritional and medicinal properties. In some countries such as Japan and China, researchers are working on the artificial production of A. mellea fruiting bodies for medical applications. Dehydrated wild fruiting bodies of this species have been reported to contain high carbohydrate content (81.25%), but low protein content, lipid, and ash contents (1.81%, 1.97%, and, 8.84% respectively) (Li et al. 2022). Mannitol is the major sugar free, whereas malic acid is the predominant organic acid. Kostić et al. (2017) reported that δ-tocopherol is the most abundant tocopherol, with 42.41 μg per hundred gram dry weight.

Ustilago maydis Ustilago maydis, commonly referred to as "huitlacoche" by the Aztecs, describes the edible, immature fruiting bodies that grow as galls on maize ears. This basidiomycete fungus (family; Ustilaginaceae order; Ustilaginales), is the causal factor of corn smut. Its life cycle has two distinct stages: the first one is unicellular filamentous form, nonpathogenic, and haploid, while the second is pathogenic and dikaryotic. When haploid cells of opposing mating types fuse, a dikaryotic filament forms, which invades the host plant and causes tumor-like gall formation. Due to its remarkable flavor profile, food applications, and use in haute cuisine, this is famous and more recognized as an alternative crop (Villagrán et al. 2023). Its nutritional composition consists of 1.6–2.3% lipids, which include necessary fatty acids like linoleic acid, 11–16% protein, which is high in lysine, and 16–23% dietary fiber. It is also a nutritious source because it contains 55–66% carbohydrates and 5–7% ash. Because it is economically significant due to its consumption and commercialization in regional markets, maize producers in Mexico's Central-South region frequently tolerate and even actively promote this infection. It is also processed into canned goods by some food companies. It offers significant financial returns because its market value is much higher than that of maize grain. Its cultivation has been optimized through different inoculation techniques, like spraying, injection, and spore sprinkling. The most successful method for causing gall formation among these is the injection of haploid mycelium, which is created by the combination of opposite mating types (heterothallic mating).

1.3.4 Psychoactive Edible Mushrooms

Historically, indigenous communities in Mexico have used psychoactive fungi, particularly psilocybin-containing species known as "magic mushrooms," for spiritual and ceremonial uses. Magic mushroom use first came to the public's notice in 1957 and has since become widely accepted. Psychoactive mushroom enthusiasts' interest and usage patterns were greatly influenced by detailed accounts of their ritual consumption by Mexican tribes and other cultures around the world (Wasson 1978). The Yoruba tribe (from Nigeria, Africa) was given credit for the first documented use of hallucinogenic mushrooms. It was discovered to have originated between 7000 and 9000 years ago during the Paleolithic period. Psilocybe species, commonly called "shrooms," are the "true" magic mushrooms out of the 180 types reported worldwide. They contain psilocybin and psilocin, psychoactive indoles of tryptamines that have low toxicity and never cause addiction other than mild-to-severe psychedelic functions (Johnson et al. 2008; Tylš et al. 2014). Psychoactive fungi from genera like Psilocybe and possibly Panaeolus have been used in traditional practices for more than 3000 years. Tryptamine derivatives are also known to be present in a number of other genera, including Gymnopilus, Conocybe, Galerina, Copelandia, Inocybe, Mycena, Lycoperdon, Panaeolina, Panaeolus, Pluteus, and Pholotina. Though it is generally thought to have a low risk profile, taking psychoactive mushrooms, especially when combined with alcohol, can have mild-to-severe side effects, such as psychological distress, erratic behavior, and long-term mental health issues (van Amsterdam et al. 2011; Carbonaro et al. 2016). In the wild, psilocybe species (Psilocybe spp.) are frequently mistaken for toxic or non-psychoactive mushrooms that share morphological similarities. Due to the illegal sale of poisonous species that resemble psilocybe, this misidentification has resulted in instances of accidental poisoning which may be fatal. The illicit growing, selling possessing, and magic mushrooms is illegal in several countries, such as America, Europe, and Australia.

Psilocybin, a unique type of hallucinogenic (neurotropic) chemical component, is the reason psilocybe has been revered. The indigenous religious people regard these mushrooms as miniature saints or the flesh of the gods. This neurotropic species, which has long been used in spiritual rituals, was identified by French mycologist Heim as Psilocybe (Guzman 2008). This demonstrates that ancient Egyptians were not the only people to use this substance in ceremonies (Guzman 2008). In India, these mushrooms are often taken with a variety of foods, such as bread with butter and jam, omelets, bread with bananas, or honey, demonstrating their integration into local dietary traditions. Even though it is a serious infraction, many law enforcement officers are unaware of it, which adds to its ubiquity. As a result, a thorough and systematic examination of its usage, particularly among youth, is urgently needed. Scientists have developed easy screening approaches to distinguish between psilocybin and non-psilocybin mushrooms. A DNA-based technique and LC/MS were utilized to detect hallucinogenic mushrooms and psychedelic chemicals in growing kits bought from the black market (Gambaro et al. 2015; Pepe et al. 2023). Psilocybin has been used by clinicians as a potential treatment for neuropsychiatric illnesses. Ongoing research investigates its usage, along with other hallucinogens, in the treatment of psychiatric diseases such as major depressive disorder, substance use disorders, and post-traumatic stress disorder, including alcohol and drug dependency.

1.3.5 Worldwide Genra of Most Common Edible Mushrooms

Of the estimated 1.5 million fungal species, approximately 14,000 have been formally described worldwide. Approximately 7000 species (50%) are classified as edible to varying degrees, and 3000 species from 31 genera are recognized as potentially edible. However, approximately 10% of these species have been documented to possess toxic or lethal properties. Macrofungi contain over 200 genera, many of which consist of species that are used as direct sources of nutrition or for their bioactive compounds that promote health. Edible genera such as Agaricus, Dictyophora, Flammulina, Auricularia, Lentinula, Hericium, Pholiota, Tremella, Volvariella, and Pleurotusm are popular globally due to their therapeutic and nutritional values (Chang and Wasser 2017; Azam et al. 2018; Gautam et al. 2022; Dávila Giraldo et al. 2023; Nawarathne and Daranagama 2024; Kumar et al. 2025). Certain mushroom genera, such as Sparassis, Cantharellus, Lactarius, Tuber, Morchella, and Suillus, are highly valued for culinary and medicinal properties. However, perceptions of edibility differ by region; species such as A. semotus, Agaricus arvensis, Lenzites elegans, Coprinus atramentarius, and Amanita gemmata are consumed in some areas but regarded as toxic or harmful elsewhere. In many developing countries, wild mushrooms are considered a good source of traditional medicine and nutrition (Gaglarirmak 2011) and give additional income. Recent advances in molecular techniques have greatly improved mushroom identification and taxonomy. Identifying a species' accurate scientific name significantly improves the accuracy of assessing its edibility. Rural communities in Mexico use approximately 300 wild edible mushroom species, with 180 species identified as potentially edible. In some cases, simply identifying the genus is enough to determine its edibility nature; For instance, all recognized species of the genus Cantharellus are considered as edible (Martins 2016; Zhang et al. 2021b; Kissanga et al. 2022). Tuber spp., also known as truffles, are among the most expensive and sought-after fungi, especially in Northern European regions. Certain desert truffle species, such as Tirmania and Terfezia are naturally adapted to arid and semiarid Mediterranean ecosystems (Farag et al. 2021; Graziosi et al. 2022). The major genera and its associated species that are used as edible mushroom globally given below. (Heim 1977; Singer 1986; Boa and Nations 2004; Aida et al. 2009; Hall et al. 2010; Thawthong 2014; Zhang et al. 2014; Ao et al. 2016a; Kalac 2016; Diego and Pardo-Giménez 2017; Zied and Pardo-Giménez 2017; Adeniyi et al. 2018; Ye et al. 2019; Kumar et al. 2021b; Li et al. 2021a; Khastini et al. 2023; Drewinski et al. 2024; Salazar-Vidal et al. 2025).

Amanita The genus amanita contains 83 mushroom species have been reported worldwide in 31 countries. Among these, Amanita caesarea is highly valued and sought after, particularly in Nepal, Turkey, and Mexico (Wang and Chen 2014). Several edible Amanita species, including *A. bingensis, A. masasiensis, A. caesarea, A. crocea, A. crassiconus, A. strobilaceovolvata, A. argentea, A. aurea, A. calyptratoides, A. perphaea, A. ceciliae, A. subviscosa, A. calopus, A. calyptroderma, A. inaurata, A. craseoderma, A. gemmate, A. flavoconia, A. flammeola, A. goossensiae, A. hemibapha, A. hovae, A. loosii, A. muscaria, A. rubescens, A. robusta, A. spissa, A. tuza, A. virgineoides, A. fulva, A. zambiana, A. vaginata, and A. umbonata*, are recognized for their edible properties in different regions of the world. Certain species, such as *A. gemmata*, are reported to be edible in specific regions (e.g., Mexico and Costa Rica), but have caused poisoning incidents in others (e.g., Guatemala), highlighting regional differences in toxicity. Similarly, reports on the edibility of *A. flavoconia and A. spissa* are different. Certain Amanita species, particularly *A. phalloides* (death cap), are well known for their high toxicity, resulting in numerous deaths worldwide when consumed. While some Amanita species are consumed and even traded, caution is advised due to the complex toxicity profiles displayed by different species (Boa and Nations 2004; Diaz 2018).

Suillus Approximately 27 edible Suillus species have been reported in more than 25 countries. *Suillus luteus* is the most widely consumed and collected species globaly, with significant harvesting taking place in Ecuador, Chile, and Argentina (Martins 2016). Suillus granulatus, another widely distributed edible species, is an excellent source of essential minerals and carbohydrates (Zhou et al. 2016). *Suillus placidus,* while edible in Russia, is considered toxic in China. In the late 1980s, Estonia and Mexico were major producers of Suillus species, with yields of around 280 kg/ha. Some most important species in this group include *Suillus acidus, S. bovinus, S. cavipes, S. grevillea, S. luteus, S. hirtellus, S. placidus, S. lactifluus, S. pungens, S. plorans, S. pseudobrevipes, S. granulatus, S. pictus, S. americanus, S. tomentosus, S. subluteus, S. abietinus, S. brevipes, S. viscidus*, and *S. variegatus*.

(Zhang et al. 2014; Saba et al. 2016; Sun et al. 2019; Demirtas et al. 2024; Legoshchina and Ufimtsev 2024; Zhao et al. 2024).

Auricularia The genus Auricularia, also known as ear fungi, and Judas ear, is found across subtropical, and temperate areas around the world (Du et al. 2011). Approximately 13 species of Auricularia have been reported in 24 countries, with Southeast Asian countries are leading globally in Auricularia cultivation. China produces approximately 3.6 million tons of Auricularia each year, accounting for roughly 6% of global production. Auricularia auricula and Auricularia polytricha are among the earliest cultivated mushrooms, dating back to approximately 600 A.D. China (Li 2012). *Auricula-judae, A. delicata, A. fuscosuccinea, A. cornea, A. tenuis,* and *A. mesenterica* are some of the other edible species in this genus that are consumed globally (Flores et al. 2002; Montoya-Alvarez 2011; Dulay et al. 2022; Jahedi et al. 2024).

Agaricus The genus Agaricus contains over 500 species worldwide, with approximately 60 of these species classified as edible in 29 countries. This genus has significant nutritional and medicinal value. The genus Agaricus contains several economically significant species, most notably Agaricus bisporus, also known as the button mushroom. This species is the most widely grown edible mushroom, contributing above 32% of global mushroom production (Largeteau et al. 2011). Worldwide some edible species of this genus includes *A. blazei, A. bingensis, A. rodmani, A. subedulis, A. erythrotrichus, A. subrutilescens, A. comptulus, A. purpurellus, A. silvaticus, A. goossensiae, A. bisporus, A. nivescens, A. micromegethus, A. arvensis, A. gennadii, A. bulbillosus, A. croceolutescens, A. semotus, A. maculatus, A. sylvicola, A. aurantiacus, A. endoxanthus, A. volvatulus, A. placomyces, A. subperonatus, and A. campestris.* Other species are less frequently acknowledged, but economically farmed species include, *A. subrufescens, A. bitorquis, A. arvensis,* and *A. campestris* (Kerrigan 2005; Glamočlija et al. 2015; Wang et al. 2015b; Chen et al. 2019; Allen et al. 2023). Some of the species have been documented to be poisonous, including *A. litoralis* and *A. xanthodermus* (Boxshall et al. 2021; Katoch et al. 2023).

Sparassis This genus also contains edible species with cosmopolitan distribution, among them Sparassis crispa is important due to multipurposes uses. Sparassis crispa, commonly called as the cauliflower mushroom, is an edible ectomycorrhizal mushroom present in coniferous forests of Eastern Asia, North America, and Europe (Adhikari et al. 2005). In addition, it is collected in China, Canada, Mexico, Russia, India, Ukraine, Turkey, and the United States (Tedder et al. 2002). It is valued for its culinary applications and is especially well-known for its medicinal properties, which have been the focus of recent research (Nowacka-Jechalke et al. 2021; Sharma et al. 2022).

Terfezia Terfezia species, also known as "desert truffles," are present in arid as well as semiarid part of the Mediterranean, parts of Asia, North Africa, Europe,

South, and North America. The most notable species like *T. arenaria* and *T. claveryi* are highly valued in Europe due to their culinary uses and high price. In Middle Eastern, Gulf, and North African countries, these truffles are frequently consumed raw (Bradai et al. 2015). Approximately seven edible Terfezia species have been identified, including T. *boudieri, T. leonis, T. claveryi, T. decaryi, T. Leptoderma T. pfeilii*, and *T. arenaria* with reports from eight countries (Kagan-Zur 2001; El Enshasy et al. 2013; Gouzi et al. 2013; Kagan-Zur and Akyuz 2014; Martínez-Tomé et al. 2014; Kıvrak 2015; Tejedor-Calvo et al. 2021; Sonji et al. 2022).

Lycoperdon The genus Lycoperdon, which contains approximately 22 edible species, is found in over 19 countries and has a worldwide distribution. Popular species include *L. perlatum, L. giganteum*, and *L. pyriforme*, also known as puffballs. Some species, such as *L. giganteum*, can reach 150 cm in diameter, ranking among the largest edible mushrooms (Boa 2004; Leffingwell and Alford 2011). These species are most commonly found in wooded areas grassy fields, and roadsides.

Moreover, *L. perlatum* resembles immature fruiting bodies of toxic Amanita species, increasing the risk of misidentification. *L. endotephrum, L. asperum, L. echinatum, L. marginatum, L. gemmatum, L. candidum, L. peckii, L. oblongisporum, L. umbrinum, L. rimulatum, L. pyriforme, L. pusillum, L. umbrinum var. floccosum*, and *L. spadiceum* are other edible species reported from various regions (Vetter 2004; Lian et al. 2007; Zeković et al. 2010; Mędyk et al. 2017; Lestari et al. 2019; Kıvrık et al. 2022).

Morchella The genus Morchella contain approximately 62 recognized species spread across 28 countries, including several well-known edible types. Morchella esculenta is a highly valued species, both for its culinary value and for its unique morphological features, but it is toxic when consumed raw (Thakur 2021). In Turkey, M. esculenta can cost up to €130 per kilogram. This genus has a global distribution, with edible species such as M. *angusticeps, M. conica, M. crassipes, M. esculenta var. umbrina, M. conica var. rigida, M. costata, M. elata, M. esculenta var. rotunda, M. intermedia M. esculenta var. vulgaris*, and *M. deliciosa*, reported in a variety of ecological conditions (Laala et al. 2020; Ranjitkar et al. 2020; Dissanayake et al. 2021; Taşkın et al. 2021; Kapoor et al. 2023; Li et al. 2024).

Macrolepiota The genus Macrolepiota contains approximately 13 species reported across 33 countries, with *M. procera* being the most frequently traded on a small scale due to its high nutritional value and global consumption (Boa 2004). A notable concern is the frequent misidentification of *M. procera* as *Chlorophyllum molybdites*, a species with conflicting toxicology reports. *M. africana, M. procera var. vezo, M. gracilenta var. goossensiae, M. dolichaula, M. excoriata, M. excoriata var. rubescens, M. gracilenta,, M. prominens, M. procera, M. puellaris, M. Rhacodes*, and *M. zeyheri* are examples of widely distributed edible species (Tedder et al. 2002; Ge et al. 2010; Lebel and Syme 2012; Kumar et al. 2017; Cho et al. 2019; Ćirić et al.

2019). *M. neomastoidea*, which is found in East Asia and Korea has been documented to be toxic (Kim et al. 2009; Lebeuf et al. 2024).

Pleurotus The genus Pleurotus includes approximately 40 edible varieties spread across 35 countries (Boa 2004). *P. ostreatus*, the oyster mushroom, is the most widely consumed and cultivated species worldwide. This species accounts for approximately 19% of global mushroom production (Stabnikova et al. 2024). Other notable Pleurotus species consumed worldwide include *P. abalonus, P. citrinopileatus, P. pulmonarius, P. concavus, P. cornucopiae, P. cystidiosus, P. djamor, P. floridanus, P. eryngii, P. spodoleucus, P. rhodophyllus, P. salignus*, and *P. sapidus* (Guzmán 2000; Shnyreva et al. 2012; Maftoun et al. 2015; Grabarczyk et al. 2019; Lin et al. 2022; Touml et al. 2022; Hassan and Muhammad 2023).

Boletus The genus Boletus includes about 72 edible species spread across more than 30 countries. Among these, *Boletus edulis*, also known as porcini, suilli, king bolete, panza, and penny bun, is important for both the environment and the economy because of its widespread consumption around the world. In South Africa, exotic tree plantations introduced *B. edulis* to native forests, but local populations were initially hesitant to consume this unfamiliar species. In China, mountainous communities have traditionally harvested *B. edulis* and sold it in local markets to help alleviate their financial problems (Boa 2004; Zhang et al. 2017). Boletus edulis, which is widely used in Asia, North America, and Europe, is important for both the environment and the economy (Águeda et al. 2008b). Some important species consumed globally are: *B. barrowsii, B. aereus, B. aestivalis, B. aurantiacus, B. emodensis, B. bulbosus, B. bouriqueti, B. citrifragrans, B. calopus, B. appendiculatus, B. elegans, B. colossus, B. atkinsonii, B. caudicinus, B. crassus, B. communis, B. cyanescens, B. bicoloroides, B. felleus, B. erythropus, B. griseus, B. frostii, B. luridus, B. impolitus, B. violaceofuscus, B. loyo, B. michoacanus, B. luridiformis, B. pseudoloosii, B. nigroviolaceus, B. pinetorum, B. pinicola, B. rubellus, B. pinophilus, B. reticulatus, B. russellii, B. regius, B. scaber, B. sulphureus, B. subtomentosus, B. truncatus, B. speciosus, B. variipes, B. variegatus, B. zelleri*, and *B. vitellinus* (Hall 1998; Hall et al. 1998; Hanuš et al. 2008; Sitta and Floriani 2008; Luki 2009; Vidović et al. 2010; Wang et al. 2015a; Dimitrijevic et al. 2016; Falandysz et al. 2017; Khanh 2022).

Cantharellus About 22 edible species of the ectomycorrhizal fungus Cantharellus spp. are found in 45 countries and are well-known for their culinary value (Arora and Dunham 2008; Buyck 2008). These fungi are widely distributed geographically and are distinguished by their unique fruity, apricot-like odor. Among these, the golden chanterelle, or Cantharellus cibarius, is a highly valued and economically important species that is only collected from its natural habitat. These mycorrhizal mushrooms have a high market value when purchased from local suppliers because, they have not been successfully mass-cultivated (Yun and Hall 2004; Hall and Zambonelli 2012). Additionally, there are no known toxic species in the genus. In different regions, they are frequently sold in mixed-species assortments in local markets. Several edible species distributed globally are: *C. densifolius, C. floccosus, C. eucalyptorum, C. ignicolor, C. incarnatus, C. formosus, C. rufopunctatus, C. isabellinus, C. platyphyllus, C. longisporus, C. infundibuliformis, C. luteopunctatus, C. luteocomus, C. lutescens,, C. miniatescens, C. pseudocibarius, C. madagascariensis, C. splendens, C. minor, C. cyanescens, C. ruber, C. platyphyllus, C. floridulus, C. odoratus, C. pseudofriesii, C. symoensii, C. cinereus, C. tenuis, C. cinnabarinus, C. subalbidus, C. congolensis, C. cyanoxanthus, C. tubaeformis, C. cibarius*, and *C. subcibarius* (Danell 1999; Flores et al. 2002; Tedder et al. 2002; Buyck 2008; Ogawa et al. 2018; Zhang et al. 2022; Kumar et al. 2025).

Termitomyces A very valuable fungal genus popular for its remarkable nutritional qualities is Termitomyces. As an example of its importance, T. clypeatus has a significant quantity of essential nutrients (Tibuhwa 2012). Approximately 27 edible species of Termitomyces, including *T. titanicus*, the largest mushroom species in the world, have been reported worldwide in over 35 countries (Boa 2004). However, there is still limited scientific data about these fungi are primarily consumed in Asia and Africa. *T. clypeatus* is regularly harvested and sold in local markets in Northern India, Nepal, and Tibet. Some edible species of this genus are: *T. albuminosus, T. letestui, T. cylindricus, T. aurantiacus, T. heimii, T. clypeatus, T. eurrhizus, T. globulus, T. microcarpus, T. entolomoides, T. titanicus T. fuliginosus, T. mammiformis, T. medius, T. robustus, T. striatus*, and *T. schimperi* (Sangvichien and Taylor-Hawksworth 2001; Kansci et al. 2003; Gunasekara et al. 2021; Paloi et al. 2023).

Podaxis They are secotioid fungi that look like a stalked-puffball within family Agaricaceae and consist of 44 mushroom species (Conlon et al. 2016). among these, *P. pistillaris* is a noteworthy edible species that is frequently found in regions like the Hawaiian Islands, Australia, and Afghanistan. Apart from their culinary application, P. pistillaris has been used historically as baby powder in West Africa and as a hair dye in Australia (Medina-Ortiz et al. 2017; Siddique et al. 2017). while This species has been found to be toxic in Niger, despite being regarded as edible in countries like India and Pakistan.

Polyporus In the genus Polyporus, approximately 30 edible mushroom species documented in 20 countries. Different species are identified for their application in food and medicine with relatively little importance (Boa 2004; Yu 2022). The following edible species reported worldwide belong to Polyporus genus are: *P. aquosus, P. alveolaris, P. arcularius, P. brumalis, P. brasiliensis, P. elegans, P. blanchettianus, P. sapurema, P. croceoleucus, P. tricholoma, P. confluens, P. grammocephalus, P. fimbriatus, P. umbellatus, P. indigenus, P. badius, P. moluccensis, P. mylittae, P. rugulosus, P. rhizomorphus, P. squamosus, P. stipitarius, P. eucalyptorum, P. tenuiculus, P. sanguineus, P. tubaeformis, P. tuberaster*, and *P. tinosus* (Prance 1984; Härkönen 2001; Zjawiony 2004; Bandara et al. 2015; Quintero-Cabello et al. 2021; Pawłowicz et al. 2024).

Ramaria The genus Ramaria contains 44 edible mushroom species documented across 18 countries. Some important mushroom species are continuous harvested and traded in the Mexico and Nepal markets, with Ramaria botrytis being among the most widely consumed (Boa 2004). Notably, regional variations in species use and perception are highlighted by the fact that Ramaria formosa is considered toxic in Bulgaria but edible in Nepal. Some reported edible species reported globaly are: *R. apiculata, R. flava, R. araiospora, R. bonii, R. mairei, R. botrytoides, R. botrytis, P. aurea, R. rubripermanens, R. cystidiophora, R. flavobrunnescens, R. obtusissima, R. ochracea, R. rosella, R. rubiginosa, R. subaurantiaca, R. stricta, R. subbotrytis R. sanguinea*, and *R. sandaracina* (Flores et al. 2002; Revisiting Ramaria species: the Coral Fungi as Food and Pharmaceuticals 2020; Elkhateeb et al. 2021; Semwal et al. 2023).

Clitocybe Clitocybe is a genus of about 1131 species, with only a few recognized as edible and others known to be poisonous or toxic. Despite the genus' overall association with toxicity, some species have medicinal properties. The species Clitocybe nebularis, Clitocybe maxima, and Clitocybe alexandri are all reported to be therapeutic (Pohleven et al. 2009; Vaz et al. 2010; Zou et al. 2012). Several species, including *C. fragrans, C. squamulosa, C. clavipes, C. gibba, C. hypocalamus, C. infundibuliformis, C. geotropa, C. suaveolens, C. nebularis*, and *C. odora*, are consumed in various regions around the world, like as Bulgaria, Hong Kong, China, Australia, India, Mexico, Russia, Chile, Ukraine, and Indonesia (Takahashi 2003; Vizzini et al. 2010; Cooper 2014; Alvarado et al. 2018).

Tricholoma The genus Tricholoma consists of 57 edible species spread over 30 different countries. Tricholoma matsutake is one extreme prized and expensive species. It is primarily exported from China, Bhutan, Russia, and Korea. Farmers in China who cultivate *T. matsutake* can earn between $5 and $6 million per year. In contrast, Tricholoma pessundatum, which is thought to be toxic, which used in Hong Kong even though its potential risks (Boa and Nations 2004). Some edible species of genus includes *T. columbetta, T. sejunctum, T. caligatum, T. georgii, T. magnivelare, T. rutilans, T. terreum, T. imbricatum, T. quercicola, T. matsutake, T. mauritianum, T. mongolicum, T. saponaceum, T. nauseosum, T. tigrinum, T. equestre, T. personatum, T. pessundatum, T. portentosum, T. flavovirens, T. russula, T. scabrum, T. sulphureum, T. vaccinum*, and *T. ustaloides* (Tedder et al. 2002; Chen 2004; Trudell et al. 2017; Aoki et al. 2022).

Tuber The genus Tuber contains approximately 18 species spread across eight countries (Boa and Nations 2004). Significant species include *T. borchii, T. aestivum, T. indicum, T. brumale, T. melanosporum*, and *T. magnatum*, which are widely studied for their economic and culinary values. These species sell for high prices, ranging from €600 to €6000 per kilogram (Luard 2006). *T. indicum*, a commercially important species from China (Yunnan), has been commonly exported since the 1980s to United States, Japan, Australia, and Europe. Truffles are highly valued for their distinct aromatic properties, which vary between species. Tuber melanosporum is known for its earthy "wet forest" aroma, which is accompanied by hints of radish and hazelnut, whereas Tuber magnatum has a complex fragrance with garlicky cheese notes and small methane undertones (Culleré et al. 2010).The species of genus *Tuber* reported worldwide, including *T. hiemalbum*, *T. sinosum T. magnatum*, *T. gibbosum*, *T. melanosporum*, *T. rufum*, *T. brumale*, *T. indicum*, *T. californicum*, *T. mesentericum*, *T. borchii*, *T. moschatum*, *T. aestivum*, and *T. oligospermum* (Leonardi et al. 2021; Zambonelli et al. 2021).

Volvariella Volvariella contains approximately 12 edible mushroom species spread among 27 countries. The most prevalent species is *V. volvacea*, also known as the paddy straw mushroom. Historically, Buddhist monks cultivated it for consumption before presenting it as a tribute to royalty. This species was first cultivated approximately 300 years ago, in the eighteenth century, and was introduced to various parts of Asia between 1932 and 1935 (Boa and Nations 2004; Ahlawat and Arora 2016). Countries that harvest or cultivate *V. volvacea* and other species include Central Africa, China, Benin, Chile, Mexico, Costa Rica, Congo, Nigeria, Ghana, Thailand, Indonesia, Madagascar, Malawi, Israel, Taiwan, Mauritius, Hong Kong, Nepal, Peru, Russia, and India. This genus includes notable edible species such as *V. esculenta, V. bakeri, V. speciosa, V. diplasia, V. terastria, V. parvispora, V. volvacea, V. earlei*, and *V. bombycina* (Ghosh 2020; Thuc et al. 2020).

Cordyceps Cordyceps is a genus of Ascomycota that parasitizes insect larvae. Approximately 37 edible species have been reported in Japan, Nepal, and China. While these species are mostly used for their reported health advantages, they are also highly valued for their medicinal properties. Most notable species include *C. liangshanensis, C. cicadicola, C. ophioglossoides, C. sinensis*, *C. gunnii*, and *C. militaris* (Hall 1998; Hall et al. 1998; Yang et al. 2009). Among these, *C. militaris* is known for its pharmacological properties, which include anti-inflammatory, immunomodulatory, antioxidant, and antitumor effects. Anthropogenic activities and habitat destruction have negatively affected *C. sinensis* populations, resulting in declining natural yields (Jędrejko et al. 2021; Phull et al. 2022).

Russula Approximately 128 edible species in the genus Russula are reported across 28 countries. Russula emetica is consumed in Russia and Mexico, but it is considered toxic if not cooked properly. Russula is a mycorrhizal fungus and difficult to cultivate due to its symbiotic relationship with plant roots, and it exhibits significant ecological diversity (Liu et al. 2024). Some worldwide reported edible species of this genus includes *R. cellulata, R. grisea, R. compressa, R. roseostriata, R. cyanoxantha, R. congoana, R. delica, R. diffusa var. diffusa, R. roseoalba, R. depallens, R. lepida, R. emetic, R. striatoviridis, R. xerampelina, R. erythropus, R. atrovirens, R. viscida R. meleagris, R. chamaeleontina, R. hiemisilvae, R. cyclosperma, R. oleifera, R. rubra, R. minutula, R. pectinata, R. atropurpurea, R. pseudostriatoviridis, R. alutacea, R. olivacea, R. pseudopurpurea, R. phaeocephala, R. sesenagula, R. vesca, R. testacea*, and *R. virescens* (Tedder et al. 2002; Liu et al. 2024).

Cortinarius The genus includes about 30 edible species distributed across more than 11 countries. Cortinarius species are unpopular in North America and Europe due to toxic species that pose risks to consumers. Only small number of edible mushroom species have been documented in countries such as China, Russia, Japan, Ukraine Rica, and Costa (de Mattos-Shipley et al. 2016). Some important edible species of this genus includes *C. armillatus, C. orichalceus, C. elatior, C. albovio-laceus, C. multiformis, C. varius, C. praestans, C. purpurascens, C. collinitus, C. armeniacus, C. rufo-olivaceus, C. crassus, C. mucosus, C. claricolor, C. clari-color, C. cornucopioides, C. largus*, and *C. glaucopus* (Montoya-Esquivel et al. 2001; Ridwan et al. 2018; Deng et al. 2024b).

Flammulina Flammulina velutipes, also known as the "golden needle mushroom," is native to Southeast Asia, specifically China, Korea, Taiwan, and Japan (Psurtseva 2005). Initially, Flammulina was considered a monotypic genus, with only F. ononides and F. velutipes. Interestingly, with further research, several species have been identified worldwide. To date, 14 species have been formally recognized under in this genus with limited data on their edibility (Pérez-Butrón and Fernández-Vicente 2007; Ge et al. 2008; Zaiwei et al. 2015).

Laccaria In the genus laccaria nine edible species have been documented in 17 countries. Laccaria laccata, the most common species, is found throughout temperate regions of the Northern Hemisphere, including Mexico, Europe, North America, and Costa Rica (Chamberlain 1996; Tedder et al. 2002; Nuytinck et al. 2013; Lee et al. 2019). This genus forms symbiotic mycorrhizal associations, making cultivation difficult; however, wild specimens are widely harvested and sold in local markets (Zhang et al. 2023). The following edible mushroom species of this genus are of global culinary significance: *L. scrobiculatus, L. laccata, L. proxima, L. amethys-tine, L. edulis, L. farinacea, L. amethysteo, L. bicolor*, and *L. amethystea* (Tedder et al. 2002; Stubbe and Verbeken 2012; Nuytinck et al. 2013; Lee et al. 2019).

Lactarius 94 edible mushroom species in the genus Lactarius have been identified in more than 39 countries. Fresh Lactarius species are distinguished by the ability to exude a distinctive fluid in response to injury or incision. This fluid's color and organoleptic properties differ between species and serve as important taxonomic indicators (Athanasakis et al. 2013). The species of this genus show cosmopolitan distribution from Europe to America and Asia (Delgat et al. 2019). Lactarius deliciosus has a high economic value in Spain, particularly in the Palencia region, where it is sold for approximately €2 per kilogram, with daily market volumes reaching 4000 kilograms during the harvest season. China is the leading producer of L. deliciosus, supplying approximately 308,000 tons annually, while Estonia produces more than 250 tons. *Lactarius rufus, Lactarius piperatus* (commonly referred as peppery milk-cap and now classified under the genus Lactifluus), and *Lactarius torminosus* are also considered edible species in Turkish cuisine. However, *L. piper-atus* is classified as toxic in China, indicating regional variation in culinary use and safety perception (Ayaz et al. 2011). Some common edible mushroom species of the genus *Lactarius include; L. carbonicola, L. akahatsu, L. chrysorrheus, L. heimii, L. luteopus, L. rubidus, L. pubescens, L. angustus, L. densifolius, L. kabansus,*

L. lapponicas, L. laevigatus, L. medusae, L. piperatus, L. indigo, L. camphoratus, L. quietus, L. hatsudake, L. laeticolor, L. rubrilacteus, L. phlebophyllus, L. rubroviolascens, L. yazooensis L. mitissimus, L. zonarius, L. princeps, L. flavidulus, L. xerampelinus, L. congolensis, L. insulsus, L. edulis, L. gymnocarpus, L. controversus, L. sesemotani, L. sanguifluus, L. necator, L. deliciosus, L. inversus, L. trivialis, L. volemus, L. vellercus, L. denigricans, L. pseudovolemus, L. subindigo, L. pyrogalus, L. resimus, L. corruguis, L. rufus, L. volemoides, L. latifolius, L. salmonicolor, L. subdulcis, L. annulatoangustifolius, L. gymnocarpoides, L. torminosus, L. tanzanicus, L. scrobiculatus, and *L. pelliculatus* (Çaglarlrmak et al. 2002; Lian et al. 2007).

Leccinum The genus Leccinum contains 14 species that are extensively harvested and consumed in countries like New Zealand and Europe (Boa and Nations 2004). *Leccinum versipelle* is particularly prized in Poland, where it is highly valued in its fresh form but less attractive when dried (Kotowski et al. 2019).The *L. holopus, L. versipelle, L. griseum, L. testaceoscabrum, L. chromapes, L. rugosiceps, L. lepidum, L. manzanitae, L. aurantiacum, L. scabrum, L. oxydabile*, and *L. extremiorientale* are examples of worldwide-known edible species from this genus.

Lentinula The genus Lentinula contains approximately three edible species that have been documented in six countries. Lentinula edodes, commonly called "shiitake," is one of the world's extensively cultivated mushrooms, particularly in Southeast Asia (Kobayashi et al. 2020). By 1986, *L. edodes* was responsible for 14% of global mushroom production (Miles and Chang 2004). *The L. lateritia, L. boryana*, and *L. edodes* are edible species that have been documented from various regions, including India, Chile, Mexico, Nepal, Papua, Thailand, and New Guinea (Hajdú et al. 2022).

Lentinus Approximately 28 edible mushroom species of Lentinus have been documented in over 24 countries (Boa and Nations 2004). While most Lentinus species are edible, some are lesser known due to their tough texture. Important edible species, including Lentinus strigosus and *Lentinus sajor-caju* are known for their antioxidant properties. Other species, such as L. brunneofloccosus, *L. glabratus, L. araucariae, L. tuber-regium, L. crinitus, L. velutinus*, and *L. squarrosulus* have been reported in Brazil, Benin, China, Central Africa, Ethiopia, Burundi, Gabon, India, Congo, and Ghana. These mushrooms have nutritional and ecological value, contributing to traditional dietary and medicinal practices in these regions (Obodai 2001).

Lepista The genus Lepista, widely referred to as the "blewit mushroom," is an edible, wild basidiomycete known for its culinary uses. The species of this genus show worldwide distribution, specifically found in different parts of Europe and North America. Wood blewit mushrooms (Clitocybe nuda) are economically significant and are grown in controlled conditions, such as the Mushroom Research Center in France. However, cultivated wood blewit mushrooms are generally thought to have a poorer taste than their wild counterparts, which are prized for their superior flavor and texture (Spencer and Palmer 2021). Wood blewits (Clitocybe nuda) are harvested for their medicinal and nutritional value. These mushrooms are high in essential macronutrients, with total protein, carbohydrates, ash, and lipids (44.2%, 41.4%, 5.4%, 9.0%), respectively. Making them an excellent source of nutrition (Colak et al. 2007). The Medicinal application is antioxidant, antimicrobial, and beriberi treatment (Dulger et al. 2002; Pinto et al. 2013). Some important species of the genus Lepista that use globaly are: *L. nuda, L. caespitosa, L. luscina, L. sordida, L. dinahouna, L. irina, L. glaucocana, L. caffrorum*, and *L. personata* (Hall 1998; Hall et al. 1998).

1.4 Identification Techniques for Edible Mushrooms

There are different methods for the identification of edible mushrooms, starting with traditional to the latest and most advanced as given below (Fig. 1.2).

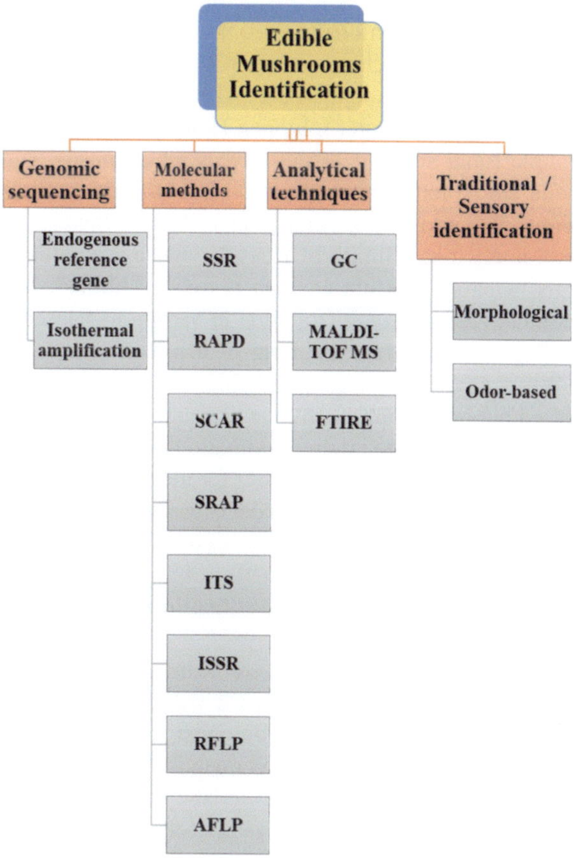

Fig. 1.2 Different methods are used in the identification of edible mushrooms in various regions of the world

1.4.1 Traditional or Sensory Identification Techniques

Sensory identification uses human sensory organs and practical experience to assess food characteristics. This method is simple, quick, cost-effective, and extremely useful, making it difficult to replace with other methods. It focuses on visual, olfactory, gustatory, tactile, and auditory assessments. In the case of edible mushrooms, identification is primarily based on morphological characteristics and odors detected through vision and smell (Cazabonne 2021; Wei et al. 2022).

1.4.1.1 Morphological Identification

Traditional morphological analysis is used to identify edible mushrooms, with a focus on the shape, color, and spore characteristics of the fruiting body. Visual examination of macroscopic features includes the shape, color, and surface markings of the cap, as well as the stipe, gills, volva, and their structural connections. Microscopic examination further evaluates spore and hyphal structures to confirm identification (Inyod et al. 2017; Wei et al. 2022; Nurhakiki and Putra 2024). Fungal taxonomy has traditionally been used to identify edible mushroom species, but this method is highly reliant on expert knowledge. Toxic mushrooms frequently resemble edible varieties in their natural habitat, making accurate identification difficult. For example, Amanita fuliginea Hongo shares morphological characteristics with edible fungi such as Termitornyces albuminosus, which is highly valued among wild edible mushrooms (Jo et al. 2014). This mushroom is renowned for its delicious flavor and high nutritional value, whereas Amanita fuliginea Hongo is extremely toxic. Similarly, Russula vinosa is a naturally occurring edible wild fungus, whereas Russula emetica is toxic. Although morphological identification is simple, it can be unreliable due to variations caused by environmental conditions and growth patterns, resulting in inconsistencies and possible misidentifications (Wei et al. 2022; Ferreira et al. 2023; Semwal 2024). Morphological identification of edible mushrooms is frequently subjective, making accurate species classification difficult due to inherent limitations in this method. As a result, depending only on morphological characteristics for identification can lead to significant errors and limited applicability (Chathurika et al. 2023).

1.4.1.2 Olfactory/Odor Identification Technique

The unique aroma of different edible mushrooms arises mostly from volatile organic compounds (Sun et al. 2020). C8 volatiles such as esters, alcohols, ketones, and aldehydes are thought to be the most important contributors to their odor profile (Aisala et al. 2019). These compounds have a variety of odor properties, and their interactions produce the characteristic fragrance of mushrooms. Additionally, only a small subset of these volatiles actively influence aroma perception, while the rest act as complementary factors. Looking into the variability in volatile compositions among edible mushroom species can help establish chemical markers for identification and classification (Aisala et al. 2020; Zhou et al. 2023; Deng et al. 2024a).

Artificial Identification Method

Artificial identification of edible mushrooms is primarily based on human olfaction, which distinguishes smells based on previous experience (Portalo-Calero et al. 2019a). While the human sense of smell is more sensitive than previously thought, accurate description of subtle odor differences remains difficult due to linguistic constraints. Odors are frequently described using broad and imprecise terms such as earthy, sweet, or fruity, which can lead to ambiguous or contradictory interpretations. Although olfaction can detect a variety of odors, verbal descriptions are insufficient for accurately identifying edible mushrooms. This method works well for species with distinct and identifiable odors, but it has limitations when dealing with mushrooms that emit similar or difficult-to-define odors (Portalo-Calero et al. 2019b).

Use of Electronic Nose Technique

The electronic nose (E-nose) technique is widely used to detect volatile compounds in food matrices. The E-nose, which mimics the mammalian olfactory system, is an effective analytical tool for odor detection and differentiation. Its main component, a sensor array, works based on the sensitivity of individual sensors to target gases. The fundamental concept of odor discrimination is based on generating distinct response patterns for different volatile compounds, allowing their identification. E-nose systems have proven effective in profiling flavors in a variety of food categories, including edible mushrooms, wines, meats, fruits, and vegetables, demonstrating their high sensitivity and broad applicability in food quality assessment (Pei et al. 2016). This E-nose technique extensively applied to assess the impact of dehydration on different volatile aromatic compounds in edible mushrooms, allowing for the differentiation of species exposed to varying drying times. Furthermore, the electronic nose has proven effective in detecting toxic toadstools and has been used to identify species in the genus Amanita (Portalo-Calero et al. 2019a). The odor profile analysis of mushrooms demonstrated the effectiveness of electronic nose technology for species identification. This technique achieved classification accuracy ranging from 80% to 100%. Electronic nose systems provide a robust approach to assessing food aroma, distinguished by rapid detection, consistent sensory assessment, operational simplicity, and cost-effectiveness. (Chen et al. 2017). The main hurdle for E-nose is to accurately collect sensor data and analyze different volatile compounds in several edible mushrooms. In general, E-nose can improve the identification of edible fungal species by means of olfactory profiles by detecting specific volatile organic compounds.

1.4.1.3 Analytical Techniques

Various analytical methods are used to identify edible mushrooms, with a focus on chemical composition. The different chromatographic techniques are used, each with unique detection capabilities and applications. To improve analytical accuracy some advanced hyphenated techniques, are used in this field as discussed below (Wei et al. 2022; Yan et al. 2023).

1.4.1.4 Use of Gas Chromatography

Each edible mushroom has a unique composition of volatile compounds, which is typically analyzed using advanced instrumental techniques. Although these measurements do not have a direct correlation with human olfactory perception or sensory experience (Zhang et al. 2018). Only a small number of volatile compounds found in edible mushrooms interact with olfactory receptors, influencing odor perception only when the concentration exceeds the olfactory threshold. While some edible mushrooms contain similar volatile compounds, with relative concentrations vary high. The distinct volatile profiles of each species, as determined by sensory evaluation and instrumental analysis, vary in composition and prioritize compounds differently in terms of sensory relevance (Aisala et al. 2019; Zhu et al. 2024). Headspace-Solid-Phase Micro-Extraction (HS-SPME) is an effective approach that help in the identification of distinctive chemicals in various fungal species, allowing for a comparative examination of flavor profiles across edible mushrooms. Analytical methods such as headspace-gas chromatography (HS-GC), gas chromatography coupled with flame ionization detection (GC-FID), and gas chromatography-mass spectrometry (GC-MS) are widely used for profiling volatile compounds in fungi, with GC-MS being the most popular due to its high sensitivity and specificity. Gas chromatography effectively separates complicated mixtures based on polarity, different boiling points, and adsorption qualities, allowing individual components to be identified as they elute off the column. The detector changes the sample's compositional data into electrical impulses, which are amplified and processed for chromatograms. Gas chromatography (GC) alone is less effective for qualitative analysis; but, when combined with mass spectrometry (MS), it gives complementing benefits, improving analytical precision and accuracy. This hybrid technique is typically combined with electronic nose technology, a unique analytical instrument for detecting and quantifying trace volatile chemicals in various food items (Zhou et al. 2015; Lin et al. 2023). Moreover, little research has used gas chromatography-mass spectrometry and E-nose technologies to investigate the olfactory characteristics of edible mushrooms. One study used headspace solid-phase microextraction (HS-SPME) in combination with a gas chromatography-ion trap mass spectrometer (GC-IT-MS) for the identification of volatile chemicals in six edible fungus species, revealing eleven significant volatiles as potential indicators for species classification. Furthermore, researchers used GC-MS and E-nose technologies to study the volatile profiles of eight edible fungus species, revealing considerable compositional differences in their volatile components (Wei et al. 2022; Yan et al. 2023). The main findings show that GC and E-nose technologies are useful for identifying edible mushrooms. When combined with chemometric approaches, these technologies enable the analysis of volatile chemicals, which aids in the successful discrimination of distinct edible mushrooms. GC-MS has advantages such as low sample requirements, excellent sensitivity, and rapid analysis, allowing for detailed characterization of individual taste components. However, its wider application is limited by the expensive cost of reagents and equipment, as well as the complexities of data processing (Yang et al. 2016b).

1.4.1.5 Use of Matrix-Assisted Laser Desorption and Ionization-Time of Flight MS (MALDI-TOF MS)

MALDI-TOF MS technique has evolved as a strong analytical technique, widely applied in the exact identification of microorganisms and taxonomic classification such as fungal species (Calderaro and Chezzi 2024). MALDI-TOF MS technique works by co-crystallizing the sample and matrix in a predetermined ratio on a target plate. Laser irradiation ionizes the proteinaceous crystals, resulting in ions with various mass-to-charge (m/z) ratios. These ions are propelled in an electric field, and their flight durations, which vary with their m/z ratios, allow for separation. The detector measures the ions, generating a mass spectrum that plots the m/z ratio against ion intensity. This spectrum mostly represents the diversity of ribosomal RNA-associated proteins. The resulting spectrum is accurately identified by comparing it to a database. This technique gives a quick and reliable method to find the protein profiles of edible mushrooms (Gopal and Muthu 2023). The study found that the MALDI-TOF-MS technique can accurately identify different edible mushrooms based on their fruiting bodies. This technology provides quick analysis comparable to morphological inspections while maintaining the same level of stability as DNA sequencing. For the first time, MALDI-TOF-MS was used to detect truffles alongside phylogenetic DNA sequence analysis. The findings demonstrated that this approach allows for the accurate and efficient identification of truffle species as well as the detection of misidentified specimens in commercial truffles, hence contributing to the prevention of market fraud (El Karkouri et al. 2019). The technique is extremely cost-effective, with the key advantage being the capacity to rapidly identify species. Edible mushrooms can be properly identified using MALDI-TOF-MS on the same day they are collected, considerably increasing efficiency in taxonomic studies (Patel 2019). To ensure precise identification, compare the mass spectrogram produced during detection to a reference spectrogram from a database. Incomplete or erroneous reference spectra in the database can cause identification mistakes or failures.

1.4.1.6 Fourier Transform Infrared (FTIR)

Edible fungi are primarily made up of proteins and carbohydrates. Fourier transform infrared (FTIR) spectroscopy, which relies on molecular vibrations and rotations, is a quick and inexpensive method for finding edible fungi. This method enables for the qualitative and quantitative investigation of fungal samples. The position of absorption peaks in the infrared spectrum makes it easier to identify particular compounds, while the intensity of these peaks can be used to calculate the relative concentration of each compound (Dian-xu et al. 2018). FTIR spectroscopy can scan the sample's whole chemical constituents in the mid-infrared range, and the spectral fingerprint reflects the sample's actual quality (Qi et al. 2017). FTIR spectroscopy, when combined with modern chemometric approaches such as principal component analysis (PCA), hierarchical cluster analysis (HCA), and partial least squares discriminant analysis (PLS-DA), allows for the identification of unique spectral fingerprints for specific samples (Wang et al. 2019d). The FTIR, in conjunction with PCA, has been successfully used to identify edible mushrooms, allowing precise categorization at the genus, species, and subspecies levels. In a study of *B. bicolor* 58 specimens from Yunnan, FTIR combined with PCA successfully distinguished the species at the regional level. FTIR has several benefits, including non-destructive analysis, minimum sample needs, and quick operation. However, accurate identification requires the creation of robust models and complicated mathematical computations (Yang et al. 2016a; Wei et al. 2022).

1.5 Identifications of Edible Mushrooms by Molecular Methods

Molecular technology is frequently employed in the identification of species due to its speed, specificity, sensitivity, and accuracy. DNA can be found in numerous developmental stages and different tissues of the organism since it contains the genetic makeup of life and is extremely stable. It remains unchanged by the environment, seasons, and different other factors (Yao et al. 2020).

1.5.1 Molecular Marker Technology

Form long time, Molecular marker approaches that use nucleic acid polymorphisms to capture genomic DNA variations have long been used in the genetic analysis of edible mushrooms (Yan et al. 2023). Molecular markers are usually shown as electrophoresis bands. They can be broadly divided into three types: PCR-based markers, hybridization-based markers, and emerging molecular marker technologies (Wei et al. 2022).

1.5.2 Use of Simple Sequence Repeats (SSR) in Mushroom Identification

The SSRs, also known as microsatellite DNA, are defined by brief tandem repeats of 1 to 6 base pairs. These repeats are prevalent in eukaryotic genomes and have a high mutation rate, which contributes to their fast evolution. Microsatellite DNA is widely dispersed across the genome, primarily in noncoding areas. The flanking regions of these repetitions are typically preserved, while the repeats themselves frequently appear in a single copy. Length polymorphisms in SSRs can be identified by PCR amplification of the core repeat sequences, followed by electrophoretic analysis to determine variation (Bhariya et al. 2024). Different studies show that the SSR technique has been effectively used to identify economically important edible mushrooms like Auricularia auricula-judae (Yan et al. 2023), Flammulina velutipes (Lallawmsanga and Carrasco 2022), Phellinus igniarius (Dong et al. 2024), and Lentinula edodes (Lee et al. 2020). Furthermore, highly polymorphic SSR markers have been developed to accurately differentiate these species(Liu et al. 2016). SSR markers show codominant inheritance and a high level of allelic diversity at every locus, making them an effective tool for species identification. However, developing traditional SSR markers is time-consuming and costly (Saito et al. 2019). The emergence of next-generation sequencing (NGS) technology has enabled the low-cost production of large-quantity SSR markers for whole-genome research (Singh et al. 2022).

1.5.3 Application of Inter-simple Sequence Repeats (ISSR) in Mushroom Identification

The ISSR approach was developed to differentiate between closely related species and is extensively used in the identification of fungi, insects, vertebrates, and plants. It is an improvement on the simple sequence repeat (SSR) technique, which uses anchored SSR primers. These primers contain random nucleotides (2–4) added to the SSR sequence at the 3′ or 5′ end, which improves the specificity and unpredictability of the amplification process (Marwal and Gaur 2020). PCR amplification specificity is improved by targeting regions that align with anchored nucleotides, ensuring that only the desired sequences are amplified. Furthermore, investigations have shown that ISSR markers may distinguish between several strains of *Pleurotus djamor*, including hybrid strains, monokaryotic strains, and dikaryotic wild-type strains used in hybrid production (Aguilar Doroteo et al. 2018). In a study, the SRAP and ISSR were used to assess the genetic diversity of 72 specimens of *Lepista nuda* from China (Du et al. 2018). The five pairs of SRAP and Six pairs of ISSR primers were shown to be successful in producing different polymorphic bands. The study found that both ISSR and SRAP markers are accurate, efficient, and highly polymorphic techniques for identifying L. nuda. Furthermore, ISSR markers demonstrated notable robustness, repeatability, and quick results, making them appropriate for genetic study of this species (Ruiz-Moyano et al. 2019). The negative aspect of ISSR is that the ideal reaction conditions must be carefully examined.

1.5.4 Random Amplified Polymorphic DNA (RAPD)

The RAPD approach is extensively used for pedigree analysis, evolutionary studies, and identification of species. This approach uses a single, randomly generated oligonucleotide primer, typically ten base pairs long, to amplify genomic DNA using PCR. The primers bind to specific locations in the genome. Variations like as insertions, deletions, or mutations within these areas can change the primer binding sites, resulting in detectable polymorphisms in amplified DNA fragments (Wei et al. 2022). Various studies show the efficacy of RAPD markers in the genetic investigation of Agaricus bisporus. The data indicate that RAPD profiles, particularly combined with morphological and physiological characterizations, are reliable techniques for identifying and differentiating A. bisporus strains (Saha et al. 2024). To accurately identify diverse Cordyceps species, researchers developed a method that uses ITS, RAPD, and SCAR markers drived-sequences. The findings showed that two DNA-based techniques are complimentary, improving the reliability and precision of species identification (Deshmukh et al. 2020). The RAPD-PCR technology has numerous advantages, including great flexibility, minimal laboratory equipment requirements, cost effectiveness, and fast execution. However, its principal disadvantage is low reproducibility, which necessitates careful primer selection and validation to provide reliable results (Felbinger et al. 2020).

1.5.4.1 Application of Internal Transcribed Spacer (ITS) in Mushroom Identification

The ITS is the sequence of nuclear ribosomal RNA that is commonly used in molecular systematics because it shows higher variability than the large subunit (LSU) and small subunit (SSU) of rRNA genes. This diversity is frequently seen within individual ribosomal DNA (rDNA) repetitions. DNA sequence-based phylogenetic analyses, notably of the ITS1-5.8S-ITS2 sections of rDNA, are a typical method for accurate fungal identification (Ghosh et al. 2019; Bradshaw et al. 2023). The ITS region has been used to study various aspects of edible mushroom diversity, such as the phylogenetic and biogeographic relationships within the Suillus genus (Pildain et al. 2021), molecular phylogenetics of Amanita species in Japan (Varga et al. 2024), phylogenetic diversity of wild edible species in the Russula genus (Tharnrat Kaewgrajang 2020), and the genetic diversity of Pleurotus eryngii (He et al. 2016).

1.5.4.2 Sequence-Related Amplified Polymorphism

Li and Quiros (2001) proposed Sequence-Related Amplified Polymorphism (SRAP), a molecular marker method widely used in genomic research. SRAP markers have been shown useful for genetic linkage mapping, diversity evaluation, gene tagging, and fungal population genetics (Du et al. 2024). This approach extensively applied in the study of genetic diversity in species such as Pleurotus pulmonarius (Vidal-Diez De Ulzurrun et al. 2021), Ganoderma spp. (Chao et al. 2018), Auricularia cornea (Du et al. 2024), Lentinula edodes strains (Dong et al. 2019), Tricholoma matsutake (Ma et al. 2010), Agaricus bisporus (Harfi et al. 2021).

1.5.4.3 Use of Restriction Fragment Length Polymorphism (RFLP) in Mushroom Authentication

RFLP technology depends on the differences between genotypes, such as base deletions, recombination, insertions, or mutations in restriction endonuclease recognition sites. When DNA from various individuals is digested through particular restriction enzymes, it yields fragments of varying length. After electrophoresis, these fragments produce distinct RFLP patterns. DNA polymorphisms are identified by hybridizing these patterns with cloned DNA probes and analyzing them using Southern blotting and radiography. This technique accurately determines the number and position of restriction sites in PCR amplification products. The size of restriction fragments, a stable genetic trait, is used as a molecular fingerprint to distinguish individuals as well as intra- and interspecific genotypes (Kordalewska et al. 2019; Wei et al. 2022). Multiple studies have shown that PCR-RFLP is a reliable molecular tool for distinguishing between Flammulina velutipes strains. This method has been successfully used to analyze ITS regions for the identification of Flammulina strains. Furthermore, PCR-RFLP has been used to distinguish strains of Agaricus species, conduct phylogenetic analyses of Pleurotus species, and perform DNA fingerprinting within the Pleurotus eryngii complex, demonstrating its broad use in phylogenetics and fungal taxonomy(Guan et al. 2008). Multiple studies have shown that PCR-RFLP is a reliable molecular tool for distinguishing between Flammulina velutipes strains. This method has been successfully used to analyze ITS regions for the identification of Flammulina strains. Furthermore, PCR-RFLP has been used to distinguish strains of Agaricus species, conduct phylogenetic analyses of Pleurotus species, and perform DNA fingerprinting within the Pleurotus eryngii complex, demonstrating its broad use in phylogenetics and fungal taxonomy PCR-RFLP, a fingerprinting-based technology, is widely used for species identification because of its simplicity, cost-effectiveness, and reproducibility. However, it takes longer than RAPD because it requires more labor-intensive procedures such as DNA digestion with multiple enzymes, membrane transfer, and probe preparation. The SCAR, SPAR, AFLP, and SNP are additional molecular techniques used to identify edible fungi species. To address the limitations of individual methods, researchers frequently combine multiple approaches (e.g. RAPD-SCAR, ISSR-RAPD, ISSR-SRAP, and AFLP-SCAR) to enhance the efficiency and accuracy of edible mushroom identification.

1.5.4.4 Sequence Characterized Amplified Region (SCAR)

The SCAR technique, which was introduced in 1991 and has been widely used since 1996, has become an important tool in basidiomycete research, including population genetics, breeding, and genetic analysis (Liu et al. 2012). SCAR markers have been used successfully to identify species in mushrooms like L. bicolor (Li et al. 2014), Flammulina velutipes (Li et al. 2014), and Lentinula edodes (Sheng-jin et al. 2022). Furthermore, the method has been successful in the authentication and strain identification of medicinal mushrooms such as Lentinula edodes (Sheng-jin et al. 2022) and Ganoderma lucidum (Su et al. 2008).

1.5.4.5 Amplified Fragment Length Polymorphism (AFLP)

The AFLP is a molecular technique for DNA fingerprinting in a different of taxa, such as mammals, microbes, and plants, due to its capacity to study DNA with different complexity. AFLP has several advantages, including specificity, reliability, and high reproducibility (Vuylsteke et al. 2007; Restrepo et al. 2018). This approach has been used to differentiate between several mushroom species, including Agaricus bisporus (Foulongne-Oriol et al. 2011), H. marmoreus (Wang et al. 2009), Armillaria gallica (Brazee et al. 2012), and Pleurotus eryngii (Pawlik et al. 2012). Mukhopadhyay et al. (2012) employed AFLP with 32 primer combinations to investigate 30 Lentinula strains, including L. novae-zelandiae, L. lateritia, and L. edodes.

1.5.4.6 Genomic Sequencing Technology

Sequencing technology identifies species through analysis of specific DNA fragment sequences. Several genetic markers are commonly employed for fungal classification and identification, including the ribosomal ITS, LSU, RNA polymerase II (RPB2), β-tubulin, and elongation factor 1-alpha (EF1-α). Fungal ribosomal DNA (rDNA) is organized in tandem repeats, with conserved coding segments and highly variable spacer regions. The ITS region, recognized for its sequence polymorphism, provides information on the evolutionary relationships of eukaryotic organisms by successfully differentiating strains, species, and genera (Liu et al. 2016, 2022). This specific region is the most commonly used as a universal DNA barcode identifier to identify mushrooms (Mohammed et al. 2018). The LSU rRNA gene is organized in tandem repeats within the genome as ribosomal gene cluster (Sonnenberg et al. 2007). Specific regions, particularly the LSU D1-D2 domain, are useful molecular markers for phylogenetics, identification, and classification of species (Antil et al. 2023). The second-largest RNA polymerase II component, RPB2, has a single-copy gene structure and develops gradually. Because of these properties, RPB2 has emerged as a valuable molecular marker for fungal taxonomy and identification. β-Tubulin, a key cytoskeletal protein and microtubule constituent, is coded by multiple genes in different organisms and varies by species (Gadadhar et al. 2017). Elongation factors are proteins that play an important role in the elongation phase of mRNA translation, allowing the polypeptide chain to extend. Three major elongation factors have been identified in eukaryotes. EF1-α regulates the binding of aminoacyl-tRNA to the A-site of the ribosome during translation. The EF1-α gene is highly conserved across species and regarded as a housekeeping gene. Fraiture et al. (2019) analyzed lethal Amanita species across the country using five gene regions: β-tubulin, ITS, EF1-α, RPB2, and LSU. Their findings showed that including more loci in the analysis improved phylogenetic accuracy. He et al. (2016) studied the phylogenetic relationships of Pleurotus eryngii, Pleurotus nebrodensis, and Bailinggu using ITS region, partial EF1-α, and RPB2 gene sequences. The ITS, RPB2, and EF1-α sequence variations differentiate the three fungi at the species level. These genetic markers are reliable tools for distinguishing between species. Earlier fungal identification methods relied on a single gene fragment, such as ITS, which had inherent flaws and could lead to misidentification (Kiss 2012). The comprehensiveness of the referenced gene bank can have an impact on the accuracy of analytical results, especially when identifying edible mushroom with minor variations in the spacer region. A study analyzed sequence variations in the and RPB2, EF1-α, and ITS regions of Ganoderma australe from the same individual using cloning and sequencing to determine if these differences affect the precision of molecular identification (Yu-Li et al. 2019). The study found that ITS, EF1-α, and RPB2 sequences vary by individual. Specifically, the ITS sequence can vary by up to 2.3% within the same organism, whereas the RPB2 sequence differs by only 0.5%. The EF1-α sequence can vary by up to 1.8% within an individual. As a result, researchers are increasingly analyzing multiple gene fragments concurrently, which improves the precision and reliability of species identification.

1.5.4.7 Endogenous Reference Gene Identification

It is a conserved species-specific, DNA sequence with a steady copy number, and is free of allelic variation. This approach of detecting endogenous reference genes is widely used in food research, particularly for the detection of food adulteration. Various crops, including soybeans (Long et al. 2022), peach (Xu et al. 2022), and corn (Gao et al. 2020), have been investigated for the identification and application of endogenous reference genes. Endogenous reference genes in edible mushroom species are now being identified more comprehensively. The first step in discovering a specific edible mushroom is to identify a good endogenous reference gene. To accomplish so, large gene sequences are obtained from GenBank and submitted to BLAST homology, and alignment analysis. Genes with high specificity but low homology are then identified as suitable endogenous reference genes for the target fungus. Specific primers are constructed based on these chosen genes, and their specificity is confirmed using qualitative and quantitative analysis. Sensitivity testing is carried out to create a reliable set of methodologies for quantitative and qualitative detection of the selected endogenous reference gene in edible mushroom. Zheng et al. (2018) used sequence alignment and BLAST to identify pyrG as the Pleurotus ostreatus endogenous reference gene. Three P. ostreatus strains and seven other fungal species were analyzed using quantitative and qualitative PCR. The findings demonstrated that pyrG is a reliable reference gene for identifying P. ostreatus. Shan et al. (2019) identified the pol gene from Tricholoma matsutake as a good endogenous reference gene. They formed a quantitative PCR technique to detect the pol gene, allowing for accurate identification of T. matsutake. Zhang et al. (2021a) revealed that the Tps2 gene is a reliable endogenous reference gene for Flammulina velutipes using both qualitative and quantitative PCR. This method, which uses an endogenous reference gene, is more cost-effective and efficient because the findings can be obtained straight from normal PCR amplification without the requirement for sequence alignment. While this technique has significant potential for wider application, it is limited by the requirement that the target species' gene sequence be publically available.

1.5.4.8 Isothermal Amplification Technology

Polymerase chain reaction (PCR)-based detection is a well-established nucleic acid analysis technology that involves amplification of target genes at a variable temperature making use of a thermal cycler. However, PCR is frequently time-consuming, complex, and requires skilled personnel, rendering it unsuitable for on-site detection in non-laboratory situations. As a result, there is increased interest in the development of isothermal amplification technologies, which provide simpler, faster, and more efficient alternatives. One such method, loop-mediated isothermal amplification (LAMP), has received a lot of attention because to its potential. LAMP employs four unique primers targeting six areas of the target gene and is aided by a DNA polymerase with high strand displacement activity. The reaction generates a huge amount of amplification products, including white magnesium pyrophosphate precipitates visible to the naked eye, which indicates successful amplification (Negi et al. 2021). LAMP products can be identified using agarose gel electrophoresis (AGE), as well as optical detection methods such as cationic polymers, SYBR Green I, calcein, or lateral flow dipsticks (LFD) and hydroxynaphthol blue (Vaagt et al. (2013). To present, research into the fast identification of mushroom species utilizing LAMP has been limited. Vaagt et al. (2013) pioneered the use of LAMP technology in conjunction with LFD for the quick identification of mushroom species, achieving identification in around 2 hours. This approach can detect harmful mushrooms in complex mushroom mixtures. He et al. (2019) used the LAMP and HRCA procedures for cost-effective as well as rapid identification of distinct dangerous Amanita species. Their findings showed that both approaches were extremely specific, sensitive, and effective for identifying these lethal fungi. Isothermal amplification technologies, such as LAMP and HRCA, enable target sequence amplification at a constant temperature, providing benefits such as reduced time requirements, high sensitivity, and increased specificity(Becherer et al. 2020). There are several major differences between isothermal amplification strategies, including reaction duration, primer design, enzyme concentration, template requirements, and temperature. For example, the LAMP approach often requires numerous primers to achieve specificity, differentiating it from other approaches in terms of primer complexity and overall sensitivity.

1.5.4.9 Geographical Distribution of Edible Mushrooms

The worldwide climate is often divided into five major regions based on temperature, precipitation, and vegetation type. The distribution and diversity of edible mushrooms, as well as the size of their harvested areas, are heavily impacted by microclimatic conditions unique to each place. Habitat variability and microclimatic variables have a significant impact on fungal communities, with studies highlighting temperature, humidity, and precipitation as critical factors of fruiting body development and species diversity across different climatic zones. These climate parameters are the best predictors of fungal species richness. As a result, the diversity of edible mushroom species varies depending on the climatic circumstances in each place (Boddy et al. 2014; Ruiz-Almenara et al. 2019; Meena et al. 2020; Alem et al. 2021; Kewessa et al. 2022). Different climatic regions of the world and the associated diversity of edible mushrooms (Table 1.3) are given below:

Table 1.3 Distribution of different edible mushroom species in various regions of the world

Sr. No.	Name of mushrooms	Distributed area/ country	References
1	*Phellorina inquinans, Lycoperdon* sp., *Agaricus rodmani, Podaxis pistillaris*	Baluchistan, Pakistan	Ahmad et al. (2014), Khan and Khan (2021), Khalid (2022, pp. 1998–2020), Raza et al. (2022)
2	*P. ostreatus, P. dryinus, Pleurotus comucopiae*	Punjab, Pakistan	
3	*Podaxis pistillaris, Phellorina inquinans*	Sindh, Pakistan	
4	*Boletus edulis, Agaricus bisporus, Morchella conica, A. Silvaticus*	Azad Kashmir, Pakistan	
5	*Agaricus moelleri, Amanita cokeri, Clavaria fragilis, Coprinus comatus, Amanita fulva*	Nagaland, India	Jones and Janardhanan (2000), Agrahar-Murugkar and Subbulakshmi (2005), Ajith and Janardhanan (2007), Dai et al. (2009), Choudhary et al. (2015), Ao et al. (2016b)
6	*Gomphus floccosus, Lactarius quiteticolor, Russula integra*	Meghalaya, India	
7	*Phellinus rimosus, Lentinus tuber-regium, Ganoderma lucidum*	Southern regions, India	
8	*Laetiporus sulphureus, Lentinus sajor-caju, Cantharellus cibarius*	Assam, India	
	Lentinus torulosus, L. conatus, L. sajor-caju, L. squarrosulus, L. cladopus	Northern part of India	
9	*P. geesteranus, P. ostreatus, Pleurotus highking*	Common to most part of Bangladesh	Jayakumar et al. (2006), Amin et al. (2007)
10	*Auricularia polytricha, Pleurotus ostreatus, Flammulina velutipes*	China	Amin et al. (2007)
11	*M. heinemannianus*	West, Africa	Antonín (1998)
12	*Boletus edulis, Ramaria sp, Russula brevipes, Lactarius indigo, Amanita caesarea*	Tlayacapan part of Mexico	Álvarez-Farias (2016)
13	*Hypholoma capnoides, Kuehneromyces mutabilis*	Asia, Japan, Australia, North America	Boa and Nations (2004), Wardley (2007), Alanne et al. (2019), IUCN (2019)
14	*T. nivea*	Europe, Spain, Algeria, arid zone of Tunisia	Alsheikh and Trappe (1983), Moreno et al. (2000), Hamza and Jdir (2016)
15	*L. rufus*	temperate zone of Northern America & Europe	Lagutina (2018)
16	*Termitomyces striatus, Psathyrella atroumbonata, Lentinus subnudus*	Different part of Nigeria	Alofe et al. (1996)

(continued)

Table 1.3 (continued)

Sr. No.	Name of mushrooms	Distributed area/ country	References
17	*Lentinus conatus*	Common to Abakaliki par of Nigeria	Afiukwa et al. (2015)
18	*H. leucopus pers, M. pulchella clowez*	Turkey &Eastern part of Antonia	Acar et al. (2021)
19	*P. eryngii*	Bingol, Turkey, Malatya, Tunceli, Adiyaman	Akyuz and Kirbag (2009)
20	*V. volvacea*	Southeast Asia	Acharya et al. (2016)
21	*Laccaria laccate, Ramaria botrytis, Grifola frondosa, Clavulinopsis fusiformis*	Katmandu valley	M.k and G (1996), Christensen et al. (2008)
22	*R. subalpina*	Subalpine forest of Sikkim Himalayas	Das (2010), Acharya et al. (2017)
23	*A. auricular judae, Calvatia giganetea*	Temperate zone of world	Akgul et al. (2017)
24	*A. phalloides*	Canada and United States	Ammirati (1985)
25	*C. caperatus*	Northern latitude, Finland	Seitz (2020)
26	*C. cornocopioides*	Europe, East Asia, *North* America	Kotlaba (1984), Seitz (2020)
27	*A. caesarea*	Italy, Northern Africa, Southern Europe	Seitz (2020)
28	*R. cyanoxantha*	Europe	Seitz (2020)
29	*A. silvaticus*	Coniferous forest—*North* America and Europe	Pegler (2001), Miller and Miller (2006), Phillips (2010), Seitz (2020)
30	*C. atramentaria*	Europe, Sydney, Australia N. America, Asia and New Zealand	Ramsbottom (1953), Seitz (2020)
31	*F. velutipes*	China, new Europe, Western and North America, East Asia	Davis et al. (2012), Zaiwei et al. (2015), Wang et al. (2018), Seitz (2020)
32	*C. squamosus*	Australia, Europe, America	Seitz (2020)
33	*A. bisporus*	Grasslands in Europe, N. America	Seitz (2020)

1.5.5 Tropical Regions

Edible mushrooms are abundant in tropical climates and play an important role in local diets and economies. Species including Auricularia auricula-judae (Phallus indusiatus), Volvariella volvacea, and Pleurotus ostreatus are common in these places (Thawthong 2014). In Southeast Asia, wild edible mushrooms are not only culinary staples but also sources of bioactive chemicals with antioxidant and antibacterial effects (Thu et al. 2020). Phallus indusiatus, also called the veiled lady or bamboo mushroom, thrives in the warm, humid environs of tropical forests in Asia, Africa, the Americas, and Australia (Cabral et al. 2019). This fungus often develops in rich soils and well-rotted woody waste, frequently prospering after heavy rains in bamboo forests (Nion and Djaya 2015). In Africa, the Nigerian savannah contains species such as Lentinus squarrosulus and Termitomyces robustus, which are vital to indigenous culture and rituals (Buba et al. 2019). In Africa, underutilized wild edible fungi contribute greatly to food security and nutrition, but their ecosystem functions are devalued (Sileshi et al. 2023). In Pakistan, researchers have documented the distribution of Agaricus campestris, a common edible mushroom, across different districts, highlighting the species' adaptability to tropical climates (Iqbal et al. 2023). Furthermore, the tropical bolete Phlebopus spongiosus, endemic to southern Vietnam, is highly valuable and demands high market prices (Kumla et al. 2020).

1.5.6 Subtropical Regions

Edible mushrooms in the world's subtropical regions have gained importance because of their diverse nutritional and medicinal properties (Valverde et al. 2015). These regions, which have warm climates with distinct wet and dry seasons, promote the growth of several kinds of mushroom species that are highly valued both as food and in traditional medicine (El Enshasy et al. 2013). Rainfall, temperature, and host substrate availability all have an impact on the distribution of edible mushrooms in subtropical regions (Liu et al. 2020; Mathur and Mathur 2023). Among the most commonly found species in subtropical climates are A. auricula-judae, L. edodes and P. ostreatus (Kamal et al. 2021). Lentinula edodes, a mushroom extensively farmed and consumed, is native to East Asia but has become popular in subtropical regions because to its resilience to diverse climates and substrates, such as rotting hardwood (Yoo et al. 2024). Similarly, Pleurotus ostreatus, distinguished for its quick growth and characteristic oyster-shaped cap, thrives in subtropical climates, particularly on decaying plant material and agricultural waste (Rahmat et al. 2020). Auricularia auricula-judae is another extensively distributed species, noted for its therapeutic values and tendency for decayed wood in subtropical forests (Sun et al. 2022b). Other notable edible species in subtropical regions include Psilocybe cubensis, Macrolepiota procera, Termitomyces titanicus, Clitocybe nuda, Coprinus comatus, Ganoderma lucidum, Agaricus bisporus, and Hericium erinaceus. Ganoderma lucidum is highly prized in traditional Chinese medicine for its immune-boosting effects. It is usually found on decaying wood in subtropical forests (Wang et al. 2020). Termitomyces titanicus, recognized for its relationship with termites, is highly valued in numerous parts of Africa and Asia, particularly for its distinct flavor and texture (Paloi et al. 2023). Coprinus comatus is a commonly recognized fungus found in subtropical grasslands, known for its tasty cap before it liquefies (Miles and Chang 2004). Clitocybe nuda, found on decaying organic materials, is a coveted delicacy in subtropical climates, although caution must be exercised due to its possible toxicity (Kuo 2006). Macrolepiota procera, the Parasol fungus, grows naturally in subtropical meadows and woodlands and is a popular eating fungus because of its big, umbrella-shaped crown (Jaworska 2013). Hericium erinaceus is prized for its medicinal advantages and culinary purposes. It thrives in subtropical forests on rotting wood (Szućko-Kociuba et al. 2023). Psilocybe cubensis, a tropical and subtropical species that grows on dung, especially in humid climates, is well-known for its hallucinogenic qualities. Agaricus bisporus, the common button mushroom, is farmed globally, with subtropical climes offering optimal conditions for mass production (Karunarathna 2016). These species are not only important for food, but also for the global mushroom market, with continuous study into their health benefits and ecological significance.

1.5.7 Temperate Regions

Edible mushrooms are found throughout temperate zones, thriving in forests, grasslands, and cities. These places, which have moderate climates and varied seasons, provide ideal conditions for a variety of mushroom species (Ocañas et al. 2019). Edible mushroom variety in temperate zones is regulated by soil composition, moisture levels, and the presence of certain host plants. (Procházka et al. 2023). According to the latest research findings, more than 300 wild edible mushroom species are consumed in Mexico, with the majority located in temperate woods in the center and southeast areas. Russula and Lactarius genera contain some of the most abundant edible mushrooms in temperate zones. Russula vesca, characterized by its brittle texture, is widely distributed and considered edible (Burrola-Aguilar et al. 2013; Stojek et al. 2022).

1.5.8 Polar and Subpolar Regions

Edible mushrooms can be found all across the world, including polar and subpolar locations, however, their distribution is influenced by harsh environmental conditions (Tsuji 2023). Fungi such as Chaga (Inonotus obliquus) grow in colder climes, especially in Siberia and Northern Europe. While edible mushroom species are fewer in these harsh conditions, the Chaga fungus is well-known and picked for its medicinal potential (Alhazmi 2018; Osono et al. 2019). It mostly grows on birch trees in cold woodlands, making it a distinctive species in polar habitats (Géry et al. 2018). These environments also sustain a variety of edible Lactarius and Russula mushrooms, which are predominantly found in boreal forests (Persad 2024). The short growth season and decreased biodiversity in these places limit the variety of mushrooms, only a few species grow (Hussain et al. 2023). Edible mushrooms like Armillaria (honey mushrooms) occur frequently in subpolar regions such as Canada, Scandinavia, and Alaska, with some species used in traditional cuisines. Another common edible species in these locations is the Boletus edulis, or king bolete, which is especially valued for its rich flavor and texture in subpolar woodlands. These mushrooms typically grow in symbiotic relationships with coniferous trees, adapting well to the nutrient-poor soils of colder climates. Despite the difficult conditions, these mushrooms play a significant role in local diets, givening essential nutrients and culinary diversity in the subpolar and polar ecosystems.

1.5.9 Arid and Semiarid Regions

Edible mushrooms are found in a variety of environments, including dry and semiarid locations, and play important roles in local diets and economics. Certain species in these places have adapted to the harsh environment, providing considerable nutritional and therapeutic benefits (Morte et al. 2021). One famous example is the desert truffle (Terfezia spp.), a hypogeous Ascomycete that lives in dry and semiarid environments all over the world (Morte et al. 2021). These truffles create ectomycorrhizal relationships with plants such as Helianthemum and Cistus, which help to cycle nutrients in their environments. They are highly appreciated for their culinary and medicinal characteristics, which improve the host plant roots' ability to withstand dry seasons (Bell et al. 2022). Another species, Agaricus campestris, also known as the field mushroom, has been reported in several districts of Punjab, Pakistan. This mushroom is recognized for its edibility and is found in a variety of environments, including desert places (Iqbal et al. 2023). In addition, Pleurotus opuntiae is an edible mushroom species that grows on the remnants of Opuntia plants in central Mexico's semiarid climate. It is considered a delicacy in Mexico's local cuisine (Morales-Flores et al. 2022).

1.6 Cultural and Economic Significance of Edible Mushrooms

Mushrooms have held enormous cultural value across diverse communities worldwide, providing not just as a food source but also as vital parts of traditional medicine, rituals, and folklore (Vunduk and Biketova 2022). Edible mushrooms were treasured in ancient Egypt as symbols of immortality and were thought to be a gift from the god Osiris himself. Because of their distinct flavor, they were considered a delicacy reserved for the nobility. Similarly, the ancient Greeks equated mushrooms with increased power, believing they gave warriors energy during combat. Mushrooms were revered as the "Food of the Gods" in Roman society. For decades, Chinese culture has regarded mushrooms as an essential component of health, frequently referring to them as a "elixir of life" due to their therapeutic capabilities (Hou et al. 2019). Mushrooms are strongly established in local traditions in various indigenous communities, where they are valued for their nutritional and therapeutic benefits. For example, among the Zapotecs of Oaxaca, Mexico, wild edible mushrooms are not only a dietary staple but also play a part in cultural identity and heritage (Garibay-Orijel et al. 2007). Mushrooms such as Lingzhi (Ganoderma lucidum) have long been valued in East Asian cultures, particularly in China and Japan, for their reported health advantages, and they are frequently connected with longevity and spiritual power (Wachtel-Galor et al. 2004; Ekiz et al. 2023). These mushrooms are commonly featured in art and essential to various ancient medical treatments. Similarly, throughout European cultures, particular mushrooms have been tied to folklore and appear in numerous cultural tales, indicating everything from magic to nourishment (Hou et al. 2019).

Edible mushrooms greatly impact the global economy, contributing to food security, income production, and sustainable farming methods (Jayaraman et al. 2024). In China, local farmers receive up to 62% of their cash revenue from the mushroom industry (export), highlighting the significant economic impact of mushroom production in rural areas (Thakur 2020a; Ullah et al. 2022). The global mushroom market has grown steadily, driven by rising consumer demand for nutritional and functional meals. This spread has led to employment creation and economic development in both urban and rural regions, with mushroom growing practiced in over 80 different countries worldwide (Fongnzossie et al. 2020; Globe Newswire 2020). In addition to their direct economic benefits (cultural, medical, and nutritional applications), mushroom production contributes to environmental sustainability by utilizing agricultural waste products as substrates, thus reducing waste while promoting recycling (Gao et al. 2023; Viriato et al. 2024).

1.7 Nutritional and Medicinal Values of Edible Mushrooms

Mushrooms have been consumed for over 2000 years, because of their dietary and therapeutic properties (Pérez-Montes et al. 2021). They are particularly beneficial in vegetarian diets because they contain almost necessary amino acids and have more protein than compared to many vegetables. Edible mushrooms have applications other than nutrition, such as pharmaceuticals, nutraceuticals, and cosmetics (Pashaei et al. 2024a). They contain bioactive compounds with anti-tumor, immunomodulatory, cardioprotective, hepatoprotective, hypocholesterolemic, antiviral, antibacterial, antiparasitic, and antidiabetic properties (Al Qutaibi and Kagne 2024).

Mushrooms are high in carbohydrates, proteins, fiber, and chitin, but less in fat contents and net calories (Pérez-Montes et al. 2021). They contain essential trace elements which makes them suitable for those with high cholesterol due to their less fat levels (Pashaei et al. 2024a, b; Słyszyk et al. 2024). Mushrooms also contain vitamins A, C, D, and B-complex (for example, B1, B2, B3, and B5), with vitamin C and D levels up to ten times higher than vitamin B3 (Tagkouli et al. 2020; Assemie and Abaya 2022). Mushrooms contain a variety of secondary metabolites (such as steroids, phenolics, quinones, terpenoids, variegatic acid, carotenoids, polyketides, and flavonoids) (Bhambri et al. 2022). Phenolics and flavonoids play important roles in antioxidant properties by acting as quenchers of singlet oxygen and strong reducing agents (Liuzzi et al. 2023). Mushrooms' medicinal value is supported by the presence of proteins, dietary fiber, carbohydrates, lipids, minerals, and trace amounts of cholesterol (Pérez-Montes et al. 2021; Ayimbila and Keawsompong 2023; Pashaei et al. 2024a).

1.7.1 Energy

Raw white mushrooms comprise about 22 kcal per 100 grams, however veal has 172 kcal, chicken has 239 kcal, and pulses have 116 kcal (Frassetto and Sebastian 2019).

1.7.2 Protein

Mushrooms are nutritionally significant primarily because they contain protein. These bioactive proteins enhance the functional properties of mushrooms and have significant medicinal properties. They produce a wide range of proteins with fundamental biological functions, including laccases, ribonucleases antibacterial proteins, lectins, ribosome-inactivating proteins, and fungal immunomodulatory proteins. These proteins are essential for immune modulation, antimicrobial defense, and enzymatic processes, which contribute to mushrooms' health benefits (Xu et al. 2011). Edible mushrooms are generally high in total protein content, which change significantly by depending on the developmental stage of the fruiting bodies as well as type of species (Kumar et al. 2021a). Edible mushrooms are an excellent protein source, with approximately 200–250 g/kg of dry biomass. They are especially high in different amino acids such as valine, glutamic acid, glutamine, aspartic acid, and leucine, which enhance their functional and nutritional value (Pashaei et al. 2024b). The proteins of Pleurotus species of mushrooms' are highly valued for their balanced profile of essential and non-essential amino acids, including gamma-aminobutyric acid (GABA), an important neurotransmitter (Thu et al. 2020). The protein content of mushrooms is approximately 2.25 g, which is comparatively lower than veal (24 g), chicken (27 g), and lentils (9 g) per 100 g (Frassetto and Sebastian 2019).

1.7.3 Lipids

Mushroom lipids (Crude fat) include a wide range of complex molecules, such as phospholipids, sterols, free fatty acids, glycerides, and sterol esters. Edible mushrooms are a low-calorie food source that contains little fat (4–6%) and no cholesterol (Kaliyaperumal et al. 2018). Agaricus bisporus has a total fat content of 1.66% to 2.2% by dry weight (Assemie and Abaya 2022). Oleic acid, palmitic acid, Linoleic acid are important fatty acids in edible mushrooms, with linoleic acid being particularly beneficial in lowering blood lipid levels and relieving arthritis symptoms (Podkowa et al. 2021). The main sterol found in mushrooms is ergosterol, which has strong antioxidant qualities (Guillamón et al. 2010), and sterol-rich diets are linked to a lower risk of cardiovascular disease (Kalač 2013). In comparison, mushrooms have 0.3 g of total fat per 100 g, which is much less than that of lentils (0.4 g), veal (8 g), and chicken (14 g) (Frassetto and Sebastian 2019).

1.7.4 Fiber

Many edible mushrooms are underutilized as potential sources of dietary fiber, even with their nutritional value. Mushrooms' crude fiber content varies significantly between species, reflecting their diverse biochemical composition. Pleurotus species, for example, have crude fiber contents that range between 7.4% to 29.75% on a dry matter basis (Ritota and Manzi 2019). In contrast, Volvariella volvacea has fiber levels ranging from 3.24% to 8.70% (Okoro and Achuba 2012). The majority of mushrooms consist of approximately 40% crude fiber on a dry matter basis, except Craterellus aureus and Sarcodon aspratus, which have relatively low fiber contents (averaging 5 g per 100 g of dry weight). The increase in the consumption of wild mushrooms is largely due to their high bioactive compound content (Fernandes et al. 2015). Comprehensive research has been carried out to check the dietary fiber composition of mushrooms. For example, an analysis of eleven mushroom species found that total dietary fiber content ranged from 27% to 44% of dry weight, with insoluble fiber comprising 25–41% and soluble fiber ranging from 0.6% to 2.8% (Hussain et al. 2020). Similarly, an examination of twenty mushroom species revealed total dietary fiber contents ranging from 27% to 36%, with insoluble fiber ranging from 12% to 21% and soluble fiber from 2% to 4% of dry weight (Nile and Park 2014).

1.7.5 Carbohydrates

Edible mushrooms are known as excellent source of carbohydrates, specifically glucose and mannitol, with little content of fructose as well as sucrose (Pk et al. 2020). They are also high in dietary fiber, specifically non-starch polysaccharides. Insoluble dietary fibers are particularly abundant in the stems. Mushrooms contain an average of soluble fibers (4–9%) and insoluble fiber (22–30%). According to reported studies, 100 g of P. ostreatus contains approximately 4.1 g of dietary fiber (Ma et al. 2020).

1.7.6 Mushroom-Derived Phenolic Compounds

Many mushrooms show anti-inflammatory effects due to specific phenolic compounds, like as hydroxycinnamic acids, hydroxybenzoic acids, flavonoids, lignans, stilbenes, tannins, polyphenols, and phenolic acids (Côté et al. 2010; D'Archivio et al. 2010). A quantitative study of phenolic and flavonoid contents in eight edible mushroom species found phenolic concentrations ranging from 1 to 6 mg/g of dry weight and flavonoid levels vary from 0.9–3.0 mg/g. The most common flavonoids were found to be myricetin and catechin. Among these species, A. bisporus, B. edulis and had the maximum phenolic content, while L. deliciosus contained the most flavonoids. P. ostreatus, C. gambosa, and A. bisporus had significantly lower phenolic levels (Palacios et al. 2011).

1.7.7 Vitamins

Edible mushroom fruiting bodies are high in vitamins, like tocopherols (E), riboflavin (B2), ascorbic acid (C), ergocalciferol (D2), and thiamine (B1). Thiamine, pyridoxine, riboflavin, pantothenic acid, folate, nicotinamide, niacin, and cobalamin are the most frequent B vitamins found. Pleurotus ostreatus is especially high in folate, thiamine, and niacin, making it a nutritionally dense species (Tagkouli et al. 2020). Wild mushrooms are naturally high in ergocalciferol, however, cultivated species needed UV-B light exposure to synthesis this vitamin because they often grow in low-light conditions (Cardwell et al. 2018). Furthermore, mushrooms, especially P. ostreatus, have 5–10 times greater niacin levels than vegetables (Pashaei et al. 2024b). They can also bioaccumulate selenium when grown in selenium-rich substrates, which increases their nutritional value (Kula et al. 2011).

1.7.8 Mineral Composition

Edible mushrooms contain essential minerals elements, including macro-elements such as calcium, phosphorus, magnesium, sodium, and potassium, in addition, trace elements are also a copper, molybdenum, iron, cadmium, and zinc (Gałgowska and Pietrzak-Fiećko 2020; Dhull 2022). The mineral content of mushrooms can range from 80 to 120 g/kg of dry matter, depending on the species, developmental stage, fruiting body size, and cultivation conditions (Temesgen 2018). In addition, mushrooms can bio-accumulate different heavy metals like mercury, silver, chromium, lead, copper, arsenic, cadmium, and nickel from contaminated environments, particularly in soils or water with elevated pollution levels (Chen et al. 2009; Ediriweera et al. 2022). While bioaccumulation is a concern for human health, the mushroom's growing conditions can influence mineral content and potential heavy metal accumulation (Esmaeili et al. 2022). Wild mushrooms typically contain more minerals than those grown under controlled conditions (Temesgen 2018).

1.8 Ecological Significance of Edible Mushrooms

The functional and structural variations of edible mushrooms have a significant impact on biodiversity and ecological processes. Fungi, as heterotrophic organisms, lack the ability to photosynthesize and therefore depend on other nutritional strategies. They evolved three main trophic modes: saprotrophic, parasitic, and symbiotic. Saprotrophic fungi break down dead organic matter from plants, animals, and other organisms.

1.8.1 Degradation of Debris and Cycling of Nutrient

These fungi perform a key role in ecosystem functions, particularly nutrient cycling between natural and man-made environments. Basidiomycetes, the most common fungi, play a significant economic and ecological role. Of the 2180 edible mushroom species documented worldwide, approximately 2012 species (92.29%) are Basidiomycetes, with the remaining 168 species (7.71%) belonging to the Ascomycetes group (Li et al. 2021a). During the Carboniferous period, around 300 million years ago, ecosystems faced a critical challenge due to the accumulation of lignocellulosic plant debris, which was primarily made up of lignin and cellulose. At the time, no organisms can degrade these complex polymers, posing a risk to ecological stability. However, a mutation occurred among Basidiomycetes fungi during early lineage are resulting in the production of peroxidase enzymes, allowing the breakdown of lignocellulosic compounds. This enzymatic activity aided in the recycling of plant biomass, thereby ensuring global ecosystem function (Floudas et al. 2012). At that time, the ancestors of Basidiomycetes underwent significant functional and structural diversification, as evidenced by their modern forms. Basidiomycetous fungi continue the lignin degradation in natural ecosystems, demonstrating a "evolutionary footprint," or a trait that remains due to old genetic modifications (Nagy et al. 2016; Ayuso-Fernández et al. 2018).

1.8.2 Mycorrhizal Function

Mycorrhizal fungi are the third trophic group, and they form mutualistic symbioses with plants, helping to shape forest ecosystems and support global biodiversity. These fungi absorb carbon compounds from their plant hosts while facilitating the uptake of essential nutrients and water (both macronutrients and micronutrients) from the soil in exchange. According to Li et al., (2021a, b), there are 2180 species of edible mushrooms reported worldwide, with 1181 being ectomycorrhizal. Mycorrhizal associations are common, affecting about 5000 different plant species, such as angiosperms and gymnosperms that dominate tropical, subtropical, temperate, and boreal biomes. Mycorrhizal fungi mycelium infiltrates root cortical cells, spreads outward to form fungal mantles, and forms extensive networks of mycelial that link trees of varying ages throughout the forest ecosystem (Beiler et al. 2010).

The networks of Mycorrhiza, also known as the *"wood wide web"* serve as an underground route for the exchange of essential resources such as water, carbon, nitrogen, and minerals (Rhodes 2017). These networks also enable inter-plant signaling, facilitate the transfer of defense compounds and allelochemicals, and increase communication between individual trees. Additionally, they support resource sharing across tree species (Angiosperms and Gymnosperms), including complex interactions between forest ecosystems (Tedersoo et al. 2020). The majority of economically important edible mushrooms are ectomycorrhizal fungi, such as chanterelles, truffles, Caesar's, porcini, and matsutake mushrooms (Reis et al. 2012a; Diego and Pardo-Giménez 2017).

1.9 Edible Mushroom Cultivation

Form millennia, mushrooms have been used as traditional remedies in different countries like China, Japan, Africa, and the Middle East. Initially gathered in the wild, large-scale cultivation began in the early twentieth century with advances in pure culture techniques. Mushroom farming is now a sustainable agricultural approach that uses agricultural, industrial, and forestry waste to produce nutritious food (Arora 2008; Chang and Wasser 2017). Mushrooms break down lignin, making cellulose and hemicellulose available to other species, which helps contribute to sustainable agriculture. Mushroom production improves food security, decreases poverty, and promotes long-term livelihoods, particularly among small farmers. It takes little land, investment, and work, making it affordable to rural and semi-urban communities. Over 110,000 fungal species have been found, with 16,000 categorized as mushrooms (Ashton 2009; Wasser 2010; Hawksworth 2012). Of these, 3000 species are edible, 700 have therapeutic characteristics, and approximately 100 are economically grown (Miles and Chang 2004; Chang 2006; Wasser 2010; Chang and Wasser 2017). Popular edible mushrooms include *Volvariella* spp., *Pleurotus* spp., *A. bisporus,* and *L. edodes*. Mushroom cultivation has expanded dramatically since World War II, with mechanization and automation driving global production at a 6–7% annual growth rate. Currently, Butten, oyster, shiitake, and other mushrooms contribute 38.8%, 29.9%, 17.9%, 11.9%, respectively, to global mushroom production (Deb 2024). The mushroom business has spread to over 100 countries, boosting economies, health, and social structures, especially in India, China, and developing countries. The cultivation and expansion of the mushroom industry have made a significant contribution to economic growth, with mushroom farming and its byproducts improving human welfare. This global phenomenon, especially in the twenty-first century, can be described as a "non-green revolution" because of its long-term impact on food security and economic development.

1.9.1 Important Steps in Edible Mushroom Cultivation

Mushroom cultivation, once a rudimentary practice, has grown into a large-scale industrial activity due to its recognized health benefits and economic importance. As a result of this transformation, many countries now produce hundreds of thousands of tons per year. However, small-scale production remains an important component of local markets in developing countries, contributing to both economic growth and food security (Thakur 2020a; Gupta et al. 2022). In either case, cultivation aims to increase yield in a short period of time. This necessitates a thorough understanding of how to select high-yielding strains and media for spawn production, as well as improved mushroom bed management, including pest and disease control (Mamiro and Royse 2008). Mushroom farming's sustainability depends on a consistent supply and effective marketing. Successful cultivation requires a thorough understanding of scientific principles, practical experience, and ongoing training to ensure optimal production results (Thakur 2020b; Olawale et al. 2024). The cultivation of mushrooms generally consists of six stages (Fig. 1.3) (Miles and Chang 2004; Kratika 2018):

For successful production of edible mushrooms, a variety of strategies must be implemented. One of the most important is to choose mushroom strains with high market demand and economic value. To maximize yield and quality, factors such as cultivation substrate availability, maintenance costs, and the impact of environmental effects on the growth of mushrooms must also be carefully considered (Gupta et al. 2018; Silva et al. 2024). Under appropriate conditions, edible mushroom cultures can produce fruiting bodies. The substrate composition and the fungal strain's genetic makeup have the greatest influence on the quality of mushroom spawn.

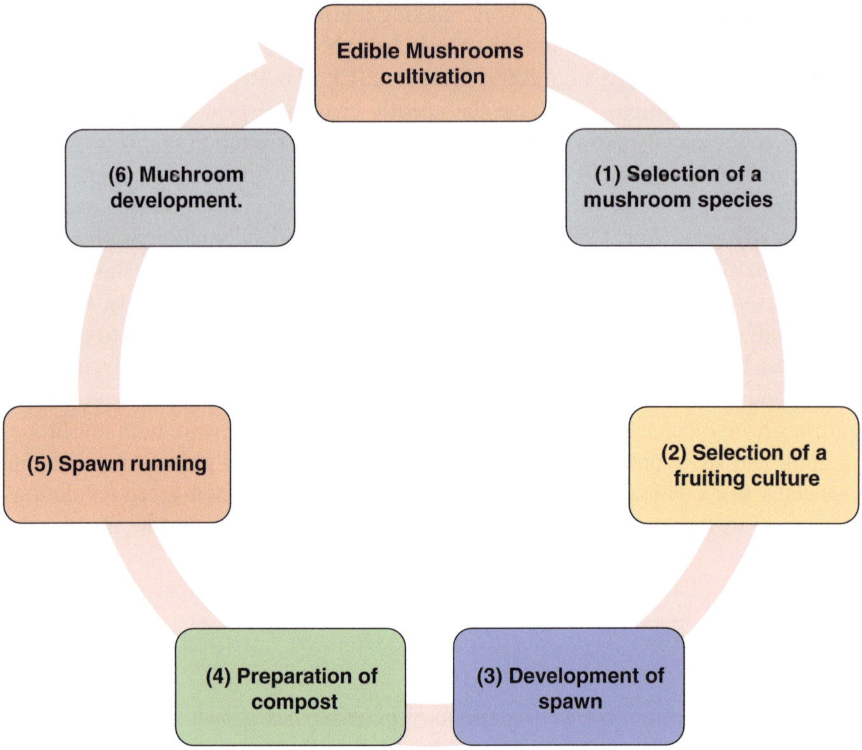

Fig. 1.3 Different phases of edible mushroom cultivation

Synthetic compost is composed of chemical and agricultural materials other than animal dung and is widely used for mushroom cultivation. While mycelium grows quickly with more spawn, production costs may rise. Temperature, humidity, pH, and aeration requirements vary by stage of mushroom production. A. bisporus, L. edodes, and different species of Pleurotus genus contributed approximately three-quarters (75%) of all edible mushrooms cultivated. Both large and small scale industrial mushroom cultivation are economically viable, and mushrooms have high nutritional value, as evidenced by their widespread cultivation in Africa, Brazil, China, and Mexico (Dimopoulou et al. 2022; Niego et al. 2023). The cultivation of ectomycorrhizal mushrooms is still evolving, with advances in techniques such as the cultivation of truffle species. Several edible mushroom species have been successfully cultivated globally including *Agaricus campestris, Pluteus cervinus, Morchella esculenta, Flammulina velutipes, Agaricus blazei, Polyporus indigenus, Coprinus comatus, Lentinula tigrinus, Volvariella volvacea, Pholiota cystidiosus, Amanita brunnescens, Pleurotus rhodophyllus, Volvariella bombycina, Agaricus augustus, Pleurotus ostreatus, Lentinula edodes, Agaricus bisporus, Morchella angusticeps, Pleurotus eryngii, Lentinus strigosus, Agaricus arvensis, Pluteus saporema, Agaricus subrufescens, Pleurotus pulmonarius, Auricularia judae,*

Pleurotus cornucopiae, Agaricus bitorquis, Laetiporus sulphureus, Auricularia fuscosuccinea, Lentinula tuber-regium, Auricularia polytricha, and *Pleurotus euosmus*. These species are widely cultivated due to their nutritional and economic value, which benefits both local and global markets (Carrasco et al. 2021; Li et al. 2021a; Balan et al. 2022; Ferraro et al. 2022; Li and Xu 2022).

1.9.2 Challenges in Edible Mushroom Cultivation

Globally, insufficient knowledge of cultivation methods and food safety problems present significant challenges to edible mushroom farming. Many small-scale farmers lack the technical expertise required for efficient mushroom cultivation, impeding the adoption of best practices. According to research conducted in 2024, approximately 45% of small-scale producers are unfamiliar with advanced cultivation methods, limiting the mushroom industry's growth and innovation potential (Kant Raut 2019; Deb 2024; Shankar et al. 2024).

1.9.3 Global Market Overview of Edible Mushrooms

The global edible mushroom market has grown significantly in recent years. According to the information provided by Food and Agriculture Organization (FAO 2024a), the mushroom and truffle production increased by more than 34% between 2012 and 2024, demonstrating the market's growing importance in these years. In 2022, the top 10 edible mushroom producers were concentrated in Asia, Europe, and North America. China led global production with over 45 million metric tons, followed by Japan, which produced around 470,000 metric tons (FAO 2024b).

1.10 Future Research Directions in Edible Mushroom

To meet the growing global food demand, with a projected population of nine billion by 2050, global food production must increase by 50%. Over the last three decades, mushroom cultivation has grown at a 10% annual rate around the world. As a result, mushrooms offer a promising opportunity to improve food security, safety, and quality. Despite their potential, research into the bioactive components of wild and cultivated edible mushrooms is still limited. Mushrooms contain numerous bioactive compounds, both traditional and emerging, that provide significant nutraceutical and health benefits. Investigating the different mechanisms of action performed by extracts of various mushroom will shed light on the roles of various phytochemicals in preventing and treating degenerative diseases. More research into cultivation practices, genetic diversity, farmer education, transportation and preservation techniques, and biotechnological advancements will help mushrooms become functional foods and sustainable agricultural products.

1.11 Conclusion

Edible mushrooms play an important role in global food systems, providing numerous nutritional, medicinal, economic, and environmental benefits. Their biodiversity ranges across ecosystems, including tropical, subtropical, temperate, arid, semiarid, and Polar Regions, with species distribution influenced by climate and environmental factors. The L. edodes, A. bisporus, Volvariella, and Pleurotus spp. are well-known species that dominate global commercial production due to their nutritional value. Morphological and molecular identification techniques have advanced significantly, allowing for more accurate species classification and safety assessments. Furthermore, the integration of genomic technologies has resulted in more accurate species identification, reducing errors in distinguishing between closely related edible mushrooms. Sustainable mushroom cultivation helps to improve food security, recycle waste, generate income, and protect the environment. As global demand rises, mushroom farming helps to alleviate poverty, improve health, and promote sustainable agricultural practices around the world.

References

Abou Raya MA, Shalaby MT, Hafez SA, Hamouda AM. Chemical composition and nutritional potential of some mushroom varieties cultivated in Egypt. J Food Dairy Sci. 2014;5:421–34.

Acar İ, Blando F, Gul B, Greco A, Mukemre M, Uzun Y, Dalar A. The phenolic profile and biological activities of the wild-edible mushrooms Helvella leucopus and Morchella pulchella. Food Measure. 2021;15:555–66.

Acharya K, Khatua S. Ganoderma: cultivation, chemistry, and medicinal applications, volume 2. CRC Press; 2024.

Acharya K, Ghosh S, Kundu I. Pharmacognostic standardization of a well known edible mushroom Volvariella volvacea. J App Pharm Sci. 2016;6:185–90.

Acharya K, Das K, Paloi S, Dutta A, Hembrom M, Khatua S, Parihar A. Exploring a novel edible mushroom Ramaria subalpina: chemical characterization and antioxidant activity. Pharmacogn J. 2017;9:30–4.

Adeniyi M, Odeyemi Y, Odeyemi O. Ecology, diversity and seasonal distribution of wild mushrooms in a Nigerian tropical forest reserve. Biodiversitas. 2018;19:285–95.

Adhikari MK, Devkota S, Tiwari RD. Ethnomycolgical knowledge on uses of wild mushrooms in Western and Central Nepal. Our Nature. 2005;3:13–9.

Afiukwa C, Ebem EC, Igwe D. Characterization of the proximate and amino acid composition of edible wild mushroom species in Abakaliki, Nigeria. 2015.

Agrahar-Murugkar D, Subbulakshmi G. Nutritional value of edible wild mushrooms collected from the Khasi hills of Meghalaya. Food Chem. 2005;89:599–603.

Águeda B, Parladé J, Fernández-Toirán LM, Cisneros Ó, De Miguel AM, Modrego MP, Martínez-Peña F, Pera J. Mycorrhizal synthesis between Boletus edulis species complex and rockroses (Cistus sp.). Mycorrhiza. 2008a;18:443–9.

Águeda B, Parladé J, Fernández-Toirán LM, Cisneros Ó, de Miguel AM, Modrego MP, Martínez-Peña F, Pera J. Mycorrhizal synthesis between Boletus edulis species complex and rockroses (Cistus sp.). Mycorrhiza. 2008b;18:443–9.

Aguilar Doroteo L, Zárate Segura PB, Villanueva Arce R, Yáñez Fernández J, Garín Aguilar ME, Guadarrama Mendoza PC, Valencia Del Toro G. Use of ITS and ISSR markers in the molecular characterisation of Pleurotus djamor hybrid strains. Rev Iberoam Micol. 2018;35:49–55.

Ahlawat OP, Arora B. Paddy straw mushroom (Volvariella volvacea) cultivation. Solan: National Research Centre for Mushroom (ICAR); 2016. p. 165–82.

Ahmad S, Chaudhry B, Chaudhry I, Chaudhry M. Fungi of Pakistan. Khalid Chaudhry; 2014.

Ahmed AF, Mahmoud GA-E, Hefzy M, Liu Z, Ma C. Overview on the edible mushrooms in Egypt. J Future Foods. 2023;3:8–15.

Aida FMNA, Shuhaimi M, Yazid M, Maaruf AG. Mushroom as a potential source of prebiotics: a review. Trends Food Sci Technol. 2009;20:567–75.

Aisala H, Sola J, Hopia A, Linderborg KM, Sandell M. Odor-contributing volatile compounds of wild edible Nordic mushrooms analyzed with HS–SPME–GC–MS and HS–SPME–GC–O/FID. Food Chem. 2019;283:566–78.

Aisala H, Manninen H, Laaksonen T, Linderborg KM, Myoda T, Hopia A, Sandell M. Linking volatile and non-volatile compounds to sensory profiles and consumer liking of wild edible Nordic mushrooms. Food Chem. 2020;304:125403.

Ajith TA, Janardhanan KK. Indian medicinal mushrooms as a source of antioxidant and antitumor agents. J Clin Biochem Nutr. 2007;40:157–62.

Akgul H, Sevindik M, Coban C, Alli H, Selamoglu Z. New approaches in traditional and complementary alternative medicine practices: <em>Auricularia auricula </em>and<em> Trametes versicolor</em>. J Tradit Med Clin Naturop. 2017;2017.

Akyuz M, Kirbag S. Antimicrobial activity of Pleurotus eryngii var. ferulae grown on various agro-wastes. Eurasia J Biosci. 2009:58–63.

Al Qutaibi M, Kagne SR. Exploring the phytochemical compositions, antioxidant activity, and nutritional potentials of edible and medicinal mushrooms. Int J Microbiol. 2024;2024:1–23.

Alanne A-L, Issakainen J, Pihlaja K, Jokioja J, Sinkkonen J. Metabolomic discrimination of the edible mushrooms *Kuehneromyces mutabilis* and *Hypholoma capnoides* (Strophariaceae, Agaricales) by NMR spectroscopy. Zeitschrift für Naturforschung C. 2019;74:201–10.

Alem D, Dejene T, Oria-de-Rueda JA, Martín-Pinto P. Survey of macrofungal diversity and analysis of edaphic factors influencing the fungal community of church forests in Dry Afromontane areas of Northern Ethiopia. For Ecol Manag. 2021;496:119391.

Alhazmi H. Extraction of phytochemicals betulin and betulinic acid from the chaga mushroom and their effect on MCF-7 Cells. Thesis. 2018.

Ali S, Yousaf N, Usman M, Javed MA, Nawaz M, Ali B, Azam M, Ercisli S, Tirasci S, Ahmed AE. Volvariella volvacea (paddy straw mushroom): a mushroom with exceptional medicinal and nutritional properties. Heliyon. 2024;10:e39747.

Allen S, Bharadwaj G, Chakma JK. Genus *Agaricus* species found in the Himalayas: *Agaricus campestris* L. and *Agaricus augustus* Fr. In: Sharma A, Bhardwaj G, Nayik GA, editors. Phytochemistry and nutritional composition of significant wild medicinal and edible mushrooms. Royal Society of Chemistry; 2023. p. 31–60.

Alofe FV, Odeyemi O, Oke OL. Three edible wild mushrooms from Nigeria: their proximate and mineral composition. Plant Foods Hum Nutr. 1996;49:63–73.

Alsheikh AM, Trappe JM. Desert truffles: The genus *Tirmania*. Trans Br Mycol Soc. 1983;81:83–90.

Alvarado P, Moreau P-A, Dima B, Vizzini A, Consiglio G, Moreno G, Setti L, Kekki T, Huhtinen S, Liimatainen K, Niskanen T. Pseudoclitocybaceae fam. nov. (Agaricales, Tricholomatineae), a new arrangement at family, genus and species level. Fungal Divers. 2018;90:109–33.

Álvarez-Farias Z. Ethnomycological knowledge of wild edible mushrooms in Tlayacapan, Morelos. Mycosphere. 2016;7:1491–9.

Amend A, Keeley S, Garbelotto M. Forest age correlates with fine-scale spatial structure of Matsutake mycorrhizas. Mycol Res. 2009;113:541–51.

Amin SR, Sarker NC, Khair A, Alam N. Detection of novel supplements of paddy straw substrate on oyster mushroom cultivation. 2007.

Ammirati JF. Poisonous Mushrooms of the Northern United States and Canada. U of Minnesota Press; 1985.

Antil S, Abraham JS, Sripoorna S, Maurya S, Dagar J, Makhija S, Bhagat P, Gupta R, Sood U, Lal R, Toteja R. DNA barcoding, an effective tool for species identification: a review. Mol Biol Rep. 2023;50:761–75.

Antonelli A, Smith RJ, Fry C, Simmonds MSJ, Kersey PJ, Pritchard HW, Abbo MS, Acedo C, Adams J, A. A.M., A. B., A. W, B. S P, Bacon K, Bárrios S, Barstow C, Battison A, Bell E, Bensusan K, Bidartondo MI, Blackhall-Miles RJ, Borrell JS, Brearley FQ, Breman E, Brewer RFA, Brodie J, Cámara-Leret R, Campostrini Forzza R, Cannon P, Carine M, Carretero J, Cavagnaro TR, Cazar M-E, Chapman T, Cheek M, Clubbe C, Cockel C, Collemare J, Cooper A, Copeland AI, Corcoran M, Couch C, Cowell C, Crous P, da Silva M, Dalle G, Das D, David JC, Davies L, Davies N, de Canha MN, de Lirio EJ, Demissew S, Diazgranados M, Dickie J, Dines T, Douglas B, Dröge G, Dulloo ME, Fang R, Farlow A, Farrar K, Fay MF, Felix J, Forest F, Forrest LL, Fulcher T, Gafforov Y, Gardiner LM, Gâteblé G, Gaya E, Geslin B, Gonçalves SC, Gore CJN, Govaerts R, Gowda B, Grace OM, Grall A, Haelewaters D, Halley JM, Hamilton MA, Hazra A, Heller T, Hollingsworth PM, Holstein N, Howes M-JR, Hughes M, Hunter D, Hutchinson N, Hyde K, Iganci J, Jones M, Kelly LJ, Kirk P, Koch H, Krisai-Greilhuber I, Lall N, Langat MK, Leaman DJ, Leão TC, Lee MA, Leitch IJ, Leon C, Lettice E, Lewis GP, Li L, Lindon H, Liu JS, Liu U, Llewellyn T, Looney B, Lovett JC, Łuczaj Ł, Lulekal E, Maggassouba S, Malécot V, Martin C, Masera OR, Mattana E, Maxted N, Mba C, Mcginn KJ, Metheringham C, Miles S, Miller J, Milliken W, Moat J, Moore PGP, Morim MP, Mueller GM, Muminjanov H, Negrão R, Nic Lughadha E, Nicolson N, Niskanen T, Nono Womdim R, Noorani A, Obreza M, O'donnell K, O'hanlon R, Onana J-M, Ondo I, Padulosi S, Paton A, Pearce T, Pérez Escobar OA, Pieroni A, Pironon S, Prescott TAK, Qi YD, Qin H, Quave CL, Rajaovelona L, Razanajatovo H, Reich PB, Rianawati E, Rich TCG, Richards SL, Rivers MC, Ross A, Rumsey F, Ryan M, Ryan P, Sagala S, Sanchez MD, Sharrock S, Shrestha KK, Sim J, Sirakaya A, Sjöman H, Smidt EC, Smith D, Smith P, Smith SR, Sofo A, Spence N, Stanworth A, Stara K, Stevenson PC, Stroh P, Suz LM, Tambam BB, Tatsis EC, Taylor I, Thiers B, Thormann I, Trivedi C, Twilley D, Twyford AD, Ulian T, Utteridge T, Vaglica V, Vásquez-Londoño C, Victor J, Viruel J, Walker BE, Walker K, Walsh A, Way M, Wilbraham J, Wilkin P, Wilkinson T, Williams C, Winterton D, Wong KM, Woodfield-Pascoe N, Woodman J, Wyatt L, Wynberg R, Zhang BG. State of the World's Plants and Fungi. Royal Botanic Gardens (Kew): Sfumato Foundation; 2020.

Antonín V. Marasmius Heinemannianus, a New Edible Species from Benin, West Africa. Belg J Bot. 1998;131:127–32.

Ao T, Seb J, Ajungla T, Deb CR. Diversity of Wild Mushrooms in Nagaland, India. OJF. 2016a;06:404–19.

Ao T, Seb J, Ajungla T, Deb CR. Diversity of Wild Mushrooms in Nagaland, India. OJF. 2016b;06:404–19.

Aoki W, Bergius N, Kozlan S, Fukuzawa F, Okuda H, Murata H, Ishida TA, Vaario L-M, Kobayashi H, Kalmiş E, Fukiharu T, Gisusi S, Matsushima K, Terashima Y, Narimatsu M, Matsushita N, Ka K-H, Yu F, Yamanaka T, Fukuda M, Yamada A. New findings on the fungal species Tricholoma matsutake from Ukraine, and revision of its taxonomy and biogeography based on multilocus phylogenetic analyses. Mycoscience. 2022;63:197–214.

Arora D. Notes on Economic Mushrooms, Xiao Ren Ren: the "Little People" of Yunnan. Econ Bot. 2008;62:540–4.

Arora D, Dunham SM. A new, commercially valuable Chanterelle Species, Cantharellus californicus sp. nov., associated with Live Oak in California, USA. Econ Bot. 2008;62:376–91.

Ashton H. Ainsworth and Bisby's Dictionary of the Fungi (10th edition). Ref Rev. 2009;23:42.

Assemie A, Abaya G. The effect of edible mushroom on health and their biochemistry. Int J Microbiol. 2022;2022:1–7.

Athanasakis G, Aligiannis N, Gonou-Zagou Z, Skaltsounis A-L, Fokialakis N. Antioxidant properties of the wild edible mushroom Lactarius salmonicolor. J Med Food. 2013;16:760–4.

Atila F, Nadhim Owaid M, Ali Shariati M. The nutritional and medical benefits of agaricus bisporus: a review. J Microbiol Biotechnol Food Sci. 2017;7:281–6.

Ayaz FA, Torun H, Özel A, Col M, Duran C, Sesli E, Colak A. Nutritional value of some wild edible mushrooms from Black Sea Region (Turkey). Turk J Biochem. 2011;36:213–21.

Ayimbila F, Keawsompong S. Nutritional quality and biological application of mushroom protein as a novel protein alternative. Curr Nutr Rep. 2023;12:290–307.

Ayuso-Fernández I, Ruiz-Dueñas FJ, Martínez AT. Evolutionary convergence in lignin-degrading enzymes. Proc Natl Acad Sci USA. 2018;115:6428–33.

Azam MS, Rahman MRT, Lou Z, Tang Y, Raqib SM, Jothi JS. Review: advancements and application of immunosensors in the analysis of food contaminants. Nusantara Biosci. 2018:6.

Balan V, Zhu W, Krishnamoorthy H, Benhaddou D, Mowrer J, Husain H, Eskandari A. Challenges and opportunities in producing high-quality edible mushrooms from lignocellulosic biomass in a small scale. Appl Microbiol Biotechnol. 2022;106:1355–74.

Bandara A. A review of the polysaccharide, protein and selected nutrient content of Auricularia, and their potential pharmacological value. Mycosphere. 2019;10:579–607.

Bandara AR, Rapior S, Bhat DJ, Kakumyan P, Chamyuang S, Xu J, Hyde KD. *Polyporus umbellatus*, an edible-medicinal cultivated mushroom with multiple developed health-care products as food, medicine and cosmetics: a review. Cryptogam Mycol. 2015;36:3–42.

Bao D, Gong M, Zheng H, Chen M, Zhang L, Wang H, Jiang J, Wu L, Zhu Y, Zhu G, Zhou Y, Li C, Wang S, Zhao Y, Zhao G, Tan Q. Sequencing and comparative analysis of the straw mushroom (Volvariella volvacea) genome. PLoS One. 2013;8:e58294.

Becherer L, Borst N, Bakheit M, Frischmann S, Zengerle R, Von Stetten F. Loop-mediated isothermal amplification (LAMP) – review and classification of methods for sequence-specific detection. Anal Methods. 2020;12:717–46.

Beiler KJ, Durall DM, Simard SW, Maxwell SA, Kretzer AM. Architecture of the wood-wide web: Rhizopogon spp. genets link multiple Douglas-fir cohorts. New Phytol. 2010;185:543–53.

Bell V, Silva CRPG, Guina J, Fernandes TH. Mushrooms as future generation healthy foods. Front Nutr. 2022;9:1050099.

Bhambri A, Srivastava M, Mahale VG, Mahale S, Karn SK. Mushrooms as potential sources of active metabolites and medicines. Front Microbiol. 2022;13:837266.

Bhariya SK, Sahu H, Premi V, Sapre B. Molecular diversity of commercially available edible oyster mushroom by SSR Marker. Int J Plant Soil Sci. 2024;36:537–43.

Boa ER. Wild edible fungi: a global overview of their use and importance to people. Rome: Food and Agriculture Organization of the United Nations; 2004.

Boa, E.R. and F. and A.O. of the U. Nations. Wild edible fungi: a global overview of their use and importance to people. Food & Agriculture Org; 2004.

Boddy L, Büntgen U, Egli S, Gange AC, Heegaard E, Kirk PM, Mohammad A, Kauserud H. Climate variation effects on fungal fruiting. Fungal Ecol. 2014;10:20–33.

Boxshall A-G, Birch JL, Lebel T, Symonds MRE, Callahan DL. A field-based investigation of simple phenol variation in Australian Agaricus xanthodermus. Mycologia. 2021;113:1123–35.

Bradai L, Bissati S, Chenchouni H, Amrani K. Effects of climate on the productivity of desert truffles beneath hyper-arid conditions. Int J Biometeorol. 2015;59:907–15.

Bradshaw MJ, Aime MC, Rokas A, Maust A, Moparthi S, Jellings K, Pane AM, Hendricks D, Pandey B, Li Y, Pfister DH. Extensive intragenomic variation in the internal transcribed spacer region of fungi. iScience. 2023;26:107317.

Brazee NJ, Marra RE, Wick RL. Genotypic diversity of *Armillaria gallica* from mixed oak forests in Massachusetts. Mycologia. 2012;104:53–61.
Buba T, Agbo V, Abdullahi A. The ecology of edible mushrooms of the Nigerian savannah: Towards their optimal exploitation. J Appl Biosci. 2019;132:13439.
Bulman SR, Visnovsky SB, Hall IR, Guerin-Laguette A, Wang Y. Molecular and morphological identification of truffle-producing Tuber species in New Zealand. Mycol Prog. 2010;9:205–14.
Bunyard BA. The Little Book of Fungi. Princeton University Press; 2024.
Burrola-Aguilar C, Garibay-Orijel R, Argüelles-Moyao A. Abies religiosa forests harbor the highest species density and sporocarp productivity of wild edible mushrooms among five different vegetation types in a neotropical temperate forest region. Agrofor Syst. 2013;87:1101–15.
Buyck B. The edible mushrooms of Madagascar: an evolving enigma. Econ Bot. 2008;62:509–20.
Cabral TS, Silva BD, Martín MP, Clement CR, Hosaka K, Baseia IG. Behind the veil – exploring the diversity in Phallus indusiatus s.l. (Phallomycetidae, Basidiomycota). MycoKeys. 2019;58:103–27.
Çaglarlrmak N, Ünal K, Ötles S. Nutritional value of edible wild mushrooms collected from the black sea region of turkey. Micol Apl Int. 2002;14:1–5.
Calderaro A, Chezzi C. MALDI-TOF MS: a reliable tool in the real life of the clinical microbiology laboratory. Microorganisms. 2024;12:322.
Cano JM, Berrocal-Lobo M, Domínguez-Núñez JA. Growth of *Amanita caesarea* in the presence of *Pseudomonas fluorescens* and *Bacillus cereus*. Fungal Biol. 2017;121:825–33.
Carbonaro TM, Bradstreet MP, Barrett FS, MacLean KA, Jesse R, Johnson MW, Griffiths RR. Survey study of challenging experiences after ingesting psilocybin mushrooms: acute and enduring positive and negative consequences. J Psychopharmacol. 2016;30:1268–78.
Cardwell G, Bornman JF, James AP, Black LJ. A review of mushrooms as a potential source of dietary vitamin D. Nutrients. 2018;10:1498.
Carrasco J, Zied DC, Navarro MJ, Gea FJ, Pardo-Giménez A. Commercial cultivation techniques of mushrooms. In: Advances in Macrofungi. CRC Press; 2021.
Cazabonne J. The use of human senses in the identification of mushrooms. Field Mycol. 2021;22:117–22.
Chamberlain M. Ethnomycological experiences in South West China. Mycologist. 1996;10:173–6.
Chang S-T. The origin and early development of straw mushroom cultivation. Econ Bot. 1977;31:374–6.
Chang S-T. The world mushroom industry: trends and technological development. IJM. 2006:8.
Chang S. Overview of mushroom cultivation and utilization as functional foods. In: Cheung PCK, editor. Mushrooms as functional foods. Wiley; 2008. p. 1–33.
Chang ST, Wasser SP. The Cultivation and Environmental Impact of Mushrooms. Oxford Research Encyclopedia of Environmental Science/Oxford University Press; 2017.
Chang ST, Buswell JA, Chiu SW. Mushroom biology and mushroom products: proceedings of the first international conference on mushroom biology and mushroom products the Chinese University of Hong Kong, Hong Kong: Chinese University Press. 1993. Available at https://www.google.com.pk/books/edition/Mushroom_Biology_and_Mushroom_Products/fMMsA_2naOEC?hl=en&gbpv=0&kptab=overview.
Chao W, Tang C, Zhang J, Yu L, Yoichi H. Development of a stable SCAR marker for rapid identification of Ganoderma lucidum Hunong 5 cultivar using DNA pooling method and inter-simple sequence repeat markers. J Integr Agric. 2018;17:130–8.
Chathurika KBAB, Siriwardena SER, B. H.r.a.c, P. G.s.m and D. K.v.o. Developing an identification system for different types of edible mushrooms in Sri Lanka using machine learning and image processing. Int J Eng Manag Res. 2023;13:54–9.
Cheek M, Nic Lughadha E, Kirk P, Lindon H, Carretero J, Looney B, Douglas B, Haelewaters D, Gaya E, Llewellyn T, Ainsworth AM, Gafforov Y, Hyde K, Crous P, Hughes M, Walker BE, Campostrini Forzza R, Wong KM, Niskanen T. New scientific discoveries: plants and fungi. Plants, People, Planet. 2020;2:371–88.

Chen Y. Song rong (Tricholoma matsutake), a valuable mushroom from China: consumption, development and sustainability. In: Kusters K, Belcher B, editors. Forest products, livelihoods and conservation: case studies of NTFP systems. Bogor, Indonesia: Center for International Forestry Research; 2004. p. 56–71.

Chen X-H, Zhou H-B, Qiu G-Z. Analysis of Several Heavy Metals in Wild Edible Mushrooms from Regions of China. Bull Environ Contam Toxicol. 2009;83:280–5.

Chen G, Wu F, Pei F, Cheng S, Muinde B, Hu Q, Zhao L. Volatile components of white *Hypsizygus marmoreus* detected by electronic nose and HS-SPME-GC-MS: influence of four drying methods. Int J Food Prop. 2017;20:2901–10.

Chen J, Callac P, Llarena-Hernández RC, Gerardo Mata G. Two species of Agaricus subg. Minoriopsis from Mexico. Phytotaxa. 2019;404:91.

Cho HJ, Lee H, Park MS, Kim C, Wisitrassameewong K, Lupala A, Park KH, Kim MJ, Fong JJ, Lim YW. Macrolepiota in Korea: new records and a new species. Mycobiology. 2019;47:368–77.

Choudhary M, Division of Soil and Crop Management, ICAR–Central Soil Salinity, R. Devi, Division of Soil and Crop Management, ICAR–Central Soil Salinity, A. Datta, Division of Soil and Crop Management, ICAR–Central Soil Salinity Research Institute, Karnal, Haryana-132001, India., A. Kumar, Division of Soil and Crop Management, ICAR–Central Soil Salinity Research Institute, Karnal, Haryana-132001, India., H. Jat, and CIMMYT, Karnal, Haryana-132001, India. Diversity of wild edible mushrooms in Indian Subcontinent and its neighboring countries. Recent Adv Biol Med. 2015;01:69.

Christensen M, Bhattarai S, Devkota S, Larsen HO. Collection and use of wild edible fungi in Nepal. Econ Bot. 2008;62:12–23.

Ćirić A, Kruljević I, Stojković D, Fernandes Â, Barros L, Calhelha RC, Ferreira ICFR, Soković M, Glamočlija J. Comparative investigation on edible mushrooms Macrolepiota mastoidea, M. rhacodes and M. procera: functional foods with diverse biological activities. Food Funct. 2019;10:7678–86.

Colak A, Kolcuoglu Y, Sesli E, Dalman O. Biochemical composition of some Turkish fungi. Asian J Chem. 2007;19:2193–9.

Conlon BH, de Beer ZW, De Fine Licht HH, Aanen DK, Poulsen M. Phylogenetic analyses of *Podaxis* specimens from Southern Africa reveal hidden diversity and new insights into associations with termites. Fungal Biol. 2016;120:1065–76.

Cooper J. New species and combinations of some New Zealand agarics belonging to Clitopilus, Lyophyllum, Gerhardtia, Clitocybe, Hydnangium, Mycena, Rhodocollybia and Gerronema. Mycosphere. 2014;5:263–88.

Côté J, Caillet S, Doyon G, Sylvain J-F, Lacroix M. Bioactive compounds in cranberries and their biological properties. Crit Rev Food Sci Nutr. 2010;50:666–79.

Culleré L, Ferreira V, Chevret B, Venturini ME, Sánchez-Gimeno AC, Blanco D. Characterisation of aroma active compounds in black truffles (*Tuber melanosporum*) and summer truffles (*Tuber aestivum*) by gas chromatography–olfactometry. Food Chem. 2010;122:300–6.

D'Archivio M, Filesi C, Varì R, Scazzocchio B, Masella R. Bioavailability of the polyphenols: status and controversies. Int J Mol Sci. 2010;11:1321–42.

Daba AS, Kabeil SS, Botros WA, El-Saadani MA. Production of mushroom (Pleurotus ostreatus) in Egypt as a source of nutritional and medicinal food. World J Agric Sci. 2008;4:630–4.

Dai Y-C, Yang Z-L, Cui B-K, Yu C-J, Zhou L-W. Species diversity and utilization of medicinal mushrooms and fungi in China (Review). IJM. 2009;11

Dai X, Sharma M, Chen J, editors. Fungi in sustainable food production. Cham: Springer International Publishing; 2021.

Danell E. Cantharellus. In: Cairney JWG, Chambers SM, editors. Ectomycorrhizal fungi key genera in profile. Berlin, Heidelberg: Springer Berlin Heidelberg; 1999. p. 253–67.

Das K. Diversity and conservation of wild mushrooms in Sikkim with special reference to Barsey Rhododendron Sanctuary. NeBIO. 2010;1:1–13.

Dávila Giraldo LR, Pérez Jaramillo CC, Méndez Arteaga JJ, Murillo-Arango W. Nutritional value and antioxidant, antimicrobial and cytotoxic activity of wild macrofungi. Microorganisms. 2023;11:1158.

Davis M, Sommer R, Menge J. Field guide to mushrooms of Western North America. University of California Press; 2012.

de Carvalho CSM, Sales-Campos C, Andrade CND. Mushrooms of the Pleurotus genus: a review of cultivation techniques. Interciencia. 2010;35(3):177–82.

de Mattos-Shipley KMJ, Ford KL, Alberti F, Banks AM, Bailey AM, Foster GD. The good, the bad and the tasty: the many roles of mushrooms. Stud Mycol. 2016;85:125–57.

Deb T. Mushroom Statistics 2024 By Varieties, Environment, Flavors. Market.us Media. 2024. Available at https://media.market.us/mushroom-statistics/.

Delgat L, Dierickx G, De Wilde S, Angelini C, De Crop E, De Lange R, Halling R, Manz C, Nuytinck J, Verbeken A. Looks can be deceiving: the deceptive milkcaps (Lactifluus, Russulaceae) exhibit low morphological variance but harbour high genetic diversity. IMA Fungus. 2019;10:14.

Demirtas N, Sengul GF, Dizeci N, Yildirim O. Exploring the chemical composition and nutritional properties of six edible mushroom species from Turkey. J Food Compos Anal. 2024;133:106477.

Deng G, Li J, Liu H, Wang Y. Volatile compounds and aroma characteristics of mushrooms: a review. Crit Rev Food Sci Nutr. 2024a;64:13175–92.

Deng X, Li M, Dai Y, Zhu X, Yan X, Wei Z, Yuan Y. The Diversity of Macrofungi in the Forests of Ningxia, Western China. Diversity. 2024b;16:725.

Dentinger BTM, Ammirati JF, Both EE, Desjardin DE, Halling RE, Henkel TW, Moreau P-A, Nagasawa E, Soytong K, Taylor AF, Watling R, Moncalvo J-M, McLaughlin DJ. Molecular phylogenetics of porcini mushrooms (*Boletus* section *Boletus*). Mol Phylogenet Evol. 2010;57:1276–92.

Deshmukh L, Sharma AK, Sandhu SS. Contrive Himalayan Soft Gold Cordyceps Species: a Lineage of Eumycota Bestowing Tremendous Pharmacological and Therapeutic Potential. Curr Pharmacol Rep. 2020;6:155–66.

Dhull SB, editor. Wild mushrooms: characteristics, nutrition, and processing. Boca Raton, FL: CRC Press; 2022.

Dian-xu MA, Gang LIU, Quan-hong OU, Hai-chao YU, Hui-mei LI, You-ming SHI. Discrimination of common wild mushrooms by FTIR and two-dimensional correlation infrared spectroscopy. Spectroscopy Spectral Anal. 2018;38:2113.

Diaz JH. Amatoxin-containing mushroom poisonings: species, toxidromes, treatments, and outcomes. Wilderness Environ Med. 2018;29:111–8.

Diego CZ, Pardo-Giménez A, editors. Edible and medicinal mushrooms: technology and applications. 1st ed. Wiley; 2017.

Dighton J. Fungi in ecosystem processes. 2nd ed. Boca Raton: CRC Press; 2018.

Dimitrijevic MV, Mitic VD, Cvetkovic JS, Stankov Jovanovic VP, Mutic JJ, Nikolic Mandic SD. Update on element content profiles in eleven wild edible mushrooms from family Boletaceae. Eur Food Res Technol. 2016;242:1–10.

Dimopoulou M, Kolonas A, Mourtakos S, Androutsos O, Gortzi O. Nutritional composition and biological properties of sixteen edible mushroom species. Appl Sci. 2022;12:8074.

Dissanayake AA, Mills GL, Bonito G, Rennick B, Nair MG. Chemical composition and anti-inflammatory and antioxidant activities of extracts from cultivated morel mushrooms, species of genus *Morchella* (Ascomycota). IJM. 2021:23.

Dong H, Shang X, Zhao X, Yu H, Jiang N, Zhang M, Tan Q, Zhou C, Zhang L. Construction of a genetic linkage map of *Lentinula edodes* based on SSR, SRAP and TRAP markers. Breed Sci. 2019;69:585–91.

Dong Y, Ma H, Sun L, He R, Ye X, Gan B, Zhang Q, Gong Z, Han X, Duan X, Yan J. Comparative transcriptome analysis of candidate genes associated with mycelia growth from a He-Ne laser with pulsed light mutant of Phellinus igniarius (Agaricomycetes). Int J Med Mushrooms. 2024;26:71–85.

Doroški A, Klaus A, Režek Jambrak A, Djekic I. Food waste originated material as an alternative substrate used for the cultivation of oyster mushroom (Pleurotus ostreatus): a review. Sustainability. 2022;14:12509.

Drewinski MP, Corrêa-Santos MP, Lima VX, Lima FT, Palacio M, Borges MEA, Trierveiler-Pereira L, Magnago AC, Furtado ANM, Lenz AR, Silva-Filho AGS, Nascimento CC, Alvarenga RLM, Gibertoni TB, Oliveira JJS, Baltazar JM, Neves MA, Vargas-Isla R, Ishikawa NK, Menolli N. Over 400 food resources from Brazil: evidence-based records of wild edible mushrooms. IMA Fungus. 2024;15:40.

Du P, Cui BK, Dai YC. Genetic diversity of wild Auricularia polytricha in Yunnan Province of South-western China revealed by sequence-related amplified polymorphism (SRAP) analysis. J Med Plants Res. 2011;5:1374–81.

Du J, Guo H-B, Li Q, Forsythe A, Chen X-H, Yu X-D. Genetic diversity of Lepista nuda (Agaricales, Basidiomycota) in Northeast China as indicated by SRAP and ISSR markers. PLoS One. 2018;13:e0202761.

Du P, He H-Y, Wu N-Y, Cao T, Cui B-K. Medicinal value, genetic diversity, and genetic relationship analysis of Auricularia cornea (Agaricomycetes) based on ITS, ISSR, and SRAP markers. Int J Med Mushrooms. 2024;26:43–57.

Duell G. The President's Report [online]. National Conference of the Australian Truffe Growers Association. 2012.

Dulay RMR, Batangan JN, Kalaw SP, Leon AMD, Cabrera EC, Kimura K, Eguchi F, Reyes RG. Records of wild mushrooms in the Philippines: a review. J App Biol Biotech. 2022; https://doi.org/10.7324/JABB.2023.110202.

Ediriweera AN, Karunarathna SC, Yapa PN, Schaefer DA, Ranasinghe AK, Suwannarach N, Xu J. Ectomycorrhizal mushrooms as a natural bio-indicator for assessment of heavy metal pollution. Agronomy. 2022;12:1041.

Ekiz E, Oz E, Abd El-Aty AM, Proestos C, Brennan C, Zeng M, Tomasevic I, Elobeid T, Çadırcı K, Bayrak M, Oz F. Exploring the potential medicinal benefits of Ganoderma lucidum: from metabolic disorders to coronavirus infections. Foods. 2023;12:1512.

El Enshasy H, Elsayed EA, Aziz R, Wadaan MA. Mushrooms and truffles: historical biofactories for complementary medicine in Africa and in the Middle East. Evid Based Complement Alternat Med. 2013;2013:620451.

El Karkouri K, Couderc C, Decloquement P, Abeille A, Raoult D. Rapid MALDI-TOF MS identification of commercial truffles. Sci Rep. 2019;9:17686.

El-Gharabawy HM. Ecology of Wind Mushrooms. In: Ecology of Macrofungi. CRC Press; 2023.

Elkhateeb W, Elnahas M, Wenhua L, Galappaththi MCA, Daba GM, Elkhateeb W, Elnahas M, Wenhua L, Galappaththi MCA, Daba GM. The coral mushrooms *Ramaria* and *Clavaria*. Stud Fungi. 2021;6:495–506.

El-Ramady H, Abdalla N, Badgar K, Llanaj X, Törős G, Hajdú P, Eid Y, Prokisch J. Edible mushrooms for sustainable and healthy human food: nutritional and medicinal attributes. Sustainability. 2022;14:4941.

Esmaeili A, Shamaei S, Aghaee EM, Akhtar ZN, Hosseini SF, Shokri S. Health risk assessment of heavy metals in edible mushrooms and their effect on anemia: a review study. J Chem Health Risks. 2022;12:597–607.

Falandysz J, Zhang J, Zalewska T. Radioactive artificial 137Cs and natural 40K activity in 21 edible mushrooms of the genus Boletus species from SW China. Environ Sci Pollut Res. 2017;24:8189–99.

FAO. Total production of mushrooms and truffles worldwide from 2012 to 2022 (in million metric tons). 2024a. https://www.statista.com/statistics/1018488/global-mushrooms-and-trufflesproduction/.

FAO. Leading producers of mushrooms and truffles worldwide in 2022 (in thousand metric tons). 2024b.

Farag MA, Fathi D, Shamma S, Shawkat MSA, Shalabi SM, El Seedi HR, Afifi SM. Comparative metabolome classification of desert truffles *Terfezia claveryi* and *Terfezia boudieri* via its aroma and nutrients profile. LWT. 2021;142:111046.

Felbinger C, Kutzsche F, Mönkediek S, Fischer M. Genetic profiling: Differentiation and identification of hazelnut cultivars (Corylus avellana L.) using RAPD-PCR. Food Control. 2020;107:106791.

Fernandes Â, Barreira JCM, Antonio AL, Morales P, Férnandez-Ruiz V, Martins A, Oliveira MBPP, Ferreira ICFR. Exquisite wild mushrooms as a source of dietary fiber: analysis in electron-beam irradiated samples. LWT Food Sci Technol. 2015;60:855–9.

Ferraro V, Venturella G, Pecoraro L, Gao W, Gargano ML. Cultivated mushrooms: importance of a multipurpose crop, with special focus on Italian fungiculture. Plant Biosyst. 2022;156:130–42.

Ferreira I, Dias T, Melo J, Mouazen AM, Cruz C. First steps in developing a fast, cheap, and reliable method to distinguish wild mushroom and truffle species. Resources. 2023;12:139.

Flores R, Bran M, Honrubia M, Hall I, Wang Y, Danell E, Zambonelli A. Edible mycorrhizal mushrooms of the west highland of Guatemala. 2002.

Floudas D, Binder M, Riley R, Barry K, Blanchette RA, Henrissat B, Martínez AT, Otillar R, Spatafora JW, Yadav JS, Aerts A, Benoit I, Boyd A, Carlson A, Copeland A, Coutinho PM, de Vries RP, Ferreira P, Findley K, Foster B, Gaskell J, Glotzer D, Górecki P, Heitman J, Hesse C, Hori C, Igarashi K, Jurgens JA, Kallen N, Kersten P, Kohler A, Kües U, Kumar TKA, Kuo A, LaButti K, Larrondo LF, Lindquist E, Ling A, Lombard V, Lucas S, Lundell T, Martin R, McLaughlin DJ, Morgenstern I, Morin E, Murat C, Nagy LG, Nolan M, Ohm RA, Patyshakuliyeva A, Rokas A, Ruiz-Dueñas FJ, Sabat G, Salamov A, Samejima M, Schmutz J, Slot JC, St F, John J, Stenlid H, Sun S, Sun K, Syed A, Tsang A, Wiebenga D, Young A, Pisabarro DC, Eastwood F, Martin D, Cullen IVG, Hibbett DS. The Paleozoic origin of enzymatic lignin decomposition reconstructed from 31 fungal genomes. Science. 2012;336:1715–9.

Fongnzossie EF, Nyangono CFB, Biwole AB, Ebai PNB, Ndifongwa NB, Motove J, Dibong SD. Wild edible plants and mushrooms of the Bamenda Highlands in Cameroon: ethnobotanical assessment and potentials for enhancing food security. J Ethnobiol Ethnomed. 2020;16:12.

Foulongne-Oriol M, Dufourcq R, Spataro C, Devesse C, Broly A, Rodier A, Savoie J-M. Comparative linkage mapping in the white button mushroom Agaricus bisporus provides foundation for breeding management. Curr Genet. 2011;57:39–50.

Fraiture A, Amalfi M, Raspé O, Kaya E, Akata I, Degreef J. Two new species of Amanita sect. Phalloideae from Africa, one of which is devoid of amatoxins and phallotoxins. MycoKeys. 2019;53:93–125.

Frassetto LA, Sebastian A. Aging, metabolic acidosis and renal failure: Interactive accelerating processes. Med Hypotheses. 2019;124:95–7.

Gadadhar S, Bodakuntla S, Natarajan K, Janke C. The tubulin code at a glance. J Cell Sci. 2017;130:1347–53.

Gaglarirmak N. Chemical composition and nutrition value of dried cultivated culinary-medicinal mushrooms from Turkey. IJM. 2011:13.

Gałgowska M, Pietrzak-Fiećko R. Mineral composition of three popular wild mushrooms from Poland. Molecules. 2020;25:3588.

Gambaro V, Roda G, Visconti GL, Arnoldi S, Casagni E, Ceravolo C, Dell'Acqua L, Farè F, Rusconi C, Tamborini L, Arioli S, Mora D. Taxonomic identification of hallucinogenic mushrooms seized on the illegal market using a DNA-based approach and LC/MS-MS determination of psilocybin and psilocin. J Analyt Bioanalyt Tech. 2015;6(6):1–6.

Gao H, Mutti J, Young JK, Yang M, Schroder M, Lenderts B, Wang L, Peterson D, St G, Clair S, Jones L, Feigenbutz W, Marsh M, Zeng S, Wagner J, Farrell K, Snopek C, Scelonge X, Sopko JD, Sander S, Betts AMC, Chilcoat ND. Complex trait loci in maize enabled by CRISPR-Cas9 mediated gene insertion. Front Plant Sci. 2020;11:535.

Gao Y, Wu Z, Li W, Sun H, Chai Y, Li T, Liu C, Gong X, Liang Y, Qin P. Expanding the valorization of waste mushroom substrates in agricultural production: progress and challenges. Environ Sci Pollut Res. 2023;30:2355–73.

Garibay-Orijel R, Caballero J, Estrada-Torres A, Cifuentes J. Understanding cultural significance, the edible mushrooms case. J Ethnobiol Ethnomed. 2007;3:4.

Gautam AK, Verma RK, Avasthi S, Sushma Y, Bohra B, Devadatha MN, Suwannarach N. Current insight into traditional and modern methods in fungal diversity estimates. J Fungi. 2022;8:226.
Ge Z, Yang Z-L, Zhang P, Matheny PB, Hibbett D. Flammulina species from China inferred by morphological and molecular data. Fungal Divers. 2008;32(2):59–68.
Ge ZW, Yang ZL, Vellinga EC. The genus Macrolepiota (Agaricaceae, Basidiomycota) in China. Fungal Divers. 2010;45:81–98.
Géry A, Dubreule C, André V, Rioult J-P, Bouchart V, Heutte N, Eldin de Pécoulas P, Krivomaz T, Garon D. Chaga (Inonotus obliquus), a future potential medicinal fungus in oncology? A chemical study and a comparison of the cytotoxicity against human lung adenocarcinoma cells (A549) and human bronchial epithelial cells (BEAS-2B). Integr Cancer Ther. 2018;17:832–43.
Ghosh K. A review on edible straw mushrooms: a source of high nutritional supplement, biologically active diverse structural polysaccharides. J Sci Res. 2020;64:295–304.
Ghosh A, Mehta A, Khan AM. Metagenomic analysis and its applications. In: Encyclopedia of bioinformatics and computational biology. Elsevier; 2019. p. 184–93.
Glamočlija J, Stojković D, Nikolić M, Ćirić A, Reis FS, Barros L, Ferreira ICFR, Soković M. A comparative study on edible Agaricus mushrooms as functional foods. Food Funct. 2015;6:1900–10.
GLOBE NEWSWIRE. Global mushroom cultivation industry (2020 to 2025) – economic viability of mushroom cultivation and trade by developing countries presents opportunities. GlobeNewswire News Room. 2020. Available at https://www.globenewswire.com/news-release/2020/04/27/2022477/28124/en/Global-Mushroom-Cultivation-Industry-2020-to-2025-Economic-Viability-of-Mushroom-Cultivation-and-Trade-by-Developing-Countries-Presents-Opportunities.html.
Gopal J, Muthu M. Matrix-assisted laser desorption ionization-time of flight mass spectrometry (MALDI-TOF MS) for the analysis of mushroom polysaccharides of clinical, nutritional, and medicinal significance-contemplating the achievements and eliciting future trends. TrAC Trends Anal Chem. 2023;164:117095.
Gouzi H, Leboukh M, Bouchouka E. Antioxidant and antiradical properties of methanolic extracts from Algerian Wild Edible Desert Truffles (*Terfezia* and *Tirmania*, Ascomycetes). Int J Med Mushrooms. 2013:15.
Grabarczyk M, Mączka W, Wińska K, Uklańska-Pusz C. Mushrooms of the Pleurotus genus – properties and application. Biotechnol Food Sci. 2019;83:13–30.
Graziosi S, Hall IR, Zambonelli A. The mysteries of the white truffle: its biology, ecology and cultivation. Encyclopedia. 2022;2:1959–71.
Grimm-Samuel V. On the mushroom that deified the Emperor Claudius. The Class Q. 1991;41:178–82.
Guan X-J, Xu L, Shao Y-C, Wang Z-R, Chen F-S, Luo X-C. Differentiation of commercial strains of Agaricus species in China with inter-simple sequence repeat marker. World J Microbiol Biotechnol. 2008;24:1617–22.
Guillamón E, García-Lafuente A, Lozano M, D'Arrigo M, Rostagno MA, Villares A, Martínez JA. Edible mushrooms: role in the prevention of cardiovascular diseases. Fitoterapia. 2010;81:715–23.
Gunasekara N, Nanayakkara C, Karunarathana S, Wijesundera R. Nutritional aspects of three Termitomyces and four other wild edible mushroom species from Sri Lanka. Chiang Mai J Sci. 2021;48(5):12361246.
Gupta S, Summuna B, Gupta M, Annepu SK. Edible mushrooms: cultivation, bioactive molecules, and health benefits. In: Mérillon J-M, Ramawat KG, editors. Bioactive molecules in food. Cham: Springer International Publishing; 2018. p. 1–33.
Gupta N, Mehta M, Singh K. Benefits, challenges and opportunities in mushroom production: a review. Pharma Innov J. 2022;11:360–4.
Guzmán G. Genus *Pleurotus* (Jacq.: Fr.) P. Kumm. (Agaricomycetideae): diversity, taxonomic problems, and cultural and traditional medicinal uses. IJM. 2000;2.

Guzman G. Diversity and use of traditional Mexican medicinal fungi. A review. Int J Med Mushr. 2008;10:209–17.
Györfi J, Geösel A, Vetter J. Mineral composition of different strains of edible medicinal mushroom *Agaricus subrufescens* Peck. J Med Food. 2010;13:1510–4.
Hajdú P, Abdalla Z, El-Ramady H, Prokisch J. Edible mushroom of Lentinula spp.: a case study of Shiitake (Lentinula edodes L.) cultivation. Environ Biodivers Soil Security. 2022;6:41–9.
Hajicek-Dobberstein S. Soma siddhas and alchemical enlightenment: psychedelic mushrooms in Buddhist tradition. J Ethnopharmacol. 1995;48:99–118.
Hall IR. Edible and poisonous mushrooms: an introduction. Christchurch, N.Z.: New Zealand Institute for Crop & Food Research; 1998.
Hall IR, Zambonelli A. Laying the foundations. In: Zambonelli A, Bonito GM, editors. Edible ectomycorrhizal mushrooms: current knowledge and future prospects. Berlin, Heidelberg: Springer; 2012. p. 3–16.
Hall IR, Zambonelli A, Primavera F. Ectomycorrhizal fungi with edible fruiting bodies 3.Tuber magnatum, tuberaceae. Econ Bot. 1998;52:192–200.
Hall IR, Stephenson SL, Buchanan PK, Cole ALJ, Yun W. Edible and poisonous mushrooms of the world. Timber Press; 2010.
Hamza A, Jdir H. Nutritional, antioxidant and antibacterial properties of Tirmania nivea, a wild edible desert truffle from Tunisia Arid Zone. Med Aromat Plants. 2016;05:2167.
Hanuš LO, Shkrob I, Dembitsky VM. Lipids and fatty acids of wild edible mushrooms of the genus Boletus. J Food Lipids. 2008;15:370–83.
Harfi T, Alireza M-A, Farzad R, Fariborz Z-N. Induced mutation in *Agaricus bisporus* by gamma ray to improve genetic variability, degradation enzyme activity, and yield. Int J Radiat Biol. 2021;97:1020–31.
Härkönen M. Mushroom collection in Tanzania and Hunan (Southern China): inherited wisdom and folklore of two different cultures. In: Watling R, Frankland JC, Ainsworth AM, Isaac S, Robinson CH, editors. Tropical mycology: volume 1, macromycetes. UK: CABI Publishing; 2001. p. 149–65.
Hassan AA, Muhammad LQ. Isolation and identification of edible mushroom Pleurotus spp. and evaluate its efficiency in the production of Pleuran. IOP Conf Ser: Earth Environ Sci. 2023;1225:012080.
Hawksworth DL. Global species numbers of fungi: are tropical studies and molecular approaches contributing to a more robust estimate? Biodivers Conserv. 2012;21:2425–33.
Hay WD. An elementary text-book of British Fungi. Lowrey: S. Sonnenschein; 1887.
He X-L, Wu B, Li Q, Peng W-H, Huang Z-Q, Gan B-C. Phylogenetic relationship of two popular edible Pleurotus in China, Bailinggu (P. eryngii var. tuoliensis) and Xingbaogu (P. eryngii), determined by ITS, RPB2 and EF1α sequences. Mol Biol Rep. 2016;43:573–82.
He X, Wang X, Fang J, Chang Y, Ning N, Guo H, Huang L, Huang X, Zhao Z. Polysaccharides in *Grifola frondosa* mushroom and their health promoting properties: a review. Int J Biol Macromol. 2017;101:910–21.
He Z, Su Y, Li S, Long P, Zhang P, Chen Z. Development and evaluation of isothermal amplification methods for rapid detection of Lethal Amanita Species. Front Microbiol. 2019;10:1523.
Heim R. (1900–1979)-mycologist A. 1977. Termites and mushrooms: the termitophilous mushrooms of black Africa and southern Asia. Société nouvelle des éditions Boubée.
Heo S, Jeon S, Lee S. Utilization of Lentinus edodes mushroom β-glucan to enhance the functional properties of gluten-free rice noodles. LWT Food Sci Technol. 2014;55:627–31.
Herzog R, Solovyeva I, Rühl M, Thines M, Hennicke F. Dikaryotic fruiting body development in a single dikaryon of Agrocybe aegerita and the spectrum of monokaryotic fruiting types in its monokaryotic progeny. Mycol Prog. 2016;15:947–57.
Hou Y-N, Deng G, Mao JJ. Practical application of "About Herbs" website: herbs and dietary supplement use in oncology settings. Cancer J. 2019;25:357.
Hussain S, Jõudu I, Bhat R. Dietary fiber from underutilized plant resources—a positive approach for valorization of fruit and vegetable wastes. Sustainability. 2020;12:5401.

Hussain S, Sher H, Ullah Z, Elshikh MS, Al Farraj DA, Ali A, Abbasi AM. Traditional uses of wild edible mushrooms among the local communities of Swat, Pakistan. Foods. 2023;12:1705.
Inyod T, Sassanarakit S, Payapanon A, Keawsompong S. Morphological characteristics and molecular identification of a wild Thai isolate of the tropical mushroom Hed Taen Rad (Macrocybe crassa). Biodiversitas. 2017:18.
Iqbal U, Usman M, Yousaf N, Awan AN, Rukhsar S, Saeed S, Khan MD, Ali S, Ibrahim S, Ghafoor A. First report and distribution of a common edible mushroom Agaricus campestris (Basidiomycota: Agaricaceae) across different districts of Punjab, Pakistan. GU J Phytosci. 2023;3:102–11.
IUCN. Kuehneromyces mutabilis: Dahlberg, A.: the IUCN Red list of threatened species 2019: e.T122090579A223015247. 2019.
Jahedi A, Safaie N, Goltapeh EM, Ahmadifar S. Auricularia (wood ear mushroom) genus: a contribution to classification and new species records for Iran and world. Plant Fungal Syst. 2024;69:205–17.
Jakopovic B, Oršolić N, Jakopovich I. Proteomic research on the antitumor properties of medicinal mushrooms. Molecules. 2021;26:6708.
Jasińska A, Siwulski M, Sobieralski K, Rogalski J, Ohga S. Morphological and qualitative features of Agrocybe aegerita (Brig.) Sing. Carpophores cultivated on agricultural and textile industry wastes. J Fac Agric Kyushu Univ. 2014;59:229–34.
Jaworska J. Macrolepiota olivascens, a new species for Poland. Acta Mycol. 2013;45:67–72.
Jayakumar T, Ramesh E, Geraldine P. Antioxidant activity of the oyster mushroom, Pleurotus ostreatus, on CCl4-induced liver injury in rats. Food Chem Toxicol. 2006;44:1989–96.
Jayaraman S, Yadav B, Dalal RC, Naorem A, Sinha NK, Srinivasa Rao C, Dang YP, Patra AK, Datta SP, Subba Rao A. Mushroom farming: a review focusing on soil health, nutritional security and environmental sustainability. Farming System. 2024;2:100098.
Jędrejko KJ, Lazur J, Muszyńska B. Cordyceps militaris: an overview of its chemical constituents in relation to biological activity. Foods. 2021;10:2634.
Jo W-S, Hossain MA, Park S-C. Toxicological profiles of poisonous, edible, and medicinal mushrooms. Mycobiology. 2014;42:215–20.
Jo K, Lee S, Jo C, Jeon HJ, Choe JH, Choi Y-S, Jung S. Utility of winter mushroom treated by atmospheric non-thermal plasma as an alternative for synthetic nitrite and phosphate in ground ham. Meat Sci. 2020;166:108151.
Johnson M, Richards W, Griffiths R. Human hallucinogen research: guidelines for safety. J Psychopharmacol. 2008;22:603–20.
Jones S, Janardhanan KK. Antioxidant and antitumor activity of Ganoderma lucidum (Curt.: Fr.) P. Karst.Reishi (Aphyllophoromycetideae) from South India. Int J Med Mushr. 2000;2:6.
Joshi M, Bhargava P, Bhatt M, Kadri S, Shri M, Joshi CG. Agaricaceae. Mushrooms of Gujarat. Singapore: Springer Nature Singapore; 2021. p. 41–84.
Kagan-Zur V. Terfezias, a amily of Mycorrhizal edible mushrooms for arid zones. In: Pasternak D, Schlissel A, editors. Combating desertification with plants. US, Boston, MA: Springer; 2001. p. 45–53.
Kagan-Zur V, Akyuz M. Asian Mediterranean Desert Truffles. In: Kagan-Zur V, Roth-Bejerano N, Sitrit Y, Morte A, editors. Desert truffles: phylogeny, physiology, distribution and domestication. Berlin, Heidelberg: Springer; 2014. p. 159–71.
Kalač P. A review of chemical composition and nutritional value of wild-growing and cultivated mushrooms. J Sci Food Agric. 2013;93:209–18.
Kalac P. Edible mushrooms: chemical composition and nutritional value. Academic Press; 2016.
Kaliyaperumal M, Kezo K, Gunaseelan S. A global overview of edible mushrooms. In: Singh BP, Lallawmsanga, Passari AK, editors. Biology of macrofungi. Cham: Springer International Publishing; 2018. p. 15–56.
Kamal S, Barh A, Sharma K, Sharma VP. Mushroom biology and advances. In: Kumar Srivastava D, Kumar Thakur A, Kumar P, editors. Agricultural biotechnology: latest research and trends. Singapore: Springer Nature Singapore; 2021. p. 661–88.

Kansci G, Mossebo DC, Selatsa AB, Fotso M. Nutrient content of some mushroom species of the genus *Termitomyces* consumed in Cameroon. Nahrung. 2003;47:213–6.

Kant Raut J. Current status, challenges and prospects of mushroom industry in Nepal. IJAE. 2019;4:154.

Kapoor P, Kumari P, Padhiary S, Katoch A, Tamang J. 189Morchella esculenta (Gucchi)/Morchella conica—Krombh/Morchella elata Fr., Family—Morchellaceae. In: Edible and medicinal mushrooms of the Himalayas. CRC Press; 2023.

Karunarathna S. Mycosphere essay 8: a review of genus Agaricus in tropical and humid subtropical regions of Asia. Mycosphere. 2016;7:417–39.

Katoch A, Paudel M, Katoch P. An overview on mushrooms. In: Sharma A, Bhardwaj G, Nayik GA, editors. Phytochemistry and nutritional composition of significant wild medicinal and edible mushrooms. Royal Society of Chemistry; 2023. p. 1–30.

Kerrigan RW. Agaricus subrufescens, a cultivated edible and medicinal mushroom, and its synonyms. Mycologia. 2005;97:12–24.

Kewessa G, Dejene T, Alem D, Tolera M, Martín-Pinto P. Forest type and site conditions influence the diversity and biomass of edible macrofungal species in Ethiopia. J Fungi (Basel). 2022;8:1023.

Khalid AN. A checklist of macrofungi of Pakistan published from 1998–2020. Pak J Bot. 2022:54.

Khan NA, Khan MA. Characterization of edible and poisonous mushrooms of pakistan. 2021. https://doi.org/10.13140/RG.2.2.32014.48969.

Khanh TD. Species diversity of Boletus Dill. ex Fr in Chu Yang Sin, National Park, Dak Lak, Vietnam. Advan Stud Biol. 2022;14:41–50.

Khastini RO, Maryani N, Lestari ID. Wild edible mushroom, a potential and valuable source for food security. In: Huda N, Jaswir I, Romdhonah Y, Alimuddin A, Ahamed T, Nasser N, editors. Proceedings of the 2nd international conference for smart agriculture, food, and environment (ICSAFE 2021). Dordrecht: Atlantis Press International BV; 2023. p. 160–8.

Kim KH, Park KM, Choi SU, Lee KR. Macrolepiotin, a new indole alkaloid from Macrolepiota neomastoidea. J Antibiot. 2009;62:335–8.

Kiss L. Limits of nuclear ribosomal DNA internal transcribed spacer (ITS) sequences as species barcodes for fungi. Proc Natl Acad Sci U S A. 2012;109:E1811; author reply E1812.

Kissanga R, Liberal Â, Diniz I, Rodrigues ASB, Baptista-Ferreira JL, Batista D, Ivanov M, Soković M, Ferreira ICFR, Fernandes Â, Barros L, Catarino L. Biochemical and molecular profiling of wild edible mushrooms from Huila, Angola. Foods. 2022;11:3240.

Kıvrak İ. Analytical methods applied to assess chemical composition, nutritional value and in vitro bioactivities of Terfezia olbiensis and Terfezia claveryi from Turkey. Food Anal Methods. 2015;8:1279–93.

Kıvrık M, Süfer Ö, Bozok F. A research on quality evaluation of eight wild edible macrofungi collected from East Mediterranean region of Turkey. Chem Biodivers. 2022;19:e202100967.

Kobayashi T, Oguro M, Akiba M, Taki H, Kitajima H, Ishihara H. Mushroom yield of cultivated shiitake (*Lentinula edodes*) and fungal communities in logs. J For Res. 2020;25:269–75.

Kordalewska M, Kalita J, Bakuła Z, Brillowska-Dąbrowska A, Jagielski T. PCR-RFLP assays for species-specific identification of fungi belonging to Scopulariopsis and related genera. Med Mycol. 2019;57:643–8.

Kostić M, Smiljković M, Petrović J, Glamočlija J, Barros L, Ferreira ICFR, Ćirić A, Soković M. Chemical, nutritive composition and a wide range of bioactive properties of honey mushroom Armillaria mellea (Vahl: Fr.) Kummer. Food Funct. 2017;8:3239–49.

Kotlaba F. Review of mushrooms and other fungi of Great Britain and Europe. Folia Geobotanica Phytotaxonomica. 1984;19:329–31.

Kotowski MA. History of mushroom consumption and its impact on traditional view on mycobiota—an example from Poland. Microbial Biosyst. 2019;4:1–13.

Kotowski MA, Pietras M, Łuczaj Ł. Extreme levels of mycophilia documented in Mazovia, a region of Poland. J Ethnobiol Ethnomed. 2019;15:12.

Kour H, Kour S, Sharma Y, Singh S, Sharma I, Kour D, Yadav AN. Bioprospecting of industrially important mushrooms. In: Abdel-Azeem AM, Yadav AN, Yadav N, Sharma M, edi-

tors. Industrially important fungi for sustainable development. Cham: Springer International Publishing; 2021. p. 679–716.
Kratika S. Mushroom: cultivation and processing. Int J Food Process Technol. 2018;V5:9–12.
Kula İ, Solak MH, Uğurlu M, Işıloğlu M, Arslan Y. Determination of Mercury, Cadmium, Lead, Zinc, Selenium and Iron by ICP-OES in mushroom samples from around thermal power plant in Muğla, Turkey. Bull Environ Contam Toxicol. 2011;87:276–81.
Kumar R, Pandey S, Rishi RR, Giri K. Macrolepiota procera (Scop.) Singer (Agaricomycetes) – a new generic record of edible mushroom for Nagaland, Northeast India. Asian J Agric. 2017;1:6–8.
Kumar H, Bhardwaj K, Sharma R, Nepovimova E, Cruz-Martins N, Dhanjal DS, Singh R, Chopra C, Verma R, Abd-Elsalam KA, Tapwal A, Musilek K, Kumar D, Kuča K. Potential usage of edible mushrooms and their residues to retrieve valuable supplies for industrial applications. J Fungi. 2021a;7:427.
Kumar K, Mehra R, Guiné RPF, Lima MJ, Kumar N, Kaushik R, Ahmed N, Yadav AN, Kumar H. Edible mushrooms: a comprehensive review on bioactive compounds with health benefits and processing aspects. Foods. 2021b;10:2996.
Kumar A, Devi R, Dhalaria R, Tapwal A, Verma R, Rashid S, Elossaily GM, Khan KA, Chen K-T, Verma T. Nutritional, nutraceutical, and medicinal potential of Cantharellus cibarius Fr.: a comprehensive review. Food Sci Nutr. 2025;13:e4641.
Kumla J, Suwannarach N, Lumyong S. A new report on edible tropical bolete, Phlebopus spongiosus in Thailand and its fruiting body formation without the need for a host plant. Mycobiology. 2020;48:263–75.
Kuo M. Clitocybe nuda (MushroomExpert.Com). Clitocybe nuda. MushroomExpertCom. 2006. Retrieved from http://www.mushroomexpert.com/clitocybe_nuda.html. Available at https://www.mushroomexpert.com/clitocybe_nuda.html.
Laala G, Raja MU, Irshad G, Gardezi SRH, Akram A. Four species of true morels (Morchella) recorded from Poonch District (Azad Jammu and Kashmir). J Hortic Sci Technol. 2020:52–5.
Lagutina T. Complete Encyclopedia of mushrooms. Book on Demand Ltd; 2018.
Lallawmsanga A, Carrasco BJ. Chap. 1. Diversity of the fungi kingdom: molecular tools to distinguish mushrooms considered safe and unsafe for human health. In: Stojkovic D, Barros L, editors. Food chemistry, function and analysis. Cambridge: Royal Society of Chemistry; 2022. p. 1–26.
Largeteau ML, Llarena-Hernández RC, Regnault-Roger C, Savoie J-M. The medicinal Agaricus mushroom cultivated in Brazil: biology, cultivation and non-medicinal valorisation. Appl Microbiol Biotechnol. 2011;92:897–907.
Lebel T, Syme A. Sequestrate species of Agaricus and Macrolepiota from Australia: new species and combinations and their position in a calibrated phylogeny. Mycologia. 2012;104:496–520.
Lebeuf R, Russell SD, Justo A. Macrolepiota macilenta and M. pallida, two new species from North America. Fungal Syst Evol. 2024; https://doi.org/10.3114/fuse.2024.14.09.
Lee H, Wissitrassameewong K, Park MS, Verbeken A, Eimes J, Lim YW. Taxonomic revision of the genus Lactarius (Russulales, Basidiomycota) in Korea. Fungal Divers. 2019;95:275–335.
Lee H-Y, Moon S, Ro H-S, Chung J-W, Ryu H. Analysis of genetic diversity and population structure of wild strains and cultivars using genomic SSR markers in *Lentinula edodes*. Mycobiology. 2020;48:115–21.
Leffingwell JC, Alford ED. Volatile constituents of the Giant Puffball Mushroom (Calvatia gigantea). 2011. https://doi.org/10.13140/RG.2.2.24877.28648.
Legoshchina O, Ufimtsev V. Fruiting characteristics of fungi of the genus *Suillus* in the artificial pine plantations of the Kedrovsky coal mine dump (Kemerovo region). BIO Web Conf. 2024;128:00016.
Leonardi M, Paolocci F, Rubini A, Simonini G, Pacioni G. Assessment of inter- and intra-specific variability in the main species of Boletus edulis complex by ITS analysis. FEMS Microbiol Lett. 2005;243:411–6.
Leonardi M, Iotti M, Mello A, Vizzini A, Paz-Conde A, Trappe J, Pacioni G. Typification of the four most investigated and valuable truffles: Tuber aestivum Vittad., T borchii Vittad, T magnatum Picco and T melanosporum Vittad. Cryptogamie Mycologie. 2021;42:149–70.

Lestari AS, Zulfiana D, Ismayati M, Zulfitri A, Krishanti NPRA, Kartika T, Himmi SK, Wikantyoso B, Prianto AH, Yusuf S. Morphological diversity of Agaricomycetes in Kuningan Botanical Garden, West Java, Indonesia. IOP Conf Ser: Earth Environ Sci. 2019;374:012016.
Li Y. Present development situation and tendency of edible mushroom industry in China. Mushroom Sci. 2012;18:3–9.
Li G, Quiros CF. Sequence-related amplified polymorphism (SRAP), a new marker system based on a simple PCR reaction: its application to mapping and gene tagging in Brassica. Theor Appl Genet. 2001;103:455–61.
Li C, Xu S. Edible mushroom industry in China: current state and perspectives. Appl Microbiol Biotechnol. 2022;106:3949–55.
Li D, Liu Y, Wang P, Ma Y, Wang S, Zhao S, Xu F. Development of SCAR markers to determine the mating types of Lepista nuda Protoplast Monokaryons. Curr Microbiol. 2014;68:536–42.
Li H, Tian Y, Menolli N, Ye L, Karunarathna SC, Perez-Moreno J, Rahman MM, Rashid MH, Phengsintham P, Rizal L, Kasuya T, Lim YW, Dutta AK, Khalid AN, Huyen LT, Balolong MP, Baruah G, Madawala S, Thongklang N, Hyde KD, Kirk PM, Xu J, Sheng J, Boa E, Mortimer PE. Reviewing the world's edible mushroom species: a new evidence-based classification system. Comp Rev Food Sci Food Safe. 2021a;20:1982–2014.
Li M, Yu L, Zhao J, Zhang H, Chen W, Zhai Q, Tian F. Role of dietary edible mushrooms in the modulation of gut microbiota. J Funct Foods. 2021b;83:104538.
Li H, Xu G, Yuan G. Effects of an Armillaria mellea Polysaccharide on learning and memory of D-galactose-induced aging mice. Front Pharmacol. 2022;13:919920.
Li Y, Yang T, Qiao J, Liang J, Li Z, Sa W, Shang Q. Whole-genome sequencing and evolutionary analysis of the wild edible mushroom, Morchella eohespera. Front Microbiol. 2024;14:1309703.
Lian B, Dong Y-R, Hou W-G, Tong L-H, Yuan S. Ectomycorrhizal fungi in Jiangsu Province, China. Pedosphere. 2007;17:30–5.
Lin P, Yan Z-F, Kook M, Li C-T, Yi T-H. Genetic and chemical diversity of edible mushroom Pleurotus species. Biomed Res Int. 2022;2022:6068185.
Lin H, Jiang H, Adade SY-SS, Kang W, Xue Z, Zareef M, Chen Q. Overview of advanced technologies for volatile organic compounds measurement in food quality and safety. Crit Rev Food Sci Nutr. 2023;63:8226–48.
Liu J-Y, Ying Z-H, Liu F, Liu X-R, Xie B-G. Evaluation of the use of SCAR markers for screening genetic diversity of Lentinula edodes strains. Curr Microbiol. 2012;64:317–25.
Liu XB, Feng B, Li J, Yan C, Yang ZL. Genetic diversity and breeding history of Winter Mushroom (Flammulina velutipes) in China uncovered by genomic SSR markers. Gene. 2016;591:227–35.
Liu Y, Li X, Kou Y. Ectomycorrhizal fungi: participation in nutrient turnover and community assembly pattern in forest ecosystems. Forests. 2020;11:453.
Liu K, Ma H, Liu M, Zhang G, Yin S, Li S. Genetic diversity of cultivated *A. auricula* by comprehensive analysis of rDNA-ITS and SRAP. Int J Mol Evol Biodivers. 2022; https://doi.org/10.5376/ijmeb.2022.12.0002.
Liu Y, Yong T, Cai M, Wu X, Guo H, Xie Y, Hu H, Wu Q. Exploring the potential of Russula griseocarnosa: a molecular ecology perspective. Agriculture. 2024;14:879.
Liuzzi GM, Petraglia T, Latronico T, Crescenzi A, Rossano R. Antioxidant compounds from edible mushrooms as potential candidates for treating age-related neurodegenerative diseases. Nutrients. 2023;15:1913.
Llarena-Hernández CR, Largeteau ML, Ferrer N, Regnault-Roger C, Savoie J-M. Optimization of the cultivation conditions for mushroom production with European wild strains of Agaricus subrufescens and Brazilian cultivars. J Sci Food Agric. 2014;94:77–84.
Long L, Yan W, He Y, Dong L, Xing Z, Li C, Xia W, Li F. Development of a duplex digital PCR method to quantify five genetically modified soybean events. Food Anal Methods. 2022;15:294–306.
Lu X, Brennan MA, Serventi L, Liu J, Guan W, Brennan CS. Addition of mushroom powder to pasta enhances the antioxidant content and modulates the predictive glycaemic response of pasta. Food Chem. 2018;264:199–209.

Luard E. Truffles. London: Frances Lincoln; 2006.
Luki N. The distribution and diversity of boletus genus. Kragujevac J Sci. 2009;31:59–68.
M.k, A. and D. G. Ethnomycologie nepalaise. Bull Trimest Soc Mycol Fr. 1996:112.
Ma D, Yang G, Mu L, Song Y. Application of SRAP in the genetic diversity of Tricholoma matsutake in northeastern China. Afr J Biotechnol. 2010;9:6244–50.
Ma Q-L, Zhu C, Morselli M, Su T, Pelligrini M, Lu Z, Jones M, Denver P, Castro D, Gu X, Relampagos F, Caoili K, Teter B, Frautschy SA, Cole GM. The novel omega-6 fatty acid docosapentaenoic acid positively modulates brain innate immune response for resolving neuroinflammation at early and late stages of humanized APOE-based Alzheimer's disease models. Front Immunol. 2020;11:558036.
Maftoun P, Johari H, Soltani M, Malik R, Othman NZ, Enshasy HAE. The edible mushroom Pleurotus spp.: I. Biodiversity and nutritional values. Int J Biotechnol Well Indus. 2015;4:67–83.
Maihara VA, Moura PLDC, Catharino MGM, Moreira EG, Castro LP, Figueira RCL. Cadmium determination in Lentinus edodes mushroom species. Food Sci Technol. 2012;32:553–7.
Mallick MAI, Mukherjee Sanyal S, Hingnara Anchal Public Institution, Ballabhpur, Chakdaha, Nadia 741223, West Bengal. Unlocking the potential: a comprehensive review of environmentally sustainable applications for agro-based spent mushroom substrate (SMS). In: Roychoudhury S, Sanyal T, Sen K, Mukherjee Sanyal S, editors. A basic overview of environment and sustainable development [volume 2]. International Academic Publishing House (IAPH); 2023. p. 434–77.
Mamiro DP, Royse DJ. The influence of spawn type and strain on yield, size and mushroom solids content of Agaricus bisporus produced on non-composted and spent mushroom compost. Bioresour Technol. 2008;99:3205–12.
Martin F, Kohler A, Murat C, Balestrini R, Coutinho PM, Jaillon O, Montanini B, Morin E, Noel B, Percudani R, Porcel B, Rubini A, Amicucci A, Amselem J, Anthouard V, Arcioni S, Artiguenave F, Aury J-M, Ballario P, Bolchi A, Brenna A, Brun A, Buée M, Cantarel B, Chevalier G, Couloux A, Da Silva C, Denoeud F, Duplessis S, Ghignone S, Hilselberger B, Iotti M, Marçais B, Mello A, Miranda M, Pacioni G, Quesneville H, Riccioni C, Ruotolo R, Splivallo R, Stocchi V, Tisserant E, Viscomi AR, Zambonelli A, Zampieri E, Henrissat B, Lebrun M-H, Paolocci F, Bonfante P, Ottonello S, Wincker P. Périgord black truffle genome uncovers evolutionary origins and mechanisms of symbiosis. Nature. 2010;464:1033–8.
Martínez-Tomé M, Maggi L, Jiménez-Monreal AM, Murcia MA, Marí JAT. Nutritional and antioxidant properties of Terfezia and Picoa. In: Kagan-Zur V, Roth-Bejerano N, Sitrit Y, Morte A, editors. Desert truffles: phylogeny, physiology, distribution and domestication. Berlin, Heidelberg: Springer; 2014. p. 261–73.
Martins A. The numbers behind mushroom biodiversity. In: Ferreira ICFR, Morales P, Barros L, editors. Wild plants, mushrooms and nuts. Wiley; 2016. p. 15–63.
Marwal A, Gaur RK. Molecular markers: tool for genetic analysis. In: Animal biotechnology. Elsevier; 2020. p. 353–72.
Mathur M, Mathur P. Ecological niche modelling of an industrially important mushroom – Ganoderma lucidum (Leys.) Karsten: a machine learning global appraisal. JSIR. 2023;82(11). https://doi.org/10.56042/jsir.v82i12.1973.
Matsushima Y, Eguchi F, Kikukawa T, Matsuda T. Historical overview of psychoactive mushrooms. Inflammat Regenerat. 2009;29:47–58.
Mattar TV, Gonçalves CS, Pereira RC, Faria MA, De Souza VR, Carneiro JDDS. A shiitake mushroom extract as a viable alternative to NaCl for a reduction in sodium in beef burgers: a sensory perspective. BFJ. 2018;120:1366–80.
Medina-Ortiz AJ, Herrera T, Vásquez-Dávila MA, Raja HA, Figueroa M. The genus Podaxis in arid regions of Mexico: preliminary ITS phylogeny and ethnomycological use. MycoKeys. 2017;20:17–36.
Mędyk M, Grembecka M, Brzezicha-Cirocka J, Falandysz J. Bio- and toxic elements in mushrooms from the city of Umeå and outskirts, Sweden. J Environ Sci Health B. 2017;52:577–83.

Meena B, Sivakumar V, Praneetha S. Prospects of biodiversity and distribution of mushroom fungi in India. GSC Biol Pharm Sci. 2020;13:156–63.

Mello A, Murat C, Bonfante P. Truffles: much more than a prized and local fungal delicacy. FEMS Microbiol Lett. 2006;260:1–8.

Miles PG, Chang S-T. Mushrooms: cultivation, nutritional value, medicinal effect, and environmental impact. 2nd ed. Boca Raton: CRC Press; 2004.

Miller OK, Miller H. North American mushrooms: a field guide to edible and inedible fungi. Falcon Guide; 2006.

Mittu AB, Bhat BA, Kaur CS, Begum DZ, Kaur EM, Neha E. Amanita Species. In: Sharma A, Bhardwaj G, Nayik GA, editors. Phytochemistry and nutritional composition of significant wild medicinal and edible mushrooms. Royal Society of Chemistry; 2023. p. 77–98.

Mohammed WS, Ziganshina EE, Shagimardanova EI, Gogoleva NE, Ziganshin AM. Comparison of intestinal bacterial and fungal communities across various xylophagous beetle larvae (Coleoptera: Cerambycidae). Sci Rep. 2018;8:10073.

Montoya-Alvarez AF. Phylogenetic relationships and review of the species of auricularia (fungi: basidiomycetes) in Colombia. Caldasia. 2011;33:55.

Montoya-Esquivel, A., A. Estrada-Torres, A. Kong and L. Juárez. 2001. Commercialization of wild mushrooms during market days of Tlaxcala, Mexico.

Morales-Flores S, Cepeda-Negrete J, Mata-Montes De Oca G, Ángel-Hernández A, Hernández-Ruiz J, Ruiz-Nieto JE. In vitro molecular identification and characterization of pleurotus spp. Strains in guanajuato, mexico. Agrociencia. 2022; https://doi.org/10.47163/agrociencia.v56i2.2780.

Moreno G, Díez J, Manjón JL. Picoa lefebvrei and Tirmania nivea, two rare hypogeous fungi from Spain. Mycol Res. 2000;104:378–81.

Morte A, Kagan-Zur V, Navarro-Ródenas A, Sitrit Y. Cultivation of desert truffles—a crop suitable for arid and semi-arid zones. Agronomy. 2021;11:1462.

Mukhopadhyay K, Haque I, Bandopadhyay R, Covert S, Porter D. AFLP based assessment of genetic relationships among shiitake (Lentinula ssp.) mushrooms. Mol Biol Rep. 2012;39:6059–65.

Mwangi RW, Macharia JM, Wagara IN, Bence RL. The antioxidant potential of different edible and medicinal mushrooms. Biomed Pharmacother. 2022;147:112621.

Mwilenga C. Improving Income of Akinamama Tujiendeleze Group through mushroom farming at ViwanjaSitini Ward in Ifakara Township. Masters, The Open University of Tanzania; 2023.

Nagy LG, Riley R, Tritt A, Adam C, Daum C, Floudas D, Sun H, Yadav JS, Pangilinan J, Larsson K-H, Matsuura K, Barry K, Labutti K, Kuo R, Ohm RA, Bhattacharya SS, Shirouzu T, Yoshinaga Y, Martin FM, Grigoriev IV, Hibbett DS. Comparative genomics of early-diverging mushroom-forming fungi provides insights into the origins of lignocellulose decay capabilities. Mol Biol Evol. 2016;33:959–70.

Navarro-Simarro P, Gómez-Gómez L, Ahrazem O, Rubio-Moraga Á. Food and human health applications of edible mushroom by-products. New Biotechnol. 2024;81:43–56.

Nawarathne IY, Daranagama DA. Bioremediation and sustainable mushroom cultivation: harnessing the lignocellulolytic power of *Pleurotus* species on waste substrates. N Z J Bot. 2024:1–19.

Negi A, Pare A, Meenatchi R. Emerging techniques for adulterant authentication in spices and spice products. Food Control. 2021;127:108113.

Niazi AR, Ghafoor A. Different ways to exploit mushrooms: a review. All Life. 2021;14:450–60.

Niego AGT, Lambert C, Mortimer P, Thongklang N, Rapior S, Grosse M, Schrey H, Charria-Girón E, Walker A, Hyde KD, Stadler M. The contribution of fungi to the global economy. Fungal Divers. 2023;121:95–137.

Niksic M, Klaus A, Argyropoulos D. Safety of foods based on mushrooms. In: Regulating safety of traditional and ethnic foods. Elsevier; 2016. p. 421–39.

Nile SH, Park SW. Total, soluble, and insoluble dietary fibre contents of wild growing edible mushrooms. Czech J Food Sci. 2014;32:302–7.

Nion YA, Djaya AA. Traditional edible mushrooms on tropical peatland. ResearchGate. 2015. Available at https://www.researchgate.net/publication/317721762_Traditional_edible_mushrooms_on_tropical_peatland.

Novakovic S, Djekic I, Klaus A, Vunduk J, Đorđević V, Tomovic V, Koćić-Tanackov S, Lorenzo JM, Barba FJ, Tomasevic I. Application of porcini mushroom (*Boletus edulis*) to improve the quality of frankfurters. J Food Process Preserv. 2020;44:e14556.

Nowacka-Jechalke N, Nowak R, Lemieszek MK, Rzeski W, Gawlik-Dziki U, Szpakowska N, Kaczyński Z. Promising potential of crude Polysaccharides from Sparassis crispa against colon cancer: an in vitro study. Nutrients. 2021;13:161.

Nowakowski P, Naliwajko SK, Markiewicz-Żukowska R, Borawska MH, Socha K. The two faces of *Coprinus comatus* —functional properties and potential hazards. Phytother Res. 2020;34:2932–44.

Nuñez Otaño NB, Bianchinotti MV, Saparrat MCN. Palaeomycology: a modern mycological view of fungal palynomorphs. In: Marret F, O'Keefe J, Osterloff P, Pound M, Shumilovskikh L, editors. Applications of non-pollen Palynomorphs: from palaeoenvironmental reconstructions to biostratigraphy. Geological Society of London; 2021.

Nurhakiki N, Putra IP. Morphological identification of wild edible mushroom (Termitomyces striatus) at ipb university campus forest. Biotik. 2024;12:1.

Nuytinck J, D'hooge E, Verbeken A. Lactarius (Russulales) in Europe and North America: some look-alikes tested molecular- and morphologically. In: Russulales 2010, Proceedings. National Botanical Garden Belgium; 2013. p. 106–16.

Obodai M. An ethnobotanical study of mushroom germplasm and its domestication in the Bia biosphere reserve. Ghana: Council for Scientific and Industrial Research (CSIR), Food Research Institute; 2001.

Ocañas FG, Martínez MQ, Ocañas LG, Parra AC, Mendoza HV, Rodríguez HG, Garza RV, Guerrero GG, Jiménez JG, Aranda MG. Some edible, toxic and medicinal mushrooms from temperate forests in the North of Mexico. In: Advances in macrofungi. CRC Press; 2019.

Ogawa W, Endo N, Fukuda M, Yamada A. Phylogenetic analyses of Japanese golden chanterelles and a new species description, Cantharellus anzutake sp nov. Mycoscience. 2018;59:153–65.

Ohiri RC, Bassey EE. Evaluation and characterization of nutritive properties of the jelly ear culinary-medicinal mushroom Auricularia auricula-judae (Agaricomycetes) from Nigeria. Int J Med Mushrooms. 2017;19:173–7.

Okoro IO, Achuba FI. Proximate and mineral analysis of some wild edible mushrooms. Afr J Biotechnol. 2012;11:7720–4.

Olawale OO, Oke OS, Adisa AS, Oloba OG, O.F. A. Assessment of constraints to mushroom production and scientific training needs of mushroom farmers in Oyo state Nigeria. UNIZIK J Eng Appl Sci. 2024;3:1317–23.

Osono T, Matsuoka S, Hobara S, Hirose D, Uchida M. Diversity and ecology of fungi in polar region: comparisons between Arctic and Antarctic plant remains. In: Fungi in Polar Regions. CRC Press; 2019.

Palacios I, Lozano M, Moro C, D'Arrigo M, Rostagno MA, Martínez JA, García-Lafuente A, Guillamón E, Villares A. Antioxidant properties of phenolic compounds occurring in edible mushrooms. Food Chem. 2011;128:674–8.

Paloi S, Kumla J, Paloi BP, Srinuanpan S, Hoijang S, Karunarathna SC, Acharya K, Suwannarach N, Lumyong S. Termite mushrooms (Termitomyces), a potential source of nutrients and bioactive compounds exhibiting human health benefits: a review. J Fungi (Basel). 2023;9:112.

Parab DN, Dhalagade JR, Sahoo AK, Ranveer RC. Effect of incorporation of mushroom (Pleurotus sajor-caju) powder on quality characteristics of Papad (Indian snack food). Int J Food Sci Nutr. 2012;63:866–70.

Park Y-J, Baek JH, Lee S, Kim C, Rhee H, Kim H, Seo J-S, Park H-R, Yoon D-E, Nam J-Y, Kim H-I, Kim J-G, Yoon H, Kang H-W, Cho J-Y, Song E-S, Sung G-H, Yoo Y-B, Lee C-S, Lee B-M, Kong W-S. Whole genome and global gene expression analyses of the model mushroom Flammulina velutipes reveal a high capacity for lignocellulose degradation. PLoS One. 2014;9:e93560.

Parkash V, Kaundal R, Paul S, Chauhan A, Aggarwal A. Some uncommon traditionally and medicinally important wild macrofungi of the Garhwal Himalayas, Uttarakhand, India. In: Wild Mushrooms and Health. CRC Press; 2023.

Pashaei KHA, Irankhah K, Namkhah Z, Sobhani SR. Edible mushrooms as an alternative to animal proteins for having a more sustainable diet: a review. J Health Popul Nutr. 2024a;43:205.

Pashaei KHA, Irankhah K, Namkhah Z, Sobhani SR. Edible mushrooms as an alternative to animal proteins for having a more sustainable diet: a review. J Health Popul Nutr. 2024b;43:205.

Patel R. A Moldy application of MALDI: MALDI-ToF mass spectrometry for fungal identification. J Fungi. 2019;5:4.

Pawlik A, Janusz G, Koszerny J, Małek W, Rogalski J. Genetic diversity of the edible mushroom Pleurotus sp. by amplified fragment length polymorphism. Curr Microbiol. 2012;65:438–45.

Pawłowicz T, Gabrysiak KA, Wilamowski K. Investigating the potential of polypore fungi as eco-friendly materials in food industry applications. Forests. 2024;15:1230.

Pegler D. Mushrooms and toadstools of Britain and Europe. Kingfisher; 2001.

Pei F, Yang W, Ma N, Fang Y, Zhao L, An X, Xin Z, Hu Q. Effect of the two drying approaches on the volatile profiles of button mushroom (Agaricus bisporus) by headspace GC–MS and electronic nose. LWT Food Sci Technol. 2016;72:343–50.

Peintner U, Schwarz S, Mešić A, Moreau P-A, Moreno G, Saviuc P. Mycophilic or mycophobic? Legislation and guidelines on wild mushroom commerce reveal different consumption behaviour in European countries. PLoS One. 2013;8:e63926.

Pepe M, Hesami M, de la Cerda KA, Perreault ML, Hsiang T, Jones AMP. A journey with psychedelic mushrooms: from historical relevance to biology, cultivation, medicinal uses, biotechnology, and beyond. Biotechnol Adv. 2023;69:108247.

Pérez-Butrón JL, Fernández-Vicente J. Una Nueva especie de *Flammulina* P. Karsten, *F. cephalariae* (*Agaricales*) encontrada en España. Revista Catalana de Micologia. 2007:81–91.

Pérez-Montes A, Rangel-Vargas E, Lorenzo JM, Romero L, Santos EM. Edible mushrooms as a novel trend in the development of healthier meat products. Curr Opin Food Sci. 2021;37:118–24.

Persad R. Movement patterns, food availability, and fungal diets of sympatric flying squirrels in the Kawartha Highlands | Digital Collections. 2024. Available at https://batadora.trentu.ca/islandora/movement-patterns-food-availability-and-fungal-diets-sympatric-flying-squirrels-kawartha.

Phillips R. Mushrooms and other fungi of North America. Richmond Hill, Ont: Firefly Books; 2010.

Phull A-R, Ahmed M, Park H-J. Cordyceps militaris as a bio functional food source: pharmacological potential, anti-inflammatory actions and related molecular mechanisms. Microorganisms. 2022;10:405.

Pildain MB, Marchelli P, Azpilicueta MM, Starik C, Barroetaveña C. Understanding introduction history: genetic structure and diversity of the edible ectomycorrhizal fungus, Suillus luteus, in Patagonia (Argentina). Mycologia. 2021;113:715–24.

Pinto S, Barros L, Sousa MJ, Ferreira ICFR. Chemical characterization and antioxidant properties of *Lepista nuda* fruiting bodies and mycelia obtained by in vitro culture: effects of collection habitat and culture media. Food Res Int. 2013;51:496–502.

Pk MU, Pervin R, Jahan J, Talukder RI, Ahmed S, Rahman M, Pk MU, Pervin R, Jahan J, Talukder RI, Ahmed S, Rahman M. Mushroom polysaccharides: chemistry and anticancer potentials. IntechOpen: An Introduction to Mushroom; 2020.

Podkowa A, Kryczyk-Poprawa A, Opoka W, Muszyńska B. Culinary–medicinal mushrooms: a review of organic compounds and bioelements with antioxidant activity. Eur Food Res Technol. 2021;247:513–33.

Pohleven J, Obermajer N, Sabotič J, Anžlovar S, Sepčić K, Kos J, Kralj B, Štrukelj B, Brzin J. Purification, characterization and cloning of a ricin B-like lectin from mushroom Clitocybe nebularis with antiproliferative activity against human leukemic T cells. Biochim Biophys Acta Gen Subj. 2009;1790:173–81.

Portalo-Calero F, Arroyo P, Suárez JI, Lozano J. Triangular test of amanita mushrooms by using electronic nose and sensory panel. Foods. 2019a;8:414.

Portalo-Calero F, Lozano J, Meléndez F, Arroyo P, Suárez JI. Identification of poisonous mushrooms by means of a hand-held electronic nose. In: The 8th GOSPEL workshop. Gas sensors based on semiconducting metal oxides: basic understanding & application fields. MDPI; 2019b. p. 33.

Power RC, Salazar-García DC, Straus LG, González Morales MR, Henry AG. Microremains from El Mirón Cave human dental calculus suggest a mixed plant–animal subsistence economy during the Magdalenian in Northern Iberia. J Archaeol Sci. 2015;60:39–46.

Prance GT. The use of edible fungi by Amazonian Indians. Adv Econ Bot. 1984;1:127–39.

Procházka P, Soukupová J, Tomšík K, Mullen KJ, Čábelková I. Climatic factors affecting wild mushroom foraging in Central Europe. Forests. 2023;14:382.

Psurtseva NV. Modern taxonomy and medicinal value of the Flammulina mushrooms. Int J Med Mushrooms. 2005:449–51.

Qi L-M, Zhang J, Liu H-G, Li T, Wang Y-Z. Fourier transform mid-infrared spectroscopy and chemometrics to identify and discriminate *Boletus edulis* and *Boletus tomentipes* mushrooms. Int J Food Prop. 2017;20:S56–68.

Qing Z, Cheng J, Wang X, Tang D, Liu X, Zhu M. The effects of four edible mushrooms (Volvariella volvacea, Hypsizygus marmoreus, Pleurotus ostreatus and Agaricus bisporus) on physicochemical properties of beef paste. LWT. 2021;135:110063.

Quintero-Cabello KP, Lugo-Flores MA, Rivera-Palafox P, Silva-Espinoza BA, González-Aguilar GA, Esqueda M, Gaitán-Hernández R, Ayala-Zavala JF. Antioxidant properties and industrial uses of edible Polyporales. JoF. 2021;7:196.

Rahi DK, Malik D. Diversity of mushrooms and their metabolites of nutraceutical and therapeutic significance. J Mycol. 2016;2016:1–18.

Rahmat IR, Putera MI, Asrul L. Growth and production of white oyster mushroom (Pleurotus ostreatus) by adding coconut water to agricultural waste as a carbon source media. IOP Conf Ser: Earth Environ Sci. 2020;575:012090.

Ramsbottom J. Mushrooms and toadstools. Proc Nutr Soc. 1953;12:39–44.

Ranjitkar S, Sujakhu NM, Kunwar RM, Sher H, Ur Rahman I, Hussain W, Bussmann RW, Paniagua-Zambrana NY. Morchella conica Pers. ex Fr. Morchella esculenta (Fr.) I.R. Hazll Morchellaceae. In: Kunwar RM, Sher H, Bussmann RW, editors. Ethnobotany of the Himalayas. Cham: Springer International Publishing; 2020. p. 1–16.

Raza M, Cai L, Abbasi MW, Hafeez R, Tariq M, Kirk PM, Hussain M, Wijayawardene NN. The first updated checklist of novel fungi in Pakistan (1947–2021). MycoAsia. 2022:1.

Reginald Buller AH. The fungus lore of the Greeks and romans. Trans Br Mycol Soc. 1914;5:21–66.

Reis FS, Barros L, Martins A, Ferreira ICFR. Chemical composition and nutritional value of the most widely appreciated cultivated mushrooms: an inter-species comparative study. Food Chem Toxicol. 2012a;50:191–7.

Reis FS, Stojković D, Soković M, Glamočlija J, Ćirić A, Barros L, Ferreira ICFR. Chemical characterization of *Agaricus bohusii*, antioxidant potential and antifungal preserving properties when incorporated in cream cheese. Food Res Int. 2012b;48:620–6.

Restrepo CM, Llanes A, Lleonart R. Use of AFLP for the study of eukaryotic pathogens affecting humans. Infect Genet Evol. 2018;63:360–9.

Revisiting Ramaria species: the coral fungi as food and pharmaceuticals. Biointerface Res Appl Chem. 2020;11:10790–800.

Rhodes CJ. The Whispering World of Plants: 'The Wood Wide Web.'. Sci Prog. 2017;100:331–7.

Ridwan AY, Wu J, Choi J-H, Hirai H, Kawagishi H. Bioactive compounds from the edible mushroom Cortinarius caperatus. Mycoscience. 2018;59:172–5.

Ritota M, Manzi P. Pleurotus spp. cultivation on different agri-food by-products: example of biotechnological application. Sustainability. 2019;11:5049.

Rodrigues DMF, Freitas AC, Rocha-Santos TAP, Vasconcelos MW, Roriz M, Rodríguez-Alcalá LM, Gomes AMP, Duarte AC. Chemical composition and nutritive value of Pleurotus citrino-

pileatus var cornucopiae, P. eryngii, P. salmoneo stramineus, Pholiota nameko and Hericium erinaceus. J Food Sci Technol. 2015;52:6927–39.
Rosemary Kinge T, Apiseh Apalah N, Mue Nji T, Neh Acha A, Mathias Mih A. Species richness and traditional knowledge of macrofungi (mushrooms) in the awing forest reserve and communities, Northwest Region, Cameroon. J Mycol. 2017;2017:2809239.
Royse DJ, Baars J, Tan Q. Current overview of mushroom production in the world. In: Diego CZ, Pardo-Giménez A, editors. Edible and medicinal mushrooms. Wiley; 2017. p. 5–13.
Ruiz-Almenara C, Gándara E, Gómez-Hernández M. Comparison of diversity and composition of macrofungal species between intensive mushroom harvesting and non-harvesting areas in Oaxaca, Mexico. PeerJ. 2019;7:e8325.
Ruiz-Moyano S, Esperilla A, Hernández A, Benito MJ, Casquete R, Martín-Vertedor D, Pérez-Nevado F. Application of ISSR-PCR as a rapid method for clustering and typing of yeasts isolated from table olives. LWT. 2019;109:250–4.
Saba M, Falandysz J, Nnorom IC. Mercury bioaccumulation by Suillus bovinus mushroom and probable dietary intake with the mushroom meal. Environ Sci Pollut Res. 2016;23:14549–59.
Saha S, Khatun L, Rima Biswas M, Islam R, Karim MM, Rahman M, Kakon AJ, Md. Abul Kalam Azad. Genetic discrimination of button mushroom Agaricus bisporus by DNA fingerprinting using RAPD marker. World J Adv Res Rev. 2024;21:1059–65.
Saito T, Sakuta G, Kobayashi H, Ouchi K, Inatomi S. Genetically independent Tetranucleotide to Hexanucleotide core motif SSR markers for identifying *Lentinula edodes* cultivars. Mycobiology. 2019;47:466–72.
Sakamoto Y, Ando A, Tamai Y, Yajima T. Pileus differentiation and pileus-specific protein expression in Flammulina velutipes. Fungal Genet Biol. 2007;44:14–24.
Salazar-Vidal V, Dibán MJ, Rugolo M, González G, Barroetaveña C, Urrutia E, Urrutia H. Diversity, phylogeny and taxonomy of forest fungal communities. In: Forest Fungi. Elsevier; 2025. p. 21–46.
Sánchez-Ramírez S, Tulloss RE, Amalfi M, Moncalvo J. Palaeotropical origins, boreotropical distribution and increased rates of diversification in a clade of edible ectomycorrhizal mushrooms (*Amanita* section *Caesareae*). J Biogeogr. 2015;42:351–63.
Sangvichien E, Taylor-Hawksworth PA. *Termitomyces* mushrooms: a tropical delicacy. Mycologist. 2001;15:31–3.
Savoie J-M, Largeteau ML. Production of edible mushrooms in forests: trends in development of a mycosilviculture. Appl Microbiol Biotechnol. 2011;89:971–9.
Seitz S. 39 different types of edible mushrooms. Clean Green Simple. 2020. Available at https://cleangreensimple.com/article/different-types-of-mushrooms/.
Semwal KC, editor. Wild mushrooms and health diversity, phytochemistry, medicinal benefits, and cultivation. S.l: CRC PRESS; 2024.
Semwal KC, Bhatt VK, Hussen A, Chauhan A. The genus Ramaria (Basidiomycota, Agaricales): diversity, edibility, and bioactivity. In: Edible and Medicinal Mushrooms of the Himalayas. CRC Press; 2023.
Shan L, Wang D, Li Y, Zheng S, Xu W, Shang Y. Identification of the pol gene as a species-specific diagnostic marker for qualitative and quantitative PCR detection of Tricholoma matsutake. Molecules. 2019;24:455.
Shankar BL, Veena TR, Pawar S, Gowda A, Kadadevaramath AR. A study on value chain of mushroom for value addition: challenges, opportunities and prospects of cultivation of mushroom. Int J Bus Syst Res. 2024;18:539–54.
Sharma N, Tapwal A, Verma R, Kumar D, Nepovimova E, Kuca K. Medicinal, nutritional, and nutraceutical potential of Sparassis crispa s. lat.: a review. IMA Fungus. 2022;13:8.
Sheng-jin WU, Xue-feng C, Zeng-liang LIU, Wen-long Z. Identification of *Lentinula edodes* (Berk.) Pegler hybrid progenies from dikaryon-monokaryon mating using SCAR molecular marker. J Southern Agric. 2022; https://doi.org/10.3969/j.issn.2095-1191.2022.10.018.

Shirur M, Barh A, Annepu SK. Sustainable production of edible and medicinal mushrooms: implications on mushroom consumption. In: Hebsale Mallappa VK, Shirur M, editors. Climate change and resilient food systems. Singapore: Springer Singapore; 2021. p. 315–46.

Shnyreva AA, Sivolapova AB, Shnyreva AV. The commercially cultivated edible oyster mushrooms Pleurotus sajor-caju and P. pulmonarius are two separate species, similar in morphology but reproductively isolated. Russ J Genet. 2012;48:1080–8.

Siddique AB, Khokon AM, Unterseher M. What do we learn from cultures in the omics age? High-throughput sequencing and cultivation of leaf-inhabiting endophytes from beech (Fagus sylvatica L.) revealed complementary community composition but similar correlations with local habitat conditions. MC. 2017;20:1–16.

Sileshi GW, Tibuhwa DD, Mlambo A. Underutilized wild edible fungi and their undervalued ecosystem services in Africa. CABI Agric Biosci. 2023;4:3.

Silva M, Ramos AC, Lidon FJ, Reboredo FH, Gonçalves EM. Pre- and postharvest strategies for Pleurotus ostreatus mushroom in a circular economy approach. Foods. 2024;13:1464.

Singer R. The Agaricales in modern taxonomy: with 88 plates, 4, fully rev. ed. Koenigstein: Koeltz Scienctif. Books; 1986.

Singh C, Vyas D, Singh C, Vyas D. Biodegradation by fungi for humans and plants nutrition. In: Biodegradation technology of organic and inorganic pollutants. IntechOpen; 2021.

Singh A, Ramakrishna G, Kaila T, Saxena S, Sharma S, Gaikwad AB, Abdin MZ, Gaikwad K. Next-generation sequencing technologies: approaches and applications for crop improvement. In: Wani SH, Kumar A, editors. Genomics of cereal crops. New York: Springer US; 2022. p. 31–94.

Sitta N, Davoli P. Edible ectomycorrhizal mushrooms: international markets and regulations. In: Zambonelli A, Bonito GM, editors. Edible ectomycorrhizal mushrooms: current knowledge and future prospects. Berlin, Heidelberg: Springer; 2012. p. 355–80.

Sitta N, Floriani M. Nationalization and globalization trends in the wild mushroom commerce of Italy with emphasis on porcini (Boletus edulis and allied species). Econ Bot. 2008;62:307–22.

Słyszyk K, Siwulski M, Wiater A, Tomczyk M, Waśko A. Biofortification of mushrooms: a promising approach. Molecules. 2024;29:4740.

Sonji GM, Assi M, Sonji NM, Boukhary R, Rahal M. Antioxidant activity, metal content, and essential oil composition of two desert truffles species: Terfezia boudieri and Terfezia claveryi. J Pharmacogn Phytochem. 2022;11:10–9.

Sonnenberg R, Nolte AW, Tautz D. An evaluation of LSU rDNA D1-D2 sequences for their use in species identification. Front Zool. 2007;4:6.

Spencer PS, Palmer VS. Direct and indirect neurotoxic potential of metal/metalloids in plants and fungi used for food, dietary supplements, and herbal medicine. Toxics. 2021;9:57.

Stabnikova O, Stabnikov V, Paredes-López O. Wild and cultivated mushrooms as food, pharmaceutical and industrial products. Ukr Food J. 2024;13:20–59.

Stamets PE. Notes on nutritional properties of culinary-medicinal mushrooms. IJM. 2005:7.

Stojek K, Gillerot L, Jaroszewicz B. Predictors of mushroom production in the European temperate mixed deciduous forest. For Ecol Manag. 2022;522:120451.

Stojek K, Czortek P, Bobrowska-Korczak B, Krośniak M, Jaroszewicz B. Fungal species and element type modulate the effects of environmental factors on the concentration of potentially toxic elements in mushrooms. Environ Pollut. 2024;353:124152.

Stojković D, Reis FS, Barros L, Glamočlija J, Ćirić A, Van Griensven LJID, Soković M, Ferreira ICFR. Nutrients and non-nutrients composition and bioactivity of wild and cultivated Coprinus comatus (O.F.Müll.) Pers. Food Chem Toxicol. 2013;59:289–96.

Stubbe D, Verbeken A. Lactarius subg. Plinthogalus: the European taxa and American varieties of L. lignyotus re-evaluated. Mycologia. 2012;104:1490–501.

Su H, Wang L, Ge Y, Feng E, Sun J, Liu L. Development of strain-specific SCAR markers for authentication of Ganoderma lucidum. World J Microbiol Biotechnol. 2008;24:1223–6.

Sudharsan MS, Rajagopal K, Banu N. An insight into fungi in forest ecosystems. In: Rashad YM, Baka ZAM, Moussa TAA, editors. Plant Mycobiome. Cham: Springer International Publishing; 2023. p. 291–318.

Sun X, Feng W, Li M, Shi J, Ding G. Phenology and cultivation of Suillus bovinus, an edible mycorrhizal fungus, in a Pinus massoniana plantation. Can J For Res. 2019;49:960–8.

Sun L, Zhang Z, Xin G, Sun B, Bao X, Wei Y, Zhao X, Xu H. Advances in umami taste and aroma of edible mushrooms. Trends Food Sci Technol. 2020;96:176–87.

Sun N, Gao C, Ding Y, Bi Y, Seglah PA, Wang Y. Five-dimensional straw utilization model and its impact on carbon emission reduction in China. Sustainability. 2022a;14:16722.

Sun X, Yang C, Ma Y, Zhang J, Wang L. Research progress of Auricularia heimuer on cultivation physiology and molecular biology. Front Microbiol. 2022b;13:1048249.

Szućko-Kociuba I, Trzeciak-Ryczek A, Kupnicka P, Chlubek D. Neurotrophic and neuroprotective effects of Hericium erinaceus. Int J Mol Sci. 2023;24:15960.

Tagkouli D, Kaliora A, Bekiaris G, Koutrotsios G, Christea M, Zervakis GI, Kalogeropoulos N. Free amino acids in three Pleurotus species cultivated on agricultural and agro-industrial by-products. Molecules. 2020;25:4015.

Takahashi H. New species of Clitocybe and Crepidotus (Agaricales) from eastern Honshu, Japan. Mycoscience. 2003;44:0103–7.

Tang Y, Li H-M, Tang Y-J. Comparison of sterol composition between Tuber fermentation mycelia and natural fruiting bodies. Food Chem. 2012;132:1207–13.

Taşkın H, Süfer Ö, Attar ŞH, Bozok F, Baktemur G, Büyükalaca S, Kafkas NE. Total phenolics, antioxidant activities and fatty acid profiles of six Morchella species. J Food Sci Technol. 2021;58:692–700.

Tedder S, Mitchell DA, Hillyer A. Property rights and the sustainable management of non-timber forest products. ResearchGate Available at. 2002; https://www.researchgate.net/publication/228868852_Property_rights_and_the_sustainable_management_of_non-timber_forest_products

Tedersoo L, Bahram M, Zobel M. How mycorrhizal associations drive plant population and community biology. Science. 2020;367:eaba1223.

Tejedor-Calvo E, Amara K, Reis FS, Barros L, Martins A, Calhelha RC, Venturini ME, Blanco D, Redondo D, Marco P, Ferreira ICFR. Chemical composition and evaluation of antioxidant, antimicrobial and antiproliferative activities of *Tuber* and *Terfezia* truffles. Food Res Int. 2021;140:110071.

Temesgen T. Application of mushroom as food and medicine. Adv Biotechnol Microbiol. 2018:11.

Thakur MP. Advances in mushroom production: key to food, nutritional and employment security: a review. Indian Phytopathol. 2020a;73:377–95.

Thakur MP. Advances in mushroom production: key to food, nutritional and employment security: a review. Indian Phytopathol. 2020b;73:377–95.

Thakur M. Ethnomedicinal aspects of morels with special reference to Morchella esculenta (Guchhi) in Himachal Pradesh (India): a review. CREAM. 2021;11:284–93.

Tharnrat Kaewgrajang SK. Morphology and molecular identification of some Lactarius and Russula species. Genomics Genetics. 2020;13:4458.

Thawthong A. Discovering and domesticating wild tropical cultivatable mushrooms. Chiang Mai J Sci. 2014;41:731–64.

Thongbai B, Rapior S, Hyde KD, Wittstein K, Stadler M. Hericium erinaceus, an amazing medicinal mushroom. Mycol Prog. 2015;14:91.

Thu ZM, Myo KK, Aung HT, Clericuzio M, Armijos C, Vidari G. Bioactive phytochemical constituents of wild edible mushrooms from Southeast Asia. Molecules. 2020;25:1972.

Thuc LV, Corales RG, Sajor JT, Truc NTT, Hien PH, Ramos RE, Bautista E, Tado CJM, Ompad V, Son DT, Van Hung N. Rice-straw mushroom production. In: Gummert M, Hung NV, Chivenge P, Douthwaite B, editors. Sustainable rice straw management. Cham: Springer International Publishing; 2020. p. 93–109.

Tibuhwa DD. Folk taxonomy and use of mushrooms in communities around Ngorongoro and Serengeti National Park, Tanzania. J Ethnobiol Ethnomed. 2012;8:36.

Touml G, Rős TH, El-Ramady H, Prokisch J. Edible mushroom of Pleurotus spp.: a case study of Oyster Mushroom (Pleurotus ostreatus L.). Environ Biodivers Soil Security. 2022; https://doi.org/10.21608/jenvbs.2022.117554.1161.

Trudell SA, Xu J, Saar I, Justo A, Cifuentes J. North American matsutake: names clarified and a new species described. Mycologia. 2017;109:379–90.

Tsai S-Y, Tsai H-L, Mau J-L. Antioxidant properties of *Agaricus blazei*, *Agrocybe cylindracea*, and *Boletus edulis*. LWT Food Sci Technol. 2007;40:1392–402.

Tsuji M. Survey on fungi in Antarctica and high Arctic regions, and their impact on climate change. Climate. 2023;11:195.

Tylš F, Páleníček T, Horáček J. Psilocybin—summary of knowledge and new perspectives. Eur Neuropsychopharmacol. 2014;24:342–56.

Ullah TS, Firdous SS, Shier WT, Hussain J, Shaheen H, Usman M, Akram M, Khalid AN. Diversity and ethnomycological importance of mushrooms from Western Himalayas, Kashmir. J Ethnobiol Ethnomed. 2022;18:32.

Vaagt F, Haase I, Fischer M. Loop-mediated isothermal amplification (LAMP)-based method for rapid mushroom species identification. J Agric Food Chem. 2013;61:1833–40.

Valverde ME, Hernández-Pérez T, Paredes-López O. Edible mushrooms: improving human health and promoting quality life. Int J Microbiol. 2015;2015:376387.

van Amsterdam J, Opperhuizen A, van den Brink W. Harm potential of magic mushroom use: a review. Regul Toxicol Pharmacol. 2011;59:423–9.

van Griensven LJLD. The Cultivation of mushrooms, 1st English ed. Rustington, Sussex: Darlington Mushroom Laboratories; 1988.

Varga D, Hanss J-M, Moreau P-A, Kovács GM, Dima B. Phylogenetic and morphological studies reveal large diversity and three new species in Amanita sect. Vaginatae (Agaricales, Basidiomycota) from Europe. Mycol Progress. 2024;23:38.

Vaz JA, Heleno SA, Martins A, Almeida GM, Vasconcelos MH, Ferreira ICFR. Wild mushrooms Clitocybe alexandri and Lepista inversa: in vitro antioxidant activity and growth inhibition of human tumour cell lines. Food Chem Toxicol. 2010;48:2881–4.

Vetter J. Arsenic content of some edible mushroom species. Eur Food Res Technol. 2004;219:71–4.

Vidal-Diez De Ulzurrun G, Lee Y-Y, Stajich JE, Schwarz EM, Hsueh Y-P. Genomic analyses of two Italian oyster mushroom *Pleurotus pulmonarius* strains. G3. 2021;11:jkaa007.

Vidović SS, Mujić IO, Zeković ZP, Lepojević ŽD, Tumbas VT, Mujić AI. Antioxidant properties of selected boletus mushrooms. Food Biophys. 2010;5:49–58.

Villagrán Z, Martínez-Reyes M, Gómez-Rodríguez H, Ríos-García U, Montalvo-González E, Ortiz-Basurto RI, Anaya-Esparza LM, Pérez-Moreno J. Huitlacoche (Ustilago maydis), an iconic Mexican fungal resource: biocultural importance, nutritional content, bioactive compounds, and potential biotechnological applications. Molecules. 2023;28:4415.

Virágh M, Merényi Z, Csernetics Á, Földi C, Sahu N, Liu X-B, Hibbett DS, Nagy LG. Evolutionary morphogenesis of sexual fruiting bodies in Basidiomycota: toward a new Evo-Devo synthesis. Microbiol Mol Biol Rev. 2021;86:e00019–21.

Viriato V, Carvalho SADD, Santoro BDL, Bonfim FPG. A business model for circular bioeconomy: edible mushroom production and its alignment with the sustainable development goals (SDGs). Recycling. 2024;9:68.

Vizzini A, Musumeci E, Murat C. Trichocybe, a new genus for Clitocybe puberula (Agaricomycetes, Agaricales). Fungal Divers. 2010;42:97–105.

Vunduk J, Biketova AY. Chapter 9. The cultural use of mushrooms. In: Stojkovic D, Barros L, editors. Food chemistry, function and analysis. Cambridge: Royal Society of Chemistry; 2022. p. 327–58.

Vuylsteke M, Peleman JD, Van Eijk MJ. AFLP technology for DNA fingerprinting. Nat Protoc. 2007;2:1387–98.

Wachtel-Galor S, Tomlinson B, Benzie IFF. Ganoderma lucidum ("Lingzhi"), a Chinese medicinal mushroom: biomarker responses in a controlled human supplementation study. Br J Nutr. 2004;91:263–9.

Wang Y, Chen YL. Recent advances in cultivation of edible mycorrhizal mushrooms. In: Solaiman ZM, Abbott LK, Varma A, editors. Mycorrhizal fungi: use in sustainable agriculture and land restoration. Berlin, Heidelberg: Springer Berlin Heidelberg; 2014. p. 375–97.

Wang M, Zhao R. A review on nutritional advantages of edible mushrooms and its industrialization development situation in protein meat analogues. J Fut Foods. 2023;3:1–7.

Wang L, Hu X, Feng Z, Pan Y. Development of AFLP markers and phylogenetic analysis in Hypsizygus marmoreus. J Gen Appl Microbiol. 2009;55:9–17.

Wang X-M, Zhang J, Li T, Wang Y-Z, Liu H-G. Content and bioaccumulation of nine mineral elements in ten mushroom species of the genus boletus. J Analyt Methods Chem. 2015a;2015:165412.

Wang Z-R, Parra LA, Callac P, Zhou J-L, Fu W-J, Dui S-H, Hyde KD, Zhao R-L. Edible species of Agaricus (Agaricaceae) from Xinjiang Province (Western China). Phytotaxa. 2015b;202:185.

Wang PM, Liu XB, Dai YC, Horak E, Steffen K, Yang ZL. Phylogeny and species delimitation of Flammulina: taxonomic status of winter mushroom in East Asia and a new European species identified using an integrated approach. Mycol Prog. 2018;17:1013–30.

Wang L, Guo H, Liu X, Jiang G, Li C, Li X, Li Y. Roles of Lentinula edodes as the pork lean meat replacer in production of the sausage. Meat Sci. 2019a;156:44–51.

Wang L, Li C, Ren L, Guo H, Li Y. Production of pork sausages using *Pleaurotus eryngii* with different treatments as replacements for pork back fat. J Food Sci. 2019b;84:3091–8.

Wang X, Xu M, Cheng J, Zhang W, Liu X, Zhou P. Effect of Flammulina velutipes on the physicochemical and sensory characteristics of Cantonese sausages. Meat Sci. 2019c;154:22–8.

Wang Y-Y, Li J-Q, Liu H-G, Wang Y-Z. Attenuated Total reflection-Fourier transform infrared spectroscopy (ATR-FTIR) combined with chemometrics methods for the classification of Lingzhi species. Molecules. 2019d;24:2210.

Wang L, Li J, Zhang J, Li Z, Liu H, Wang Y. Traditional uses, chemical components and pharmacological activities of the genus Ganoderma P. Karst.: a review. RSC Adv. 2020;10:42084–97.

Wardley B. American dietetic association complete food and nutrition guide. Top Clin Nutr. 2007;22:307.

Wasser SP. Medicinal mushroom science: history, current status, future trends, and unsolved problems. Int J Med Mushrooms. 2010:12.

Wasson RG. The Soma of the Rig Veda: what was it? J Am Orient Soc. 1971;91:169–87.

Wasson RG. The road to Eleusis: unveiling the secret of the mysteries. New York: Harcourt, Brace, Jovanovich; 1978.

Wei Y, Li L, Liu Y, Xiang S, Zhang H, Yi L, Shang Y, Xu W. Identification techniques and detection methods of edible fungi species. Food Chem. 2022;374:131803.

Winkler D. Fruits of the forest: a field guide to Pacific northwest edible mushrooms. Seattle, Washington: Skipstone; 2023.

Xie C, Gong W, Yan L, Zhu Z, Hu Z, Peng Y. Biodegradation of ramie stalk by Flammulina velutipes: mushroom production and substrate utilization. AMB Express. 2017;7:171.

Xu X, Yan H, Chen J, Zhang X. Bioactive proteins from mushrooms. Biotechnol Adv. 2011;29:667–74.

Xu Z, Dai J, Su W, Wu H, Shah K, Xing L, Ma J, Zhang D, Zhao C. Selection and validation of reliable reference genes for gene expression studies in different genotypes and TRV-infected fruits of peach (Prunus persica L. Batsch) during ripening. Genes. 2022;13:160.

Yan Z, Liu H, Li J, Wang Y. Application of identification and evaluation techniques for edible mushrooms: a review. Crit Rev Anal Chem. 2023;53:634–54.

Yang FQ, Feng K, Zhao J, Li SP. Analysis of sterols and fatty acids in natural and cultured *Cordyceps* by one-step derivatization followed with gas chromatography–mass spectrometry. J Pharm Biomed Anal. 2009;49:1172–8.

Yang T, Zhang J, Li T, Wang Y, Liu H. Discrimination of Boletus Tomentipes from different regions based on infrared spectrum combined with principal component analysis and cluster analysis. Guang Pu Xue Yu Guang Pu Fen Xi. 2016a;36:1726–30.

Yang W, Yu J, Pei F, Mariga AM, Ma N, Fang Y, Hu Q. Effect of hot air drying on volatile compounds of Flammulina velutipes detected by HS-SPME–GC–MS and electronic nose. Food Chem. 2016b;196:860–6.

Yao L, Lu J, Qu M, Jiang Y, Li F, Guo Y, Wang L, Zhai Y. Methodology and application of PCR-RFLP for species identification in tuna sashimi. Food Sci Nutr. 2020;8:3138–46.

Ye L, Karunarathna SC, Li H, Xu J, Hyde KD, Mortimer PE. A survey of Termitomyces (Lyophyllaceae, Agaricales), including a new species, from a Subtropical Forest in Xishuangbanna, China. Mycobiology. 2019;47:391–400.

Yoo S, Moon S, Hong CP, Park S-G, Shim D, Ryu H. Genome sequencing of Lentinula edodes revealed a genomic variant block associated with a thermo-tolerant trait in fruit body formation. J Fungi (Basel). 2024;10:628.

Yu X. Six groups of poisonous mushrooms: classified according to clinical symptoms. Highlights Sci Eng Technol. 2022;19:216–22.

Yuan B, Zhao L, Yang W, McClements DJ, Hu Q. Enrichment of bread with nutraceutical-rich mushrooms: impact of *Auricularia auricula* (mushroom) flour upon quality attributes of wheat dough and bread. J Food Sci. 2017;82:2041–50.

Yu-Li HAN, Di W, Jie Z, Xiao-Lan HE, Bao-Kai CUI. Intragenomic polymorphisms of the ITS, EF1-α and RPB2 sequences of Ganoderma australe. Mycosystema. 2019;38:679–85.

Yun W, Hall IR. Edible ectomycorrhizal mushrooms: challenges and achievements. Can J Bot. 2004;82:1063–73.

Zaiwei G, Xiaobin L, Kuan Z, Zhuliang Y. Species diversity of Flammulina in China: new varieties and a new record. Junwu Xuebao. 2015;34:589–603.

Zambonelli A, Bonito GM. Edible ectomycorrhizal mushrooms: current knowledge and future prospects. Springer Science & Business Media; 2013.

Zambonelli A, Leonardi P, Iotti M, Hall I. Ecological and genetic advances in the cultivation of Tuber Spp. Rev Fitotec Mex. 2017;40:371–7.

Zambonelli A, Iotti M, Puliga F, Hall IR. Enhancing white truffle (Tuber magnatum Picco and T. borchii Vittad.) cultivation through biotechnology innovation. In: Al-Khayri JM, Jain SM, Johnson DV, editors. Advances in plant breeding strategies: vegetable crops: volume 10: leaves, flowerheads, green pods, mushrooms and truffles. Cham: Springer International Publishing; 2021. p. 505–32.

Zeković Z, Vidović S, Mujić I. Selenium and zinc content and radical scavenging capacity of edible mushrooms Armilaria mellea and Lycoperdon saccatum. Croat J Food Sci Technol. 2010;2:16–21.

Zhang Y, Geng W, Shen Y, Wang Y, Dai Y-C. Edible mushroom cultivation for food security and rural development in China: bio-innovation, technological dissemination and marketing. Sustainability. 2014;6:2961–73.

Zhang C, He M, Liu J, Xu X, Yang C, Gao F, Fang Y, Wang W, Wang Y. Brief introduction to a unique edible Bolete—Phlebopus portentosus in Southern China. JAST-B. 2017;7:386–94.

Zhang H, Pu D, Sun B, Ren F, Zhang Y, Chen H. Characterization and comparison of key aroma compounds in raw and dry porcini mushroom (Boletus edulis) by aroma extract dilution analysis, quantitation and aroma recombination experiments. Food Chem. 2018;258:260–8.

Zhang C, Lei Z, Li Y, Yi L, Shang Y. Identification of Tps2 used as an endogenous reference gene in qualitative and real-time quantitative PCR detection of Flammulina velutipes. Food Anal Methods. 2021a;14:2152–60.

Zhang Y, Mo M, Yang L, Mi F, Cao Y, Liu C, Tang X, Wang P, Xu J. Exploring the species diversity of edible mushrooms in Yunnan, Southwestern China, by DNA barcoding. J Fungi. 2021b;7:310.

Zhang M, Wang C-Q, Gan M-S, Li Y, Shao S-C, Qin W-Q, Deng W-Q, Li T-H. Diversity of Cantharellus (Cantharellales, Basidiomycota) in China with description of some new species and new records. J Fungi. 2022;8:483.

Zhang M, Gao X-L, Mu L-Q, Deng W-Q. Morphology and molecular phylogeny reveal five new species of Laccaria (Hydnangiaceae, Agaricales) from Southern China. J Fungi. 2023;9:1179.

Zhao H, Xiong M, Yang X, Yao L, Wang Z, Wang L, Li Z, Zhang J, Lv J. Six new polyphenolic metabolites isolated from the Suillus granulatus and their cytotoxicity against HepG2 cells. Front Nutr. 2024:11.

Zheng S, Shan L, Zhuang Y, Shang Y. Identification of *pyrG* used as an endogenous reference gene in qualitative and real-time quantitative PCR detection of *Pleurotus ostreatus*. J Food Sci. 2018;83:750–5.

Zhou J, Feng T, Ye R. Differentiation of eight commercial mushrooms by electronic nose and gas chromatography-mass spectrometry. J Sens. 2015;2015:1–14.

Zhou F, Yan S, Chen S, Gong L, Su T, Wang Z. Optimization extraction process of polysaccharides from Suillus granulatus and their antioxidant and immunological activities in vitro. Pharmacogn Mag. 2016;12:S277–84.

Zhou T, Hu W, Yang Z, Li J, Zeng X. Study on nutrients, non-volatile compounds, volatile compounds and antioxidant capacity of oyster mushroom cultivated with corn distillers' grains. LWT. 2023;183:114967.

Zhu R, Wen Y, Wu W, Zhang L, Salman Farid M, Shan S, Wen J, Farag MA, Zhang Y, Zhao C. The flavors of edible mushrooms: a comprehensive review of volatile organic compounds and their analytical methods. Crit Rev Food Sci Nutr. 2024;64:5568–82.

Zied DC, Pardo-Giménez A. Edible and medicinal mushrooms: technology and applications. John Wiley & Sons; 2017.

Zjawiony JK. Biologically active compounds from Aphyllophorales (Polypore) fungi. J Nat Prod. 2004;67:300–10.

Zou Y-J, Wang H-X, Ng T-B, Huang C-Y, Zhang J-X. Purification and characterization of a novel laccase from the edible mushroom Hericium coralloides. J Microbiol. 2012;50:72–8.

Chapter 2
Domesticated Mushrooms: Techniques and Technologies

Nasir Ahmed Rajput, Muhammad Atiq, Muhammad Wahab, Muhammad Usman, Ayesha Sadiqa, Anosh Majeed, and Ayesha Shahzadi

Abstract This chapter delves into the techniques and technologies of domesticated mushroom cultivation, emphasizing its significance in sustainable agriculture, nutrition, and environmental conservation. Mushrooms are recognized for their ability to transform agricultural waste into protein-rich food, addressing malnutrition and reducing nutrient wastage. The chapter highlights the global importance of mushrooms, their nutritional and medicinal benefits, and their role in sustainable forestry and agriculture. It explores advancements in controlled environment agriculture (CEA), substrate preparation, spawn innovations, and postharvest technologies that optimize yields and quality. The cultivation of popular species such as button, oyster, and shiitake mushrooms is discussed, alongside the utilization of agricultural waste for substrate preparation. Furthermore, emerging preservation techniques like freeze-drying, high-pressure processing, and innovative packaging methods are examined to enhance shelf life and maintain nutritional value. This comprehensive overview underscores the potential of mushroom cultivation as a biotechnological solution to global food challenges and environmental sustainability.

Keywords Mushroom cultivation · Controlled environment agriculture (CEA) · Spawn innovations · Biotechnological · Sustainability

2.1 Introduction

The kingdom of fungi, which encompasses mushrooms, is fundamentally distinct from the kingdoms of microbes, plants, and animals. Mushrooms have gained global recognition due to their nutritional and therapeutic properties (Gupta et al.

N. A. Rajput (✉) · M. Atiq · M. Wahab · M. Usman · A. Sadiqa · A. Majeed · A. Shahzadi
Department of Plant Pathology, University of Agriculture, Faisalabad, Pakistan
e-mail: nasir.ahmed@uaf.edu.pk

I. U. Haq et al. (eds.), *Mushrooms*, https://doi.org/10.1007/978-981-95-2539-3_2

2022). Their cultivation represents a valuable bioconversion method, transforming organic waste into useful resources. This addresses waste management and contributes to a more sustainable and environmentally friendly way of farming (Aditya et al. 2024). By using agricultural waste as a growing medium, mushroom cultivation facilitates the conversion of nutrient-rich wastes into edible forms, preventing nutrient wastage and the eutrophication of water bodies. The spent substrate from mushroom production can then be utilized in horticultural activities, as organic fertilizer, or even as animal feed. This approach holds promising potential for sustainable forestry and agriculture, particularly in regions like India, which generate abundant crop residues (Kamthan and Tiwari 2017).

Approximately 15,000 mushroom species have been identified globally, of which around 2000 are edible for humans and over 700 are recognized for their medicinal benefits. However, fewer than 100 species are cultivated commercially (Giusti et al. 2021). Mushroom cultivation is an economically viable process for the bioconversion of agricultural and agro-industrial wastes into protein-rich food, offering a potent solution against malnutrition in developing countries. The diverse applications of mushrooms highlight their ecological importance and economic potential, making them a focal point for advancements in agricultural practices and biotechnological innovations. Furthermore, mushroom production can improve the nutritional quality of lignocellulosic wastes, making them suitable as animal feed (Sganzerla et al. 2022).

2.2 Global Impact of Mushrooms: Significance in Agriculture and Nutrition

Mushrooms offer a promising solution for converting agricultural waste into protein-rich food without composting. Their production is easy and cost-effective, boasting high biological efficiency and significant nutritional and medicinal benefits (Mane et al. 2007). Medicinal mushrooms are increasingly used as natural bioagents due to their insecticidal, fungicidal, bactericidal, herbicidal, nematocidal, and antiphytoviral properties, effectively protecting against plant diseases (Chang and Wasser 2018). Beyond their culinary appeal, mushrooms are valued for their nutritional advantages, being rich in proteins, carbohydrates, fibers, vitamins, minerals, and fats while having low cholesterol levels, calories, and sodium (Shen et al. 2014). This nutritional profile makes mushrooms an excellent choice for health-conscious individuals. Incorporating mushrooms into diets can help reduce the risk of cancer, heart disease, obesity, diabetes, and other conditions while supporting overall immune function. Oyster mushrooms, in particular, are renowned for their numerous health benefits, including anticancer, antihypertensive, antioxidant, antidiabetic, antimicrobial, anti-inflammatory, and antitherogenic properties (Sganzerla et al. 2022). The economic significance of mushrooms lies in their value as a food and therapeutic resource for humans. Their ability to be cultivated economically on various substrates, including waste materials, positions them as a potential solution to global food challenges. Annually, the Earth generates vast amounts of organic matter through photosynthesis, underscoring the potential for sustainable mushroom cultivation (Zhang 2008).

2.3 Overview of Button, Oyster, and Shiitake Mushrooms in Controlled Environments

Controlled mushroom cultivation offers optimal growth conditions, resulting in higher yields and improved quality. The three most widely cultivated basidiomycetes globally are button mushrooms, oyster mushrooms, and shiitake (Kaiser and Ernst 2019). As mushroom gardening gains popularity, mushroom production continues to expand. Commercially grown mushroom strains are valued for their appealing appearance, texture, and substantial yield (Potocnik et al. 2015). The process of mushroom production involves several systematic steps designed to ensure optimal growth and yield: preparing the growth medium, sterilizing it, inoculating the medium, incubating the inoculated material, fruiting (the actual mushroom production), harvesting mature mushrooms, and marketing.

The oyster mushroom is the second most commonly cultivated species, following the white button mushroom, and holds a significant position among the many varieties grown worldwide (OECD 2006). Button mushrooms can be grown economically in various regions due to the simplicity of maintaining the necessary environmental conditions and the easy availability of substrates and supplements. The ideal temperature range for spawn-run (vegetative growth) is between 22 °C and 25 °C, and mushrooms require an environment with a relative humidity of 85–90% (Munshi et al. 2010). Oyster mushrooms are typically grown at temperatures ranging from 10 °C to 30 °C, with species such as winter oyster, *Pleurotus cystidiosus*, *Pleurotus ostreatus*, and *Pleurotus pulmonarius* often cultivated at temperatures close to or slightly below 30 °C. For oyster mushroom mycelial growth, substrate preparation methods include pasteurization or preheating, as well as sterilization (Hoa and Wang 2015). Shiitake mushrooms (*Lentinula edodes*) are cultivated at temperatures between 5 °C and 35 °C, with substrate preparation methods including wood logs, pasteurized or preheated substrates, and sterilized substrates (Oduro 2025).

2.4 Evolution of Mushroom Cultivation

Mushroom cultivation has undergone a profound transformation, evolving from traditional methods to cutting-edge techniques. Historically, mushroom cultivation relied on outdoor settings, such as open fields or basic growing conditions, which offered limited control over critical factors like temperature, humidity, and contamination (Dong et al. 2022). This conventional approach was labor-intensive and time-consuming and faced challenges in achieving consistent yields and high-quality harvests.

However, with advancements in scientific understanding and technology, the landscape of mushroom cultivation has been revolutionized. The advent of controlled environment agriculture (CEA) has been a pivotal turning point. CEA allows for precise regulation of conditions such as temperature, humidity, light, and air quality (Kavaliauskas et al. 2022). This level of control has resulted in higher yields, reduced contamination risks, and accelerated growth cycles. In practice, mushrooms are now grown in enclosed spaces like greenhouses or indoor facilities where variables can be fine-tuned to mimic optimal natural conditions. For instance, the cultivation of specialty mushrooms like shiitake, which require specific temperature and humidity ranges, has been greatly enhanced through CEA methods (Simos 2024).

The evolution of substrates for mushroom cultivation has also significantly advanced through the utilization of various agricultural wastes. This approach enhances mushroom yield and promotes sustainable farming practices by recycling organic materials. Common substrates include wheat straw, sawdust, sugarcane trash, and corn husks, which have shown high biological efficiency in mushroom production (Kavaliauskas et al. 2022). This shift toward sustainable practices underscores the potential of mushroom cultivation to contribute to a more environmentally friendly food system.

2.5 Substrate and Spawn Innovations

Mushroom production is a specialized agricultural sector where the selection and preparation of substrates play a crucial role. Substrates, which support fungal growth and provide essential nutrients, significantly influence mushroom cultivation's yield, quality, and success. Using various substrates in production involves describing their composition and preparation techniques, directly affecting the growth and productivity of different mushroom species (Suwannarach et al. 2022). Substrates impact several characteristics, including physical, chemical, biological, and sensory traits https://www.24high.com/en/blog/76/what-types-of-substrate-are-there-for-growing-magic-mushrooms. Agricultural wastes are increasingly used as substrates due to their accessibility, value, environmental impact, and potential benefits.

In mushroom cultivation, the "seed" or planting material is known as spawn, which consists of fungus mycelium grown on suitable substrates such as cereal grains. Since mycelium cannot be propagated directly, it is first cultivated on a carrier material like cereal grains to facilitate multiplication. Spawn is a vital factor in all crop production systems. However, one of the main challenges in mushroom production is the unavailability of high-quality spawn. Good-quality spawn ensures greater yield potential, no contaminants, and better economic returns (Çağlarırmak 2024).

2.6 Utilization of Agricultural Waste for Sustainable Substrate Preparation

Mushrooms are a nutritious, protein-packed food source globally, and their cultivation has been a long-standing practice. Over the years, research has focused on utilizing agricultural waste and agroforest residues as substrates for mushroom production. Commonly used materials include wheat straw, paddy straw, bran, coffee grounds, banana leaves, and sawdust. However, the availability of these substrates varies across different cultivating regions. Ongoing research aims to develop new methods and identify optimal substrate and supplement combinations to enhance mushroom production. Agricultural by-products such as areca husk, sugarcane bagasse, paddy straw, and maize stem are frequently employed in mushroom cultivation (Doroški et al. 2022).

Mushroom cultivation represents a significant biotechnological approach to reduce and repurpose agro-industrial waste. This process involves transforming low-value by-products into valuable resources that can be used to produce high-value products (Pala et al. 2021). Numerous examples of agricultural waste are used as substrates for mushroom cultivation (Krishnappa et al. 2022). Sawdust, a by-product from logging and timber mills, is commonly used due to its suitability as a growth medium and cost-effectiveness. However, increasing concerns about deforestation have made it challenging to obtain sawdust, leading to rising demand and higher production costs for mushroom growers (Chutimanukul et al. 2023). This has created a pressing need to find alternative sustainable sources of substrates.

2.7 Advances in High-Quality Spawn Development for Optimal Yields

Recent advancements in high-quality spawn development have significantly enhanced the efficiency of mushroom cultivation. By emphasizing optimal substrate preparation, innovative inoculation techniques, nutritional optimization, and rigorous quality control measures, growers can maximize yields while reducing the risks associated with contamination and environmental factors. This integrated approach ensures that mushroom cultivation is both productive and sustainable, contributing to improved overall outcomes in the industry.

2.8 Technologies for Cultivation and Processing of Domesticated Mushrooms

The mushroom industry has experienced significant advancements in developing novel methods to enhance flavor and maintain freshness (Rai and Arumuganathan 2008). Techniques such as high-pressure processing (HPP), freeze-drying, nuclear magnetic resonance (NMR) imaging, spectroscopy, and advanced packaging technologies have been introduced to improve mushroom preservation (Verma et al. 2022). Efficient postharvest processing and preservation methods help maintain mushrooms' nutritional content and sensory appeal (Dawadi et al. 2022; Kumar et al. 2021). Further research is needed to fully explore emerging techniques like ozone treatment, edible coatings, electron beam radiation, and innovative packaging solutions (Fernanda De Sousa et al. 2015).

Mushroom production techniques are environmentally friendly and crucial for sustainable agriculture. Future cultivation methods must be efficient to increase food supply, reduce costs, and enhance integrated pest management productivity. Mushroom cultivation requires sterile conditions and involves several steps: selecting mushroom spores or strains, maintaining mycelia cultures, creating inoculums, preparing growing media, inoculating spawn, colonizing substrate, and managing crops (Kumari and Baskaran 2015). The fundamental requirements for mushroom growth include compost, manure, spawning, and optimal humidity and temperature conditions. Ideal temperatures range from 20 °C to 28 °C during the spawn run and from 12 °C to 18 °C during reproductive growth, with relative humidity between 80% and 90% (Kratika 2018).

Traditional processing techniques such as drying, pickling, freezing, sterilizing, and canning are essential for extending mushroom products' shelf life and value. These strategies help stabilize prices throughout the year by reducing seasonal fluctuations (Rai and Arumuganathan 2008). Mushrooms are preserved using various methods, including blanching, pickling, canning, solar drying, hot air-drying, and shade drying (Jayaraman and Gupta 2020). Blanching is a pretreatment technique used before several food processing procedures and is vital for fruit and vegetable processing (Arroqui et al. 2003; Negi and Kumar Roy 2001).

2.9 Postharvest Technologies

Drying is one of the oldest and most effective techniques for preserving food and extending its shelf life. Indigenous communities have traditionally relied on drying and pickling to preserve wild mushrooms, with drying being the most common method (Hoyle 2018). Mushrooms can be dried using either sun drying or mechanical drying methods. Sun drying is the most affordable and widely used option, although it often results in a darker final product. Mechanical drying, however, is quicker and includes air-drying, vacuum drying, microwave oven-drying, tray drying, and freeze-drying. Rehydrated dried mushrooms can be used in various recipes, including stews, pickles, and soups (Piskov et al. 2020).

Freezing is another effective preservation technique, especially since mushrooms are over 90% water. To minimize adverse effects during freezing, mushrooms undergo various pretreatments. Cryogenic freezing at temperatures between −80 °C and −100 °C for 5–6 min can extend the shelf life of mushrooms to up to a year (Zito 2021). Sterilization methods include steam, chemicals, or radiation. Canning involves preserving food in marinades or brines, cooling, heat sterilization, labeling, and packaging for consumption or storage (Rai and Arumuganathan 2008). Pickling is a traditional preservation technique that uses spices like turmeric, red chili, garlic, clove, and salt combined with vinegar and oil, resulting in a subtle flavor through fermentation (Bone 2023).

Conventional processing techniques are crucial for reducing postharvest losses and increasing financial gains for growers and processors (Rai and Arumuganathan 2008). Hot air-drying is a popular method in the edible mushroom sector due to its ease of use and affordability, helping to overcome the challenges posed by complex processes and low investment (Guo et al. 2021).

2.10 IoT-Based Environmental Monitoring and Automation for Optimized Mushroom Growth

Integrating Internet of things (IoT) technologies in mushroom cultivation has revolutionized the industry by providing real-time monitoring and control of environmental conditions crucial for optimal growth. IoT-based systems enable precise regulation of temperature, humidity, light intensity, and carbon dioxide levels, essential for maximizing yield and maintaining product quality (Islam et al. 2022). These systems typically consist of sensor nodes, central control units, and cloud-based data management platforms, allowing remote monitoring and automated adjustments to maintain ideal conditions (Chong et al. 2023).

IoT-based systems have been successfully implemented in oyster mushroom farming, demonstrating significant enhancements in crop yield and quality by providing real-time data on environmental parameters. Microcontrollers like ESP8266 or Raspberry Pi facilitate the integration of sensors and actuators, enabling automated control of mist makers, ventilation systems, and other equipment to maintain optimal conditions (Rukhiran et al. 2023).

Adopting IoT technologies in mushroom cultivation enhances productivity and reduces labor costs and environmental impact by optimizing resource usage (Elijah et al. 2018). However, implementing these systems can be challenging due to initial setup costs and ongoing operational expenses. Despite these challenges, IoT-based monitoring and automation systems hold considerable potential for transforming the mushroom industry by offering a more efficient, sustainable, and profitable cultivation approach than traditional methods (Singh et al. 2020).

2.11 Quality Control and Analytical Methods for Domesticated Mushrooms

Mushroom production offers an affordable agribusiness solution to address significant challenges such as resource utilization, circular economy, sustainability, nutritional security, and adverse weather conditions (Kumar Verma et al. 2022). Advanced technologies, including spectroscopy and nuclear magnetic resonance (NMR), are transforming postharvest mushroom processing by enhancing efficiency while maintaining nutritional value and sensory quality. Automation is also revolutionizing mushroom processing, making it more scalable and efficient (Dawadi et al. 2022; Kumar et al. 2021).

A novel analytical approach for domesticated mushrooms has been developed using chemometric tools and a data fusion strategy. This method integrates ultraviolet-visible (UV-vis) spectroscopy with attenuated total reflectance Fourier transform infrared (ATR-FTIR) techniques. It incorporates multivariate analysis techniques such as random forest (RF), partial least squares-discriminant analysis (PLS-DA), and t-distributed stochastic neighbor embedding (t-SNE) to improve accuracy and reliability (Zhang et al. 2021). Low- and mid-level data fusion strategies are also applied to enhance the robustness of these analytical methods.

Postharvest preservation methods are crucial to ensure high-quality mushrooms. These include physical, chemical, and thermal treatments that help maintain postharvest quality and extend shelf life (Rai and Arumuganathan 2008; Sharma et al. 2024). Cooling mushrooms immediately after harvest to around 4–5 °C is essential to reduce metabolic activities and prevent microbial growth, thereby maintaining freshness and quality (Guo et al. 2021).

2.12 Ensuring Product Quality with Spectroscopy and Magnetic Resonance

The quality control of mushrooms has been significantly enhanced by applying advanced analytical tools such as spectroscopy and nuclear magnetic resonance (NMR). These techniques are crucial for determining the chemical composition of mushrooms and accurately differentiating between species, storage duration, and geographic origin (Wang et al. 2010). Spectroscopy, particularly infrared (IR) spectroscopy, is widely used for food testing and quality control. IR spectroscopy utilizes light absorption, scattering, transmission, and reflection to analyze samples quantitatively or qualitatively. It identifies specific compounds based on their vibrational and rotational energy levels (Pizzoferrato et al. 2000).

Chemometrics plays a vital role in analyzing data using these analytical techniques. It involves similarity analysis, classification algorithms, exploratory analysis, and multivariate calibration algorithms to reveal important multivariate data (Wang et al. 2010). Techniques like thin-layer chromatography (TLC) are also used for quick analysis speed and high sensitivity (Liu et al. 2022). NMR spectroscopy is another powerful tool that provides nuclear information by transforming the energy of atomic nuclei in a magnetic field. Although NMR studies often yield qualitative results and require a large number of samples, they are repeatable, stable, and quick (Wang et al. 2010).

Recent studies have demonstrated the effectiveness of combining spectroscopic techniques with chemometric methods for mushroom species identification. For instance, Fourier transform near-infrared (FT-NIR) spectroscopy combined with machine learning algorithms has been used to accurately identify different porcini mushroom species with high accuracy rates (Liu et al. 2023). Similarly, FT-IR analysis followed by statistical methods has been proposed as a simple approach to differentiate wild mushroom species (Feher et al. 2021).

2.13 Integration of Sensory Evaluation Techniques for Market Readiness of Domesticated Mushrooms

The integration of sensory evaluation techniques is essential for ensuring that domesticated mushrooms are market ready. This process involves assessing sensory attributes such as flavor, aroma, texture, and appearance to align products with consumer preferences and enhance marketability. Sensory evaluation plays a critical role in determining the acceptability of mushroom products among consumers, which is vital for expanding market share and promoting new varieties (Oh et al. 2024).

2.14 Importance of Sensory Evaluation in Mushroom Cultivation

Consumer tests focus on taste, texture, appearance, and purchase intent. Improving sensory qualities and promoting new mushroom varieties can enhance appeal and expand the market (Oh et al. 2024). Sensory testing is frequently used to establish a food product's consumer acceptability and contributes to the creation of quality systems, serving as technical assistance for quality assurance throughout food manufacturing (Mihafu et al. 2020). Beyond their protein content, mushrooms are valued for their numerous health-promoting properties, making them an important component of dietary supplements (Sangeeta et al. 2024).

2.15 Key Sensory Evaluation Techniques

The sensory-diet tool is a pioneering approach that integrates sensory food characteristics into a composition database to assist in research into sensory characteristics, food composition, and diet (Lease et al. 2016). Affective tests, such as hedonic scaling or paired comparisons, gauge consumer acceptance and preference for mushroom varieties or products (Du et al. 2021; Guinard et al. 2016). This feedback is critical for market readiness. Additionally, electronic devices employing sensors and pattern recognition methodologies are being explored to discriminate between mushroom species based on volatile compounds (Portalo-Calero et al. 2019). Understanding mushroom flavor qualities benefits producers, product developers, chefs, sensory scientists, researchers, and consumers (Chun et al. 2020).

2.16 Recent Research Insights: Sensory Assessment and Electronic Nose Technology

Sensory evaluation is a traditional method for assessing the taste and quality of food items, including mushrooms. However, it has several drawbacks, such as the high cost of training panelists, subjectivity, inability to monitor online, and time-consuming nature (Zhou et al. 2015). As an alternative, electronic nose (e-nose) technology combined with GC-MS has emerged as a powerful tool for odor analysis. This approach offers both qualitative and quantitative assessments of trace volatile components in food samples, providing a rapid and cost-effective method for distinguishing between different mushroom species (Wang et al. 2010; Zhou et al. 2015).

Recent studies have demonstrated the effectiveness of e-nose technology in differentiating between various mushroom species based on their volatile compounds. For instance, research on *Amanita* mushrooms used a triangular sensory test and e-nose analysis to compare human olfaction with electronic detection methods, highlighting the potential for widespread application in mycology (Portalo-Calero et al. 2019). Another study differentiated eight commercial mushroom species using e-nose and GC-MS, showing that e-nose sensors can accurately respond to different mushrooms with similar fingerprint chromatograms (Zhou et al. 2015).

Furthermore, e-nose technology has been validated in studies involving cultivated and freeze-dried edible mushrooms. These studies employed sensory panels and GC-MS to elucidate the odor profiles of different species, such as *Agaricus bisporus*, *L. edodes*, and *Grifola frondosa* (Gómez et al. 2022; Chen et al. 2024). The results indicate that e-nose technology is comparable to sensory panels but offers faster analysis, making it a valuable tool for preventing fraud and evaluating medicinal mushroom species (Chen et al. 2024).

2.17 Challenges in Domesticated Mushroom Farming

Mushroom cultivation is gaining popularity among farmers, but several obstacles impede its progress. Key challenges include the lack of access to modern cultivation technologies, insufficient funding for mushroom farming, and high costs associated with raw materials like rice straw (Spark 2023). Additionally, the use of low-quality spawn and ineffective methods for harvesting and managing pests, particularly those from the Diptera order, and diseases are significant concerns (Sapkota et al. 2022).

Furthermore, there is a dearth of scientific research and studies by private organizations and government entities. The absence of specific policies and laws governing mushroom cultivation exacerbates these issues. Horticulturists have yet to classify mushrooms as a commodity, currently categorizing them as fungi at the National Agriculture Research Centre (NARC) and Plant Pathology Division. Poor transportation networks pose a major challenge to market access, a critical requirement for mushroom production. The market is largely confined to urban areas (Poudel and Bajracharya 2011). Moreover, climate changes and seasonal variations can negatively impact mushroom growth, making it difficult for farmers to adapt. Maintaining optimal growth conditions, such as substrate pH, water content, relative humidity, and temperature, also presents significant challenges (Utomo et al. 2023).

2.18 Addressing Pests and Insects in Mushroom Cultivation

Mushroom cultivation faces significant challenges from various pests and insects, leading to substantial crop losses and economic impacts. Among the key insect pests are cecids, phorids, houseflies, sciarids, and stable flies (Rijal et al. 2021). These flies, belonging to the order Diptera, are particularly problematic, contributing to yield losses of up to 60–70% globally (Marcha and Fiegalan 2020). Their larvae damage the development of mushrooms and mycelium, often making them unmarketable.

Sciarid flies, also known as fungus gnats, are notably detrimental, causing yield losses of up to 40% (Coles et al. 2021). Important species include *Sciara multiselta*, *Sciara agaris*, and *Lyeoriella mali*. Cecid flies, or gall midges, belong to Cecidomyiidae and include *Mycophila speyeri*, *Mycophila borresi*, and *Heteropeza pygmaea*. Phorid flies, part of the family Phoridae, include species such as *Megaselia nigra* and *Megaselia halterata*, which can act as vectors for pathogens.

Other pests include mites found in straw and manure, such as small mushroom mites (*Tarsonemus* sp.) and straw or hay mites (*Tyrophagus* sp.), and springtails like *Isotoma simplex*, *Lepidocyrtus* sp., and *Cynaneus* sp. (GCM 2025).

In addition to insects, microorganisms like fungi, bacteria, nematodes, and viruses cause significant losses in commercial mushroom farms worldwide. Various molds and pathogenic fungi infect mushroom species, including dry bubble disease caused by *Verticillium fungicola*, wet bubble disease caused by *Mycogone perniciosa*, cobweb or dectylim mildew caused by *Cladobotryum dendroides* (*Hypomyces rosellus*), and green mold caused by *Trichoderma* (Prasad and Singh 2020). Viral diseases, such as dieback, can result in losses of up to 100% (Biovision 2025). The presence of these pests reduces the quality, quantity, marketability, and palatability of mushrooms (Tulasi et al. 2024).

2.19 Overcoming Barriers to Technology Adoption Among Small-Scale Farmers

Facilitating the adoption of advanced agricultural technologies among small-scale farmers requires addressing a complex array of socioeconomic, institutional, technological, and educational factors. These technologies enhance productivity, sustainability, and profitability, but several barriers hinder their widespread adoption.

2.20 Socioeconomic and Financial Barriers

Small-scale farmers often face significant financial constraints, including limited access to credit and high initial investment costs for technologies like precision agriculture (Pandeya et al. 2025). Off-farm income sources can help alleviate these financial burdens by providing additional resources for technology adoption (Bilal and Jaghdani 2024).

2.21 Educational and Knowledge Barriers

A lack of awareness and educational resources is a major impediment to technology adoption. Enhancing digital literacy and technical skills through targeted farmer education and training programs is essential. Improving digital communication competence, particularly among younger farmers, can enhance technology utilization, decision-making, and productivity (FAO 2023).

2.22 Institutional and Infrastructure Barriers

Inadequate infrastructure, such as limited rural broadband connectivity, significantly hinders the integration of digital technologies in agriculture (Fragomeli et al. 2024). Agricultural extension services are vital in promoting technology adoption by providing necessary support and information (Bilal and Jaghdani 2024).

2.23 Technological and Psychological Barriers

The complexity and perceived benefits of technologies greatly influence adoption decisions. Simplifying technologies and demonstrating clear benefits can encourage adoption. Psychological factors, such as risk perception and attitudes toward change, also significantly impact technology adoption (Rukhiran et al. 2023).

2.24 Economic and Environmental Impacts of Domesticated Mushrooms

2.24.1 Utilization of Agricultural Waste

Mushroom cultivation is pivotal in circular agriculture, as it utilizes forestry and agricultural by-products such as hardwood logs and sawdust (Grimm and Wösten 2018). The absence of pesticides and artificial fertilizers in mushroom farming minimizes environmental impact, making it an exemplary model of sustainable agriculture (Okuda 2022). The biological conversion of agricultural waste into valuable products offers environmental and economic benefits (Kamthan and Tiwari 2017).

2.24.2 Soil Health and Fertility Enhancement

Mushroom farming significantly enhances soil health and fertility. Organic fertilizers, including spent mushroom substrate (SMS), improve soil fertility by increasing the number of microorganisms and enzyme activities, particularly in biodynamic and organic soils (Rajavat et al. 2022). SMS is rich in nutrients and includes nitrogen associated with high molecular weight components like lignin and humus, providing a sustainable alternative to synthetic fertilizers (Hu et al. 2021).

2.24.3 Role in Food Security

Mushrooms are a high-protein, low-calorie food source, making them a promising solution for addressing malnutrition and contributing to food security (Mishra and Shankar 2025). Mushroom cultivation effectively removes unwanted agricultural waste while providing cost-effective nutrition. It also benefits the local economy by offering employment opportunities and significant income sources for regional and national trade (Khan et al. 2024). The focus on biotransformation and clean technology maximizes output and reduces waste creation, aligning with sustainable practices (Singh et al. 2022).

2.25 Financial Viability for Growers and Enterprises

The global mushroom industry has experienced significant growth due to increasing consumer awareness of mushrooms' nutritional and therapeutic benefits. This trend has made mushroom farming an attractive venture for small-scale farmers and entrepreneurs, particularly in developing countries, due to its high-value harvests and low initial investment requirements (Dey et al. 2024). Mushroom cultivation promotes local economic development by creating job opportunities and encouraging efficient use of agricultural waste (Jones 2021).

2.25.1 Small-Scale Growers

Small-scale mushroom farming is economically viable and offers substantial benefits for local communities. It provides a secondary income source and helps utilize agricultural waste effectively, contributing to sustainable agriculture practices. The minimal initial investment makes it accessible to many farmers, fostering economic growth at the grassroots level (Dean 2021).

2.25.2 Large-Scale Enterprises

Large-scale mushroom cultivation offers significant returns on investment due to the scalability of the operation (Sujatanagarjuna et al. 2023). The global demand for mushrooms, especially gourmet varieties like shiitake and oyster, presents opportunities for industrial-scale growers. Large-scale cultivation also involves advanced marketing strategies, technology diffusion, and bio-innovation, positioning mushrooms as a crucial component of future sustainable agriculture and forestry practices (Higgins et al. 2017; Zhang et al. 2014).

2.25.3 Value-Added Products

The profitability of mushroom farming can be further enhanced by producing value-added products such as extracts, vitamins, and powders (Singh et al. 2010). A promising approach is the biotechnological conversion of lignocellulosic organic waste into edible mushrooms. Mushrooms are rich in proteins, vitamins, and minerals, making them a nutritious and safe food option for all age groups. Diversifying products increases profitability and strengthens market resilience (Kotasthane 2021).

2.26 Environmental Benefits of Mushroom Cultivation

Mushroom cultivation offers several environmental benefits, making it a sustainable agricultural practice.

2.26.1 Reduction of Agricultural Waste

The disposal of agricultural waste poses significant environmental concerns due to its high nutrient content, which can lead to pollution if not properly managed (Kumar and Chandra 2013). Cultivating mushrooms on agricultural waste is an environmentally friendly solution, as it reduces nutrient levels to a range suitable for use as manure, exemplifying sustainable farming practices (Kamthan and Tiwari 2017).

2.26.2 Carbon Sequestration

Mushrooms are a carbon-neutral crop, contributing to climate change mitigation by producing minimal greenhouse gases during their growth (Jung and Son 2021). Using agricultural waste as a substrate further reduces carbon footprints (Francis 2016). Integrating mushroom cultivation with agroforestry systems enhances carbon sequestration efforts.

2.26.3 Biodiversity and Ecosystem Support

Sustainable mushroom farming supports biodiversity conservation by protecting natural habitats and restoring ecological balance (Saldanha et al. 2023). Collaborative efforts between mushroom farmers and conservation projects highlight the positive relationship between agriculture and environmental management. Mushrooms form symbiotic relationships with trees and other plants, improving ecosystem health (Benchawattananon 2016).

2.27 Future Perspectives in Mushroom Cultivation

The mushroom industry is undergoing significant transformations driven by emerging technologies and innovative cultivation techniques.

2.27.1 Innovations in Cultivation Techniques

Advancements in AI-driven monitoring systems and precision agriculture are revolutionizing mushroom farming (Jareanpon et al. 2023). These technologies enhance yields, conserve resources, and promote sustainable practices, ensuring consistent crop quality. Smart farming technologies, including real-time monitoring of environmental parameters like temperature, humidity, and CO_2 levels, are becoming integral to modern mushroom cultivation (Mahamad et al. 2021). Precision agriculture leverages AI and data analytics to improve decision-making, while vertical farming is gaining traction in urban settings, maximizing space efficiency (Sangeeta et al. 2024).

2.27.2 Policy and Funding Opportunities

The growth of sustainable mushroom farming heavily relies on government funding programs and policies. Subsidies for eco-friendly practices and investments in innovative cultivation methods encourage wider adoption. Some countries have implemented policies like increased import charges to protect local growers, fostering a more competitive market for domestic producers. Collaborations between the public and private sectors can facilitate industry growth (Okuda 2022).

2.27.3 Expanding Global Reach

Mushrooms can improve food security, particularly in resource-limited areas. Expanding mushroom farming in developing countries can address nutritional deficiencies while generating income (Bajpai et al. 2021). Establishing international supply chains and raising public awareness about mushroom cultivation can help expand global markets.

Bibliography

Aditya N, Jarial RS, Jarial K, Bhatia JN. Comprehensive review on oyster mushroom species (Agaricomycetes): morphology, nutrition, cultivation and future aspects. Heliyon. 2024;10:e26539.

Arroqui C, Lopez A, Esnoz A, Virseda P. Mathematical model of heat transfer and enzyme inactivation in an integrated blancher cooler. J Food Eng. 2003;58:215–25.

Bajpai D, Verma S, Khare YR, Sharma A, Shrivastava AK. Income generation and livelihood security through mushroom production. Int J Curr Microbiol App Sci. 2021;10:3317–9.

Benchawattananon R. Biodiversity of mushrooms in conservative forest in Dansai District of Loei Province, Thailand. Trop Life Sci Res. 2016;27:103–9.

Bilal M, Jaghdani TJ. Barriers to the adoption of multiple agricultural innovations: insights from Bt cotton, wheat seeds, herbicides and no-tillage in Pakistan. Int J Agric Sustain. 2024;22:1–19.

Biovision I. Die-back diseases | Infonet Biovision home. Infonet-biovision.org. 2025. Available at https://www.infonet-biovision.org/PlantHealth/MinorPests/die-back-diseases.

Bone E. Pickling food preservation techniques. Mother Earth News. 2023. Available at https://www.motherearthnews.com/real-food/preservation-techniques-pickling-ze0z10zhir/.

Çağlarırmak N. Mushrooms from kinds of aspects: spawn, growing, nutrients and biocompounds. Ann Food Process Preserv. 2024;2(1):1012.

Chang ST, Wasser SP. Current and future research trends in agricultural and biomedical applications of medicinal mushrooms and mushroom products (review). Int J Med Mushrooms. 2018;20:1121–33.

Chen Q, Yang X, Hong P, Liu M, Li Z, Zhou C, Zhong S, Liu S. GC-MS, GC-IMS, and E-nose analysis of volatile aroma compounds in wet-marinated fermented Golden Pomfret prepared using different cooking methods. Foods. 2024;13:390.

Chong JL, Chew KW, Peter AP, Ting HY, Show PL. Internet of things (IoT)-based environmental monitoring and control system for home-based mushroom cultivation. Biosensors. 2023;13:98.

Chun S, Chambers E, Han I. Development of a sensory flavor lexicon for mushrooms and subsequent characterization of fresh and dried mushrooms. Foods. 2020;9:980.

Chutimanukul P, Phatthanamas W, Thepsilvisut O, Chantarachot T, Thongtip A, Chutimanukul P. Commercial scale production of Yamabushitake mushroom (Hericium erinaceus (Bull.) Pers. 1797) using rubber and bamboo sawdust substrates in tropical regions. Sci Rep. 2023;13:13316.

Coles PS, Mazin M, Nogin G. The association between mushroom Sciarid flies, cultural techniques, and green mold disease incidence on commercial mushroom farms. J Econ Entomol. 2021; https://doi.org/10.1093/jee/toaa322.

Dawadi E, Magar PB, Bhandari S, Subedi S, Shrestha S, Shrestha J. Nutritional and post-harvest quality preservation of mushrooms: a review. Heliyon. 2022;8:e12093.

Dean Z. Is mushroom growing profitable? Costs and margins. 2021. Available at https://www.mushroom-corner.com/posts/profitable-mushroom-farming.

Dey B, Habib A, Masum M, Ferdous J, Halim MA, Uddin MB, Ahmed R. Strategic insights for sustainable growth of mushroom farming industry in Bangladesh: a comprehensive evaluation using SWOT-AHP and TOPSIS frameworks. Heliyon. 2024;10:e36956.

Dong Y, Miao R, Feng R, Wang T, Yan J, Zhao X, Han X, Gan Y, Lin J, Li Y, Gan B, Zhao J. Edible and medicinal fungi breeding techniques, a review: current status and future prospects. Curr Res Food Sci. 2022;5:2070–80.

Doroški A, Klaus A, Režek Jambrak A, Djekic I. Food waste originated material as an alternative substrate used for the cultivation of oyster mushroom (Pleurotus ostreatus): a review. Sustainability. 2022;14:12509.

Du X, Muniz A, Sissons J, Wang W, Juma S. Consumer acceptance of egg white partially substituted with mushrooms and mushroom–egg white flavor pairing. Food Sci Nutr. 2021;9:1410–21.

Elijah O, Rahman TA, Orikumhi I, Leow CY, Hindia MN. An overview of Internet of Things (IoT) and data analytics in agriculture: benefits and challenges. IEEE Internet Things J. 2018;5:3758–73.

FAO. What factors shape small-scale farmers' and firms' adoption of new technologies? | Support to Investment | Food and Agriculture Organization of the United Nations. 2023. Available at https://www.fao.org/support-to-investment/news/detail/en/c/1652579/.
Feher I, Floare-Avram CV, Covaciu F-D, Marincas O, Puscas R, Magdas DA, Sârbu C. Evaluation of mushrooms based on FT-IR fingerprint and chemometrics. Appl Sci. 2021;11:9577.
Fernanda De Sousa M, António O, Martins De Oliveira A, Vicente S, Luís M, Soares R, Leite Da Cunha A, De F, Supervisor C. Edible coatings as a means to improve the quality and shelf-life of Agaricus bisporus (white button) mushroom Coorientador. 2015.
Fragomeli R, Annunziata A, Punzo G. Promoting the transition towards agriculture 4.0: a systematic literature review on drivers and barriers. Sustainability (Basel). 2024;16:2425.
Francis C. The carbon farming solution: a global toolkit of perennial crops and regenerative agriculture practices for climate change mitigation and food security, by Eric Toensmeier. Agroecol Sustain Food Syst. 2016;40:1039–40.
GCM. Mushroom cultivation and the problem with pests | GCM. Garden Culture Magazine. 2025. Available at https://gardenculturemagazine.com/mushroom-cultivation-and-the-problem-with-pests/.
Giusti A, Ricci E, Gasperetti L, Galgani M, Polidori L, Verdigi F, Narducci R, Armani A. Molecular identification of mushroom species in Italy: an ongoing project aimed at reinforcing the control measures of an increasingly appreciated sustainable food. Sustainability. 2021;13:238.
Gómez I, González RL, Tejedor-Calvo E, Clavijo MP, Carrasco J. Odor profile of four cultivated and freeze-dried edible mushrooms by using sensory panel, electronic nose and GC-MS. J Fungi. 2022;8:953.
Grimm D, Wösten HAB. Mushroom cultivation in the circular economy. Appl Microbiol Biotechnol. 2018;102:7795–803.
Guinard J-X, Myrdal Miller A, Mills K, Wong T, Lee SM, Sirimuangmoon C, Schaefer SE, Drescher G. Consumer acceptance of dishes in which beef has been partially substituted with mushrooms and sodium has been reduced. Appetite. 2016;105:449–59.
Guo L, Lan N, Li H, Xiang P, Kan H. Effect of hot air drying temperature on the quality and antioxidant activity of *Boletus edulis* Bull.: Fr. J Food Process Preserv. 2021;45 https://doi.org/10.1111/jfpp.15540.
Gupta N, Mehta M, Singh K. Benefits, challenges and opportunities in mushroom production: a review. Pharma Innov J. 2022;11(3):360–4.
Higgins C, Margot H, Warnquist S, Obeysekare E, Mehta K. Mushroom cultivation in the developing world: a comparison of cultivation technologies. In: 2017 IEEE Global Humanitarian Technology Conference (GHTC); 2017. https://doi.org/10.1109/ghtc.2017.8239314.
Hoa HT, Wang C-L. The effects of temperature and nutritional conditions on mycelium growth of two oyster mushrooms (Pleurotus ostreatus and Pleurotus cystidiosus). Mycobiology. 2015;43:14–23.
Hoyle EH. Drying foods. Home & Garden Information Center | Clemson University, South Carolina. 2018. Available at https://hgic.clemson.edu/factsheet/drying-foods/.
Hu Y, Mortimer PE, Hyde KD, Kakumyan P, Thongklang N. Mushroom cultivation for soil amendment and bioremediation. Circ Agric Syst. 2021;1:1–14.
Islam MA, Islam A, Ullah S, Bhowmik A. An automated monitoring and environmental control system for laboratory-scale cultivation of oyster mushrooms using the Internet of Agricultural Thing (IoAT). In: Proceedings of the 2nd International Conference on Computing Advancements, Dhaka, Bangladesh, vol. 10; 2022. p. 207–12.
Jareanpon C, Khummanee S, Sriputta P, Scully P. Developing an intelligent farm system to automate real-time detection of fungal diseases in mushrooms. Curr Appl Sci Technol. 2023:e0255708.
Jayaraman KS, Gupta DD. Drying of fruits and vegetables. Handbook of industrial drying, CRC Press. 2020 p. 643–690.
Jones C. From local to global: engaging in the world's food challenges through a mushroom cultivation case study. -ORCA. Cardiff.ac.uk. 2021. https://orca.cardiff.ac.uk/id/eprint/143209/1/Cynan%20Final%20thesis.pdf.

Jung D-H, Son J-E. CO2 utilization strategy for sustainable cultivation of mushrooms and lettuces. Sustainability. 2021;13:5434.
Kaiser C, Ernst M. Shiitake and oyster mushrooms. CCD-CP-82. Lexington: Center for Crop Diversification, University of Kentucky College of Agriculture, Food and Environment; 2019. Available at http://www.uky.edu/ccd/sites/www.uky.edu.ccd/files/mushrooms.pdf.
Kamthan R, Tiwari I. Agricultural wastes- potential substrates for mushroom cultivation. Eur J Exp Biol. 2017;7 https://doi.org/10.21767/2248-9215.100031.
Kavaliauskas Ž, Šajev I, Gecevičius G, Čapas V. Intelligent control of mushroom growing conditions using an electronic system for monitoring and maintaining environmental parameters. Appl Sci (Basel). 2022;12:13040.
Khan A, Murad W, Ali S, Shah SS, Halim SA, Khalid A, Kashtoh H, Khan A, Al-Harrasi A. Contribution of mushroom farming to mitigating food scarcity: current status, challenges and potential future prospects in Pakistan. Heliyon. 2024;10:e40362.
Kotasthane T. Oyster Mushroom and its value added products. 2021. Available at https://www.researchgate.net/publication/350488177_Oyster_Mushroom_and_its_value_added_products.
Kratika S. Mushroom: cultivation and processing. Int J Food Process Technol. 2018;V5:9–12.
Krishnappa M, Kantharaja R, Puneeth M. Mushroom cultivation using agricultural wastes. Bentham Science Publishers Ebooks; 2022. p. 134–48.
Kumar S, Chandra R. Bioconversion of agricultural wastes for production of milky mushroom ( ) Calocybe indica. J Sci Res. 2013;57:65–76.
Kumar H, Bhardwaj K, Sharma R, Nepovimova E, Cruz-Martins N, Dhanjal DS, Singh R, Chopra C, Verma R, Abd-Elsalam KA, Tapwal A, Musilek K, Kumar D, Kuča K. Potential usage of edible mushrooms and their residues to retrieve valuable supplies for industrial applications. J Fungi. 2021;7:427.
Kumari A, Baskaran P. An overview on mushroom technology: cultivation, harvesting, post-harvest management and marketing. 2015. Available at https://www.researchgate.net/publication/281632073_An_Overview_on_Mushroom_Technology_Cultivation_Harvesting_Post-Harvest_Management_and_Marketing.
Lease H, Hendrie GA, Poelman AAM, Delahunty C, Cox DN. A sensory-diet database: a tool to characterise the sensory qualities of diets. Food Qual Prefer. 2016;49:20–32.
Liu H, Liu H, Li J, Wang Y. Review of recent modern analytical technology combined with chemometrics approach researches on mushroom discrimination and evaluation. Crit Rev Anal Chem. 2022;54:1560–83.
Liu H, Liu H, Li J, Wang Y. Rapid and accurate authentication of porcini mushroom species using Fourier transform near-infrared spectra combined with machine learning and chemometrics. ACS Omega. 2023;8:19663–73.
Mahamad K, Mansuriwong P, Petlamul W. Application of smart farm system to enhance phoenix oyster mushroom production. ASEAN J Sci Technol Rep. 2021;24:47–57.
Mane VP, Patil SS, Syed AA, Baig MMV. Bioconversion of low quality lignocellulosic agricultural waste into edible protein by Pleurotus sajor-caju (Fr.) singer. J Zhejiang Univ Sci B. 2007;8:745–51.
Marcha S, Fiegalan E. Diversity of dipterans associated with mushroom, Volvariella volvacea bull ex Fr. Singer cultivation. CLSU Int J Sci Technol. 2020;4:38–57.
Mihafu FD, Issa JY, Kamiyango MW. Implication of sensory evaluation and quality assessment in food product development: a review. Curr Res Nutr Food Sci J. 2020;8:690–702.
Mishra A, Shankar S. Edible mushrooms for improved human health, food security and environmental sustainability: A critical review. Sci Total Environ. 2025;995:180093. https://doi.org/10.1016/j.scitotenv.2025.180093. Epub 2025 Jul 24.
Munshi NA, Dar GhH, Ghani MY, Kausar S, Mughal N. (PDF) Button mushroom cultivation. ResearchGate. 2010. Available at https://www.researchgate.net/publication/236011864_Button_Mushroom_Cultivation.
Negi PS, Kumar Roy S. The effect of blanching on quality attributes of dehydrated carrots during long-term storage. Eur Food Res Technol. 2001;212:445–8.

Oduro S. Small scale mushroom. Scribd. 2025. Available at https://www.scribd.com/document/140480238/Small-Scale-Mushroom.

OECD. Safety assessment of transgenic organisms, Volume 1. In: OECD consensus documents, harmonisation of regulatory oversight in biotechnology. Paris: OECD Publishing; 2006.

Oh M, Ju J-H, Ju S. What are the sensory attributes associated with consumer acceptance of yellow oyster mushrooms (Pleurotus citrinopileatus)? Foods. 2024;13:2061.

Okuda Y. Sustainability perspectives for future continuity of mushroom production: the bright and dark sides. Front Sustain Food Syst. 2022;6 https://doi.org/10.3389/fsufs.2022.1026508.

Pala SA, Singh DV, Wani AH, Bhat RA, Ganai BA. Mushroom cultivation technology for conversion of agro-industrial wastes into useful products. Apple Academic Press eBooks; 2021. p. 275–84.

Pandeya S, Gyawali BR, Upadhaya S. Factors influencing precision agriculture technology adoption among small-scale farmers in Kentucky and their implications for policy and practice. Agriculture. 2025;15:177.

Piskov S, Timchenko L, Grimm W-D, Rzhepakovsky I, Avanesyan S, Sizonenko M, Kurchenko V. Effects of various drying methods on some physico-chemical properties and the antioxidant profile and ACE inhibition activity of oyster mushrooms (Pleurotus Ostreatus). Foods. 2020;9:160.

Pizzoferrato L, Manzi P, Bertocchi F, Fanelli C, Rotilio G, Paci M. Solid-state ^{13}C CP MAS NMR spectroscopy of mushrooms gives directly the ratio between proteins and polysaccharides. J Agric Food Chem. 2000;48:5484–8.

Portalo-Calero F, Arroyo P, Suárez JI, Lozano J. Triangular test of amanita mushrooms by using electronic nose and sensory panel. Foods. 2019;8:414.

Potocnik I, Stepanovic M, Rekanovic E, Todorovic B, Milijasevic-Marcic S. Disease control by chemical and biological fungicides in cultivated mushrooms: button mushroom, oyster mushroom and shiitake. Pesticidi i fitomedicina. 2015;30:201–8.

Poudel S, Bajracharya A. Prospects and challenges of mushroom cultivation in Nepal: a case study of Lakuri Bhanjyang, Lalitpur. 2011;1–4. Available at www.forestrynepal.org.

Prasad D, Singh R. Diseases, moulds, insect-pests and mites of mushroom. Int J Plant Protect. 2020;13(2):211–26.

Rai RD, Arumuganathan T Post harvest technology of mushrooms technical bulletin. 2008. Available at https://www.researchgate.net/publication/262725706_POST_HARVEST_TECHNOLOGY_OF_MUSHROOMS_Technical_Bulletin.

Rajavat AS, Mageshwaran V, Bharadwaj A, Tripathi S, Pandiyan K. Spent mushroom waste: an emerging bio-fertilizer for improving soil health and plant productivity. Elsevier eBooks; 2022. p. 345–54.

Rijal R, Maity P, Kumar A. Pests of mushroom and their ecological management strategies: a review. Biolo Forum -Int J. 2021;13:375.

Rukhiran M, Sutanthavibul C, Boonsong S, Netinant P. IoT-based mushroom cultivation system with solar renewable energy integration: assessing the sustainable impact of the yield and quality. Sustainability. 2023;15:13968.

Saldanha A, Gomes LC, Pinela J, Coimbra MA, Barros L, Dias MI, Pereira C. Sustainable cultivation of edible mushrooms: preserving biodiversity and ensuring product quality. Biol Life Sci Forum. 2023;27:6.

Sangeeta D, Sharma S, Ramniwas R, Mugabi JU, Nayik GA. Revolutionizing mushroom processing: innovative techniques and technologies. Food Chem X. 2024;23:101774.

Sapkota R, Joshi D, Bogati S, Malla S. Scope and challenges of mushroom production and their mitigations in Nepal: a review. Arch Agric Environ Sci. 2022;7:272–7.

Sganzerla WG, Todorov SD, da Silva APG. Research trends of edible mushrooms: nutritional properties and health benefits. Int J Med Mushrooms. 2022; https://doi.org/10.1615/intjmedmushrooms.2022043738.

Sharma V, Singh P, Singh A. Shelf-life extension of fresh mushrooms: from conventional practices to novel technologies—a comprehensive review. Future Postharvest Food. 2024;1(3):317–33. https://doi.org/10.1002/fpf2.12029.

Shen Y, Gu M, Jin Q, Fan L, Feng W, Song T, Tian F, Cai W. Effects of cold stimulation on primordial initiation and yield of Pleurotus pulmonarius. Sci Hortic. 2014;167:100–6.

Simos N. Mushroom farming thrives with advanced climate control. Agritecturecom. 2024; Available at https://www.agritecture.com/blog/mushroom-farming-thrives-with-advanced-climate-control

Singh RB, Bishnoi DK, Singh A. Cost benefit analysis and marketing of mushroom in Haryana. Agric Econ Res Rev. 2010;23:165–71.

Singh S, Simran SA, Sushma SJ. Smart mushroom cultivation using IoT. Int J Eng Res Technol. 2020;8

Singh A, Rajput VD, Rawat S, Kumar P, Minkina T. Mushroom: key to food security, human health, and expunging environmental pollution. 2022. Available at https://www.researchgate.net/publication/360631036_Mushroom_Key_to_Food_Security_Human_Health_and_Expunging_Environmental_Pollution.

Spark G. Prospects and challenges of mushroom cultivation in Godavari municipality. Nepal Ssrncom. 2023. Available at https://ssrn.com/abstract=4585741.

Sujatanagarjuna A, Kia S, Briechle DF, Leiding B. MushR: a smart, automated, and scalable indoor harvesting system for gourmet mushrooms. Agriculture. 2023;13:1533.

Suwannarach N, Kumla J, Zhao Y, Kakumyan P. Impact of cultivation substrate and microbial community on improving mushroom productivity: a review. Biology. 2022;11:569.

Tulasi B, Bhuvaneswari K, Manideep S, Kumar S, Chellamuthu S, Nagarani C, Durgam T. A comprehensive review of pests, pesticides and residue detection techniques in mushroom. Arch Curr Res Int. 2024;24:26–41.

Utomo MS, Sukur M, Nurraharjo E, Nofiyanto E. The training in temperature and humidity conditioning equipment utilization for the oyster mushroom cultivation business at Pongangan Gunungpati. Abdi Dosen Jurnal Pengabdian Pada Masyarakat. 2023;7:470.

Verma KD, Patel AR, Billoria S, Kaushik G, Kaur M. Microbial biotechnology in food processing and health. CRC Press; 2022.

Wang B, Xu S, Sun D-W. Application of the electronic nose to the identification of different milk flavorings. Food Res Int. 2010;43:255–62.

Zhang Y-HP. Reviving the carbohydrate economy via multi-product lignocellulose biorefineries. J Ind Microbiol Biotechnol. 2008;35:367–75.

Zhang Y, Geng W, Shen Y, Wang Y, Dai Y-C. Edible mushroom cultivation for food security and rural development in China: bio-innovation, technological dissemination and marketing. Sustainability. 2014;6:2961–73.

Zhang X, Yu H, Yu H-Q, Wang Z, Xia R, Feng D, Qu Y, Tan R, Shi Y, Xiang P, Zhang S, Li C. A forensic detection method for hallucinogenic mushrooms via High-Resolution Melting (HRM) analysis. Genes. 2021;12:199.

Zhou J, Feng T, Ye R. Differentiation of eight commercial mushrooms by electronic nose and gas chromatography-mass spectrometry. J Sens. 2015;2015:1–14.

Zito BB. How to freeze mushrooms: cooked or raw. Bus Insid. 2021. Available at https://www.businessinsider.com/guides/kitchen/can-you-freeze-mushrooms.

Chapter 3
Mushrooms That Cannot Be Farmed

Tauseef Tariq Kisana, Zobia Jabeen, Muhmmad Mughees ud Din, Waqar Ahmad, Nasir Ahmad Khan, and Hayssam Mohamed Ali

Abstract Nonfarmed mushrooms have been desired for their ecological roles, bioactive compounds, and special nutritional value. They constitute a vital part of the human diet, abundant in essential nutrients such as proteins, vitamins, and minerals. The bioactive compounds, including polysaccharides, β-glucans, polyphenols, and terpenoids, give health benefits such as antioxidant, anti-inflammatory, antitumor, and cardiovascular benefits. Cultivation of nonfarmed mushrooms is more complicated than commercial cultivation of mushrooms because of their complex ecological relationships. Furthermore, these mushrooms often need soil composition, moisture level, and microbial interactions, which are hard to regulate in farmed cultivation. This chapter briefly describes the importance of nonfarmed mushrooms, their types, ecology, and importance for human health (medicinal and nutritional aspects), and their reason for complexity. It emphasizes their significance, the challenges associated with their production, and the problems related to their commercial cultivation. In addition, it describes key limitations, including substrate specificity, obligate symbiotic associations, and strict environmental requirements that inhibit commercial cultivation.

Keywords Nonfarmed mushrooms · β-glucans · Anti-inflammatory · Antitumor · Cardiovascular · Obligate symbiotic · Cultivation

---

T. T. Kisana · N. A. Khan
Department of Plant Pathology UAF, Faisalabad, Pakistan

Z. Jabeen (✉) · M. M. u. Din
Department of Plant Pathology, Balochistan Agriculture College, Quetta, Pakistan
e-mail: aridagriculture@gmail.com

W. Ahmad
Institute of Soil and Environmental Sciences UAF, Faisalabad, Pakistan

H. M. Ali
Department of Botany and Microbiology, College of Science, King Saud University, Riyadh, Saudi Arabia

I. U. Haq et al. (eds.), *Mushrooms*, https://doi.org/10.1007/978-981-95-2539-3_3

3.1 Introduction

Global food production faces many challenges, including climate change, a water crisis, land degradation, and desertification. These challenges require research into nontraditional sources of human foods (El-Ramady et al. 2022). Edible and noncultivated mushrooms are considered important next-generation healthy food sources. The global production of mushrooms increases annually at a rate of about 7%. The global consumption of mushrooms is around 12.74 million tonnes, and it is predicted that the worldwide production of mushrooms will scale up to 20.84 million tonnes by 2026 (Shirur et al. 2021). Mushrooms can be classified into some categories, such as (1) cultivated mushrooms (which can be grown commercially by farmers using various strategies to produce for sale in supermarkets; these include Enoki and Oyster mushrooms), (2) wild mushrooms (which can grow on the root systems of trees in forests and are harvested by mushroom hunters—many wild mushrooms are poisonous), (3) medicinal mushrooms (some mushrooms have medicinal benefits but are not always pleasant to consume), and (4) poisonous mushrooms (El-Ramady et al. 2022). A mushroom is a large, complex, and fleshy fruiting structure of fungi, particularly those in the basidiomycete group. It usually has a stem (known as the stipe) that supports the visible fruiting body, though some species may not have a prominent stipe. According to Boa (2004a, b), poisonous species are rare, while fatal ones are a tiny minority. The recognition of edible and inedible mushrooms largely depends on folk taxonomy, with traditional knowledge serving as a communication and information transmission tool from generation to generation (Tibuhwa 2012). Wild edible mushrooms have been a part of human diets and cultures for centuries, especially in rural and indigenous communities (Rout et al. 2020; Virágh et al. 2021; Kumar et al. 2022a, b). Proteins, fibers, and a variety of vital minerals and vitamins are among the many nutrients these mushrooms provide (Assemie and Abaya 2022a, b). Wild edible mushrooms continue to be a resource that is underutilized despite their importance, and their potential advantages are frequently disregarded (Bell et al. 2022; Sethi et al. 2024). Wild edible mushrooms are significant for reasons other than just their nutritional content. Additionally, they are essential for preserving the equilibrium of ecosystems and fostering biodiversity (Kewessa et al. 2022). Numerous symbiotic interactions between tree species and wild edible mushrooms promote the resilience and health of forests (Manjula and Kumar 2024; Pena and Tibbett 2024). Additionally, local communities may benefit greatly from the sustainable collection of wild edible mushrooms as a source of revenue. People have used wild edible mushrooms for millennia for culinary and therapeutic purposes. They are valued for their distinct flavors and textures, making them a delicacy in many cuisines (Schunko et al. 2022; Mishra et al. 2024). Wild edible mushrooms have been utilized in traditional medicine to cure various conditions, including immune system illnesses and digestive problems, in addition to their gastronomic value (Assemie and Abaya 2022a, b). Numerous wild edible mushrooms have been shown to have bioactive substances, such as antibacterial, anti-inflammatory, and antioxidant components, according to recent studies (Sundar

2024). Even though wild edible mushrooms are important, documentation and conservation measures are desperately needed. Deforestation, urbanization, and climate change threaten the habitats of numerous undiscovered wild edible mushroom species. Promoting the conservation and sustainable use of wild edible mushroom species requires thoroughly documenting their habitats, traits, and applications (Kumar et al. 2022a, b). This chapter aims to review nonfarmable mushrooms, their possible compositions, uses, and ecological and economic importance.

3.2 Nutritional Value

Wild mushrooms have been part of human diets and traditional medicine for many years, offering unique nutritional profiles and health benefits. These fungi are valued for their distinctive flavors, textures, and bioactive compounds, contributing to their nutritional and therapeutic importance. This chapter explores the diverse dietary benefits of wild mushrooms and underscores their role in promoting health and well-being. Edible mushrooms form easily digestible and very nourishing meals. Mushrooms are taken for their nutritional, flavor, and taste values.

Wild edible mushrooms are rich in essential nutrients such as proteins, dietary fibers, vitamins, and minerals (Cheung 2008). They are also a good source of high-quality protein and an excellent nutritional supplement for protein-deficient populations. The amino acid profile of mushroom protein is comparable to that of animal proteins, thus offering a plant-based alternative for meeting dietary protein requirements. Moreover, wild mushrooms are low in fat and calories, making them ideal for weight management diets (Manzi et al. 2001).

One of the notable features of wild mushrooms is their micronutrient content. They are abundant in vitamins such as riboflavin (B2), niacin (B3), and ergocalciferol (provitamin D2). Exposure to ultraviolet light increases the vitamin D content of mushrooms, an essential nutrient for bone health and immune function. Wild mushrooms are also rich in minerals like potassium, phosphorus, and selenium. Selenium is a powerful antioxidant that protects cells from oxidative stress (Barros et al. 2008).

Based on fresh weight, mushrooms contain higher protein content than any or all fruits and vegetables, and in meat and dairy products, protein content is higher than in mushrooms. In 100 g of produce, mushrooms contain 16 calories, 91.1% moisture, 0.3% fat, 4.4% carbohydrates, and 36.9% protein on a dry weight basis (Jaiswal et al. 2024). Mushrooms contain a high-protein, low-calorie diet, and all the essential amino acids required for an adult. The nutritional content differs depending on the species and substrates for their growth requirement. Lysine and tryptophan are also present in higher concentrations than methionine and cysteine. Edible mushrooms contain low amounts of carbohydrates and fat. Edible mushroom is also called vegetable meat. Mushrooms are abundant in vitamins C and D, and vitamin B3, 5–10 times more than vegetables. They have essential minerals such as calcium, phosphorus, zinc, manganese, chlorine, sodium, iron, copper, and trace amounts of bromine. Mushrooms are also a source of various B vitamins, including thiamine, riboflavin, pantothenic acid, and nicotinic acid. The high vitamin content is believed to have a soothing and calming effect on the human nervous system (Thatoi and Singdevsachan 2014). Mushrooms can be arranged into different delicious dishes and flavors for other dishes. Some vegetarians use mushrooms in their diet to supplement meat. Mushrooms are also a rich source of antioxidants.

Wild mushrooms have tremendous impacts on sustainable nutrition. They grow wild in diverse ecosystems and do not need extensive agricultural inputs. They can be produced sustainably as a food source by utilizing forests and other natural habitats. Additionally, wild mushrooms are highly nutrient-dense and versatile and, therefore, make an excellent ingredient for increasing the nutritional value of traditional and modern cuisines (Barros et al. 2008) (Table 3.1).

3.2.1 Protein

Wild edible mushrooms are greatly appreciated for their high nutritional value. Wild mushrooms are a good source of protein, low in fat, and high in moisture. Mushrooms are often called "the meat of the forest" because they contain significant protein

Table 3.1 Proximate composition of wild edible mushroom (% dry wt.)

SR. No.	Mushroom	Fat	Proteins	Ash
1.	*Craterellus aureus*	3.0	15.5	8.1
2.	*Heimiella retispora*	6.0	21.1	11.6
3.	*Lactarius glaucescens*	9.0	18.3	8.2
4.	*C. gigantea*	1.3	27.4	6.4
5.	*Russula brevispora*	1.3	24.1	9.10
6.	*Volvopluteus gloiocephalus*	3.98	18.78	13.98
7.	*Astraeus hygrometricus*	4.34	13.78	14.2
8.	*Cantharellus cibarius*	2.25	20.57	9.14
10.	*Ramaria largentii*	5.56	28.24	6.57
11.	*Russula integra*	4	21	11
12.	*Russula alboareolata*	9.0	21.0	17.0
13.	*C. cinerea*	2.4	27.4	13.8
14.	*Lactarius quieticolor*	2.8	18.9	6.5
15.	*Clitocybe subconnexa*	1.90	6.67	4.89
16.	*Russula lepida*	5.02	18.05	7.03
17.	*Turbinellus floccosus*	5.03	20.33	7.9
18.	*Lactarius deliciosus*	3.3	19.21	11.2
19.	*Russula nigricans*	4.02	22.61	6.04
20.	*Saecodon imbricatus*	4.2	17.1	4.2
21.	*Russula virescens*	4.0	20.12	11.3
22.	*Tricholoma protentosum*	3.2	14.3	8.3

content. Therefore, they are an important part of those diets in areas where traditional protein sources are scarce. On a dry weight basis, wild mushrooms normally contain 20–25% protein, the same or better than in vegetables, cereals, and legumes (Kalač 2013).

Nonfarmed mushroom protein content can strongly depend upon environmental conditions and mushroom species. Wild edible mushroom *Laetiporus sulphureus* was measured from 20% to 32% protein in its dry mass. The digestibility of mushroom proteins is relatively high because digestive enzymes easily break them down. Nonfarmed mushrooms have fungal immunomodulatory proteins (FIPs), hydrophobins, sulfur-rich aromatic proteins, ribosome-inactivating proteins (RIPs), ergothioneine-binding proteins, and polyketide synthase-associated proteins. Sulfur-rich aromatic proteins have specific aromas and tastes, making them relatively unique to wild edible mushrooms. In contrast, nonfarmed mushrooms have one more protein named ribosome-inactivating proteins (RIPs), which have antimicrobial properties that are not common in farmed mushrooms (Valverde et al. 2015).

Nonfarmed mushrooms are especially noteworthy for their strong amino acid content. All essential amino acids for human health are found in them—lysine, leucine, isoleucine, valine, threonine, methionine, tryptophan, and phenylalanine. Lysine and Leucine are found in high concentration in morels and truffles. Yet, methionine is frequently a limiting amino acid. They make their nutritional quality more complete, like aspartic acid, serine, proline, and glutamic acid, being of substance. In particular, glutamic acid is vital for mushrooms' unique (umami) taste. Wild mushrooms also have bioactive proteins—lectins that show antibiotic properties such as immunomodulation (Shen et al. 2022).

3.2.2 *Fiber and Carbohydrates*

Wild edible mushrooms (WEF) are rich in carbohydrates and comprise nondigestible carbohydrates. These mushrooms contain chitin, β-glucans, mannans, hemicellulose, and glycogen in the carbohydrate profile and differ markedly between species. For example, the carbohydrate content of WEF is within the range of 35–70% dry weight (DW), and some species, e.g., *Agaricus campestris*, contain up to 70% DW. Digestible carbohydrates such as mannitol, glucose, and glycogen represent small amounts (<1% DW) and provide negligible energy to the organism (Alzand et al. 2019), whereas the nondigestible carbohydrates include oligosaccharides such as trehalose and polysaccharides, including chitin and β-glucans, which dominate the carbohydrate content. Alcohol sugars and oligosaccharides, represented by mannitol and trehalose, are present at 28.9 and 39.2 g/kg DM, respectively. Up to 80–90% of the mushroom cell wall is chitin, which is indigestible to humans but is thought to contribute to the absorption and removal of toxins and heavy metals through the body (Borthakur and Joshi 2019). Specifically, immune-modulating and anticancer effects are attributed to the presence of β-glucans, e.g., in lentinan and pleura in mushrooms, where health-promoting roles of these mushrooms have been attributed.

Carbohydrates in WEF are mostly insoluble dietary fiber, such as chitin and β-glucans. Insoluble fiber contents of up to 220–300 g/kg DW have been reported in some species of mushrooms (Manzi et al. 2001). Soluble fibers, which are added to these fibers, are also important in gut health because they improve digestion and decrease the level of cholesterol. For example, *Termitomyces heimii* has 7.70% crude fiber content (Renthlei et al. 2024), whereas cultivated mushrooms, e.g., *Schizophyllum commune*, have 6.18% crude fiber content (Renthlei et al. 2024). With their high fiber content and bioactive polysaccharides, WEF is an important part of a healthy diet, even though their carbohydrate composition does not make it a main energy source. Wild edible mushrooms are an excellent source of high-fiber carbohydrates that bring various health benefits through their fiber and bioactive polysaccharides. Essentially, these mushrooms are a source of essential nutrients that help with physiological functions, especially immune support and detoxification. Despite their relatively low digestibility and energy contribution from their carbohydrate content, they are more valuable for their functional food properties than as a food source (Kalač 2013).

Wild mushrooms' diverse and complex carbohydrates are major in their nutritional and functional value relative to cultivated varieties. Complex polysaccharides, including β-glucans, mannans, galactans, hemicelluloses, and pectic substances, have high levels of insoluble dietary fibers, such as chitin. Prebiotics in these compounds help the growth of good gut bacteria in the gut, provide better digestion, improve colon health, and increase satiety. In addition, glycoproteins (carbohydrate-protein complexes) produced by wild mushrooms are unique and have potent antioxidant and antimicrobial properties (Wang 2002). High carbohydrate and low fat, making them superior in nutritional profile, are certain species, such as *Tricholoma matsutake*, *Lactarius matsutake*, and *Boletus aereus*; wild mushrooms, on the other hand, contain higher levels of insoluble dietary fibers like chitin prevalent in the fungal cell wall. Wild mushrooms have a much higher beta-glucan profile, a type of soluble fiber associated with immune modulation, cholesterol-decreasing effects, and improved blood sugar control.

3.2.3 Lipids and Fatty Acids

Wild edible mushrooms also contribute to human dietary intake as a rich source of numerous bioactive compounds. Wild edible mushrooms are rich in a nutritional profile; they contain carbohydrates, dietary fibers, vitamins, minerals, and unsaturated fatty acids. Because they are high in protein and low in fat and calories, they are ideal for low-energy diets. For example, 100 g dried weight of *Agaricus bisporus* contains 1.5–4% total fat. Cholesterol is lacking in mushrooms, and they do contain beneficial fatty acids, including linoleic, oleic, and palmitic, but linoleic acid is particularly useful for reducing blood lipids and relieving arthritis (Hanuš et al. 2008).

Lipids are essential to human physiology. They function as hormones or their precursors, facilitate digestion, and serve as metabolic energy sources. Additionally, they play critical roles in forming structural and functional components of biomembranes, constituting the myelin sheath, and providing thermal insulation. Betaine lipids and arseno lipids have also been found in wild fungi (Ribeiro et al. 2009).

Lipid profiling of wild mushrooms reveals three major fractions: glycolipids, neutral lipids, and phospholipids. Triglycerides, free sterols, and sterol esters are the predominant neutral lipids. In some species, neutral lipids are composed mostly of free fatty acids. Steryl glycosides and monogalactosyl diglycerides predominate among the glycolipids and are present in smaller quantities. Principal phospholipids are phosphatidylcholine, phosphatidylethanolamine, and phosphatidylinositol. In wild mushrooms, seven fatty acids are present. These are linoleic, oleic, palmitic, stearic, caproic, myristoleic, and butyric acids, but it was found that wild mushrooms had a low amount of oleic acid. Total lipids, neutral lipids, glycolipids, phospholipids (total nernst), and phospholipids (neutral) generally constitute linoleic acid as the predominant fatty acid (Hong et al. 1990).

Chemical sciences also pay attention to mushrooms because they are rich in diverse bioactive compounds such as polyunsaturated, hydroxy, and halogenated fatty acids, as well as unique lipids like betaine and arsenolipids. The bioactive enzymes, namely peroxidases and haloperoxidases, isolated from mushrooms, have many industrial applications. In addition, the genus Boletus (24 globally distributed species of which at least three are valued for their nutritional and therapeutic properties) is used. Extracts from wild edible mushrooms of the *Boletus scaber* species showed potential for antiulcer and antitumor activities with little acute toxicity and support for the functional and medicinal potential of wild collected edible mushrooms (Ribeiro et al. 2009).

3.2.4 Vitamins

Edible wild mushrooms also contribute to the nutritional profile with their high protein and fiber content and are a valuable source of vitamins. By promoting health benefits, the bioactive compounds found in them, such as phenolic compounds, are also very significant in dietary considerations. Wild edible mushrooms are excellent sources of several essential vitamins and possess great value for human nutrition. Depending on the animal species and environment, the diet is rich in vitamins of the B complex group and vitamins D, E, and C (García-Melo et al. 2022).

Vitamin D is essential to absorb calcium, maintain bone health, and prevent osteomalacia (a condition that causes bones to be soft and weak) in adults and rickets (softening of the bones) in children. Wild edible mushrooms exposed to ultraviolet (UV) radiation from sunlight can induce the production of vitamin ergocalciferol

D2 and trace amounts of vitamins cholecalciferol D3 and vitamin 22-dihydroergocalciferol D4. Vitamin D is important for bone health, but it also helps with muscle function, lowers fall risk in the elderly, and may prevent chronic diseases including cardiovascular and neurodegenerative diseases, diabetes, and some cancers, in wild mushrooms like funnel chanterelles, which produce 3–30 μg vitamin D2 per 100 g of the fresh weight, which makes them a critical nonanimal source of vitamin D, especially for vegetarians and vegans (Cardwell et al. 2018).

Vitamin C (l-ascorbic acid) is an important antioxidant essential in cells and plasma, and it acts as an electron donor to restrict damage by pro-oxidative species such as hydrogen peroxide. It is important for collagen biosynthesis in humans. Vitamin B complex is an important group of nutrients with preoxidant and antioxidant values. Vitamin B is more abundant in mushrooms than in plants, particularly B1, B2, B3, B6, and B12. Fresh mushrooms can provide up to 2–34% per serving of daily vitamin B requirements. *Lentinula edodes* (shiitake) and *Pleurotus ostreatus* (oyster mushroom) are rich in B1, B2, B3, and folic acid B9 sources. B9 content is up to 90 μg/100 g f.w. in *L. edodes* and 9 μg/100 g f.w. in *P. ostreatus*. B vitamins are significantly reduced with drying, however. For example, dried *Boletus edulis* contains 22.89 mg B3/100 g DW. *Imleria badia* contains 38.12 mg B3/100 g DW. This is important because fresh mushrooms are a nutrient source of B vitamins necessary for energy metabolism and production of red blood cells and neurological health. Wild mushrooms are rich in B vitamins, especially riboflavin B2 and niacin B3, being the most important for supporting energy metabolism and neurological function (Podkowa et al. 2021).

Fat-soluble antioxidants such as vitamin E are found in sufficient amounts in wild edible mushrooms. Eight chromanol compounds constitute the form of vitamin E, of which four are tocopherols (alpha, beta, gamma, delta) and their corresponding tocotrienols, with alpha tocopherol being the most biologically active and most abundant in mammalian tissues. Wild mushrooms contain mainly tocopherols and tocotrienols as vitamin E. It is a key fat-soluble antioxidant, protecting cell membranes, plasma lipoproteins, and phospholipids from oxidative damage by scavenging reactive oxygen species (ROS). Wild edible mushrooms contain tocopherols, but the levels vary by species and location and tend to be lower than in more traditional vitamin E-rich foods (Crescenzi et al. 2023).

Notable among wild mushrooms for their alpha-tocopherol content, the most biologically active form of vitamin E, are chanterelles and morels spp. Wild mushrooms usually provide a small but important source of vitamin E, but the levels vary widely among species. Wild mushrooms often contain more vitamin E than commercially cultivated ones. This includes wild species having higher concentrations (0.4–4.3 mg per serving), which result in greater dietary vitamin E intake. In comparison, white button mushrooms generally have negligible amounts (~0.1 mg per serving). Alpha-tocopherol has antioxidant properties that fight cell oxidative damage, potentially lowering the risk of chronic disease and promoting skin health by inhibiting inflammation and repairing tissue (Fogarasi et al. 2022).

3.3 Medicinal Properties of Nonfarmed Mushroom

Mushrooms are known to contain various chemical compounds that may offer therapeutic benefits. For example, the lipid extracts from Grifola species have demonstrated antioxidant properties and the ability to inhibit enzymes linked to chronic diseases and cancer (Inoue et al. 2002). The constituents of these mushrooms inhibit disease progression through direct cytotoxic effects on tumor cells and engage other mechanisms that suppress nonimmune responses (Chang 1996). Compounds derived from Ganoderma species exhibit antibacterial properties (Smania et al. 1999) and have effectively inhibited the growth of bacteria such as Staphylococcus (Mothana et al. 2000). The steroids from these mushrooms have also revealed a certain level of activity against microorganisms (Kuznetsov et al. 2005). Different tests done on HIV/AIDS patients with medicinal mushrooms have yielded positive results (Chang 2006). The mycelial extracts from Lentinula have been found to possess antiprotozoal activity against Paramecium (Badalyan 2004). In addition, mushrooms harbor antiviral potential, and isolates from Ganoderma fungi have demonstrated efficacy against HIV-1 (Ichimura et al. 1998) and also have the potential to combat the influenza virus type 1 (Table 3.2).

Edible mushrooms benefit heart health due to their low fat content, high levels of unsaturated fatty acids, and no cholesterol, making them an excellent choice for individuals with cardiovascular issues (Ganesan and Xu 2018). Their minimal sodium and rich potassium content help maintain salt balance and support healthy blood circulation, making them suitable for those with high blood pressure (Guillamón et al. 2010). Mushrooms are low in calories, starch, fat, and sugars, making them an ideal food for people with diabetes and those looking to lose weight, as their lean proteins aid in cholesterol reduction (De Silva et al. 2012). White button mushrooms contain compounds that may prevent cancer by inhibiting the activity of 5-alpha-reductase and aromatase, which are linked to tumor growth. Polysaccharide-K (Kresin) from *Trametes versicolor* is recognized as a leading cancer treatment (Manikandan 2011). The antioxidant properties of mushroom polysaccharides help combat aging by scavenging free radicals, with ergothioneine found in *Flammulina velutipes* being essential for maintaining healthy eyes, kidneys, and skin (Mishra and Srivastava 2024). Mushrooms also support digestive health through fermentable fibers and oligosaccharides that act as prebiotics, promoting beneficial gut bacteria (Hess et al. 2018). They enhance immune function due to various polysaccharides (beta-glucans) and minerals that stimulate innate and acquired immune responses, activating multiple immune cells (Sanodiya et al. 2009).

Among mushrooms, only a few species of fungi are commercially cultivated out of the 300 genera of mushrooms and allied fleshy basidiomycetes (Jaiswal et al. 2024). Even today, many mushroom species are foraged from the wild during their seasonal peaks. These fungi play a crucial role as decomposers, facilitating the nutrient cycling process within ecosystems. As macroscopic organisms

Table 3.2 Medicinal Values of Some Important Wild Edible Mushrooms

Sr. No.	Mushroom spp.	Medical Properties	References
1.	*Hericium erinaceus*	Therapeutic potential for gastrointestinal disorders	Gravina et al. (2023)
2.	*Craterellus cornucopioides*	Anti diabetic, Liver protection & Cholesterol-lowering properties,	Kosanić et al. (2019)
3.	*Hericium erinaceus*	Anti-depressants & antimicrobial	Chong et al. (2020)
4.	*Cantharellus cibarius*	Anti-inflammatory and wound-healing activity	Nasiry et al. (2017)
5.	*Cantharellus cibarius*	Anthelmintic, iron-chelating & anti hypoxia properties	Kumar et al. (2024)
6.	*Sarcodon imbricatus*	Anti Tumor and immunomodulatory	Tan et al. (2020)
7.	*Amanita muscaria*	Neuroprotection, Cardioprotection & Hepatoprotection,	Voynova et al. (2020)
8.	*Craterellus cornucopioides*	Anti-bacterial, anti-hyperglycemic, anti-cancer	Adamska and Felisiak (2024)
9.	*Hericium erinaceus*	Neurotrophic and Neuroprotective effects	Szućko-Kociuba et al. (2023)
10.	*Suillus luteus*	Antioxidant compounds are also good sources of Vitamin B	Roszczenko et al. (2024)
11.	*Lactarius Delicious*	Anti-inflammatory, Antihyperglycemic, immune boosting, anti-tumor	Su and Xu (2024)
12.	*Boletus reticulatus Schaeff*	Antioxidant, polyunsaturated fatty acids	Heleno et al. (2011)
13.	*Craterellus cornucopioides*	Cytotoxic activity against selected cell lines of human epithelial cervical cancer cells (HeLa)and colorectal cancer cells	Radović et al. (2022)

belonging to the kingdom Fungi (Mycota), mushrooms represent a fascinating yet underexplored biodiversity area. Of the estimated 2.2–3.8 million fungal species, only a mere 3–8% have been formally identified, highlighting fungi as one of the least understood kingdoms of life (Hawksworth and Lücking 2017). Naranjo-Ortiz and Gabaldon (2019) have divided fungi into nine phyla based on phylogenetic studies carried out in the recent past. For instance, true mushrooms fall under Basidiomycota, while truffles and morels are categorized under Ascomycota, all of which are in the subkingdom Dikarya, according to de-Miguel et al. (2014).

3.3.1 Anticancer/Antitumor Properties

Due to their rich bioactive compounds, wild edible mushrooms have received much attention for their potent anticancer and antitumor properties. Such action includes yeast polysaccharides (β-glucans), triterpenoids, phenolics, and flavonoids, which all engage in diverse cancer cell killing mechanisms as they are present within 0.21–0.530 g of dry weight per 100 g dry weight of mushroom. They act as nonself molecules, stimulate cytokine production, and promote immune-mediated apoptosis. The β-glucans also reduce cancer cell proliferation, disturb cell cycles, lower cholesterol, and have anti-inflammatory effects that contribute to cancer prevention (Assemie and Abaya 2022a, b).

The other key group of bioactive compounds, triterpenoids, is cytotoxic due to the disruption of mitochondrial membrane potential and increased reactive oxygen species (ROS) production. Therefore, this causes intrinsic apoptosis in cancer cells. One example is astrakurkurone, a unique triterpene found in the wild edible mushroom *Astraeus hygrometricus*, that kills MOLT-4 cells (T cell acute lymphoblastic leukemia) by inducing apoptosis with selective killing via inhibition of mitochondrial function, increasing oxidative stress, and modulation in the level of apoptotic proteins (Pal et al. 2021).

Wild mushrooms contain major antioxidant compounds such as phenolic compounds and flavonoids. Targeting cancer-specific molecular pathways, they mitigate oxidative stress and inhibit tumor progression. These compounds are efficient in protecting from DNA damage, disrupting cell cycle progression, and inducing apoptosis (Popovic et al. 2013). Hispolon from nonedible wild mushroom *Lactarius zonarius* has been identified as a bioactive compound that induces apoptosis by regulating apoptosis proteins, resulting in tumor growth inhibition. Furthermore, *L. zonarius* possesses high antiproliferative effects on various cancer cell lines. Wild mushrooms also benefit from a superior bioactive profile compared to their cultivated varieties due to environmental factors.

Wild mushrooms also benefit from a superior bioactive profile compared to their cultivated varieties due to environmental factors. Wild mushrooms grow in nutrient-rich natural habitats such as forests and mountainous areas, which are subjected to multiple stresses that encourage the production of secondary metabolites, namely terpenoids, polysaccharides, and phenolic compounds. For example, *Astraeus hygrometricus* contains particularly rich concentrations of these compounds, and its anticancer efficacy is elevated (Pal et al. 2021).

Another noteworthy component of mushrooms is selenium, present as selenoproteins, including selenocysteine, but primarily as selenium. As selenium itself contains antioxidant properties and can decrease cancer risk, studies indicate that eating 18 g (about two medium mushrooms) of edible mushrooms (especially wild edible mushrooms) daily can reduce cancer risk by up to 45% (Li et al. 2023).

The anticancer effects of mushrooms are mediated by inducing apoptosis, promoting autophagy, arresting cell cycles, and stimulating immune system function. They also inhibit angiogenesis and metastasis, which suppress cancer progression together with these properties (Panda et al. 2022). Thus, wild edible mushrooms such as *Astraeus hygrometricus* and *Phellinus linteus* are promising sources of novel anticancer therapies, whose bioactive compounds showed both preventative and therapeutic potential.

3.3.2 Cardiovascular Benefits

Nonfarmed mushrooms provide essential nutrients like proteins, vitamins B1, B2, C, and D, and necessary minerals potassium, magnesium, and iron (Krittanawong et al. 2021). Potassium helps control blood pressure levels; potassium lowers your risk of developing hypertension and heart problems. Nonfarmed mushrooms naturally contain small amounts of fat and sodium while providing cardiovascular benefits. Since wild mushrooms have minimal fat and contain almost no sodium, they contribute to better cholesterol numbers, which promote heart health. Studies show wild mushrooms offer cardiovascular protection because they contain beta-glucans, ergothioneine, and sterols (Guillamón et al. 2010).

Soluble fiber beta-glucans are abundant in various mushroom species. Research shows these components help decrease LDL cholesterol numbers and enhance fat profiles to protect heart health. Nonfarmed mushrooms contain ergothioneine, which protects against heart disease development by fighting oxidative stress and inflammation. Sterols from mushrooms act as bioactive compounds that limit cholesterol absorption in the digestive tract, so the body can better control its lipid processes. Research shows these biologically active substances work together to protect heart health by fighting inflammation and lowering cholesterol levels (Cerletti et al. 2021).

According to research, wild mushrooms help control blood pressure levels. Research has shown that wild edible mushrooms reduce your blood pressure, which helps prevent cardiovascular problems. The wild mushroom *Termitomyces heimii* reduces blood pressure while improving immune protection and protecting against oxidants due to its high beta-glucan availability for cardiovascular diseases. Research shows that *Termitomyces heimii* mushrooms help control high blood pressure and improve immune defense through antioxidant properties. According to research, Wild edible mushrooms can help reduce blood pressure, which helps prevent cardiovascular problems. The wild mushroom *Termitomyces heimii* helps control high blood pressure (reduce blood pressure) and improve immune defense through antioxidant properties due to its high beta-glucan availability for cardiovascular diseases (Anusiya et al. 2021).

Research shows wild mushrooms positively impact cardiometabolic health by lowering blood pressure and cholesterol levels while reducing oxidative stress. Their actions to control lipid metabolism and decrease inflammation make wild mushrooms important in protecting against heart disease. The health benefits of wild mushrooms come from their natural compounds like ergothioneine, beta-glucans, and sterols that specifically protect your heart and stop cardiovascular diseases (Fogarasi et al. 2022). Several scientific works show that including wild mushrooms in our food schedule promotes heart protection.

3.3.3 Antioxidant Properties

Wild edible mushrooms contain diverse bioactive compounds, like phenolics, flavonoids, tocopherols, ergosterol, and polysaccharides; these are rich sources of antioxidants. These antioxidants are important in fighting against diseases like cancer, rheumatoid arthritis, atherosclerosis, and many aging-related degenerative conditions. Wild edible mushrooms have free radical scavenging, lipid peroxidation inhibition, and membrane stabilization, which support both enzymatic and nonenzymatic systems (Zeb and Lee 2021).

The main contributors to the antioxidant activity of mushrooms are phenolic compounds, mainly phenolic acids like gallic acid, caffeic acid, and quercetin. Substitution of the hydroxyl group at ortho or para positions greatly increases their efficacy by improving free radical scavenging and lipid peroxidation inhibition. For example, polyphenol levels of *Hygrocybe conica* were reported to be 29.916 mg GAE/100 g of polyphenolic content (Wong and Chye 2009). Flavonoids are rare in mushrooms; flavonoids are present in some wild mushrooms, such as *Lactarius piperatus*, which protect against reactive oxygen species (ROS) and stabilize cell membranes effectively. The antioxidant properties of vitamin E include components known as tocopherols and tocotrienols, which demonstrate better performance in α-tocotrienol to scavenge singlet oxygen and protect cell membranes against oxidative damage (Khatua et al. 2013).

Wild edible mushrooms contain more essential plant minerals, such as selenium, zinc, and manganese, than other farmed mushrooms, and these are all important nutrients for our bodies to protect us from oxidative stress. They also have critical roles in maintaining the integrity of antioxidant systems (Kabuyi et al. 2017).

Wild edible mushrooms possess higher enzymatic antioxidant activity involving enzymes like superoxide dismutase (SOD), catalase, and glutathione peroxidase, and have an important role in scavenging reactive oxygen species (ROS) and protection against oxidative stress. However, environmental factors induce these enzymes and significantly contribute to the mushrooms' antioxidant defense systems. Comparatively lower levels of these enzymes occur in cultivated mushrooms due to relatively little environmental stress (Liuzzi et al. 2023).

Natural environmental stressors to wild mushrooms include ultraviolet (UV) radiation, extreme temperatures, and pathogens. These secondary metabolites, including bioactive compounds such as phenolic acids, flavonoids, tocopherols, and polysaccharides, are produced in response to these stress factors. These biological compounds are made due to environmental challenges, increasing the antioxidant capacity of mushrooms. Farmed mushrooms cultivated under controlled conditions with reduced exposure to these environmental stresses typically have fewer bioactive compounds (Tellez-Tellez 2024).

3.4 Types of Nonfarmed Mushroom

3.4.1 *Saprophytic Fungi*

Saprophytic fungi thrive on decaying organic material. These fungi secrete enzymes that decompose substances like straw, mulch, and wood chips. Replicating this natural decomposition process is relatively straightforward, which is why many commonly cultivated mushrooms, such as button mushrooms, oysters, and enoki, belong to this group. The practice of mushroom cultivation can be traced back to approximately 600 AD in China, where the *Auricularia auricula-judge* was first cultivated (Chang 1977). This was followed by cultivating the *Agaricus bisporusin* in France during the 1600s (Gayley 1938). Early methods of cultivation required minimal alterations to the environment, often utilizing natural settings like caves, tunnels, and quarries, which resulted in limited control over both yield and quality (Biswas and Biswas 2017).

As time progressed, advancements in scientific cultivation emerged, culminating in domesticating the *Volvariella volvacea* (paddy straw mushroom) in China in 1822 (Chang 1972). By the end of the twentieth century, the number of cultivated mushroom species escalated to include shiitake (*Lentinula edodes*), oyster mushroom (*Pleurotus* sp.), and reishi (*Ganoderma lucidum*), among other saprotrophic species (Fernandes et al. 2021). Early domestication undertakings whereby important food sources were established are credited for ensuring that these provisions were readily available for the Far East, Europeans, and Americans. In the past few years, there has been a growing interest in cultivating various rare parasitic species, including *Hericium erinaceus* (lion's mane), *Inonotus obliquus* (chaga), *Trametes versicolor* (turkey tail), for their medicinal properties (Rathod et al. 2021).

Farming over the years possessed the potential to domesticate the paddy straw mushroom in China in 1822. By the close of the twentieth century, cultivated mushrooms had increased considerably, like shiitake, oyster mushrooms, reishi, and several other saprotrophic species (Fernandes et al. 2021). In the last few years, more emphasis has been placed on farming rare parasitic mushrooms such as lion's mane, chaga, and turkey tail for their medicinal properties (Rathod et al. 2021). Mushrooms were first cultured using bat dung in specific caves and mine quarries because such an environment promotes the formation and growth of mycelial mass and fruiting. However, these early cultivated techniques are associated with several technical problems like humidity control, exposure to light, and contamination, which in turn lead to poor production (Xie et al. 2020). Joseph Lister changed the course in the production of mushrooms through aseptic ways; the result was more hygienic and easier to cultivate mushrooms. While contributing to greater yields, it also helped set the image blueprint for modern techniques of farming mushrooms. The techniques, such as the mushroom grain spawn method that was invented by James W. Sinden, boosted the production output of mushrooms as well as the quality of the mushrooms (Shungchang et al. 2019). Therefore, due to such invention in all types of industrialized mushroom farming, a variety of kinds of saprotrophic species has been produced.

3.4.2 *Parasitic Fungi*

Some of the parasitic fungi can infect plants, animals, and even other fungi, for instance, *Cordyceps sinensis* infects the ghost moth. As much as they present challenges to cultivation, parasitic mushrooms play a role in the interaction of host organisms and contribution to the balance of ecosystems. These fungi can be found in terrestrial and aquatic habitats, and their association with hosts can be mutualistic, parasitic, or commensal (Lu et al. 2015). Chytridiomycota and Basidiomycota are examples of large groups of parasites in many ecosystems. Parasitic fungi are important in controlling the population of the host and thus play an important role in the conservation of species diversity. Some parasitic fungi are responsible for economically important diseases that impact agriculture and natural resources, but they are also important in food webs because they affect nutrient cycling and energy flow in ecosystems. Molecular analysis of fossils shows that parasitic fungi have been in existence for millions of years, and parasitic associations have been recorded in ancient ecosystems. However, gaps in the fossil record persist, which complicates their evolutionary history analysis. Parasitic mushrooms are generally considered as being undesirable because of their impact on hosts, but they are also beneficial to the ecosystem. However, some parasitic fungi have been used in the production of medicine, thus resulting in the formation of antiparasitic drugs from some of the extracts of mushrooms.

3.4.3 *Mycorrhizal Fungi*

Mycorrhizal fungi, which include various species of mushrooms, engage in symbiotic relationships with plants. This interaction is characterized by a reciprocal exchange between the plant's root system and the mycelium of the fungi. Specifically, the fungi facilitate the uptake of water and essential nutrients for the plant, while the plant, in turn, provides carbohydrates that are vital for the growth and development of the fungi. The complexity of these relationships is profound and not yet fully elucidated, presenting significant challenges in the cultivation of mycorrhizal fungi. Notably, many of the most esteemed gourmet mushrooms fall into this mycorrhizal category, which limits their potential for large-scale cultivation compared to other mushroom varieties. As a result, replicating the conditions necessary for the successful cultivation of these mushrooms in controlled environments remains a formidable task, if not an outright impossibility. Till now, a total of 92 saprotrophic edible and medicinal fungi have been successfully cultivated globally (Boa 2004a, b), yet mycorrhizal species are notably absent from this list. For the fungi that can be cultivated, there are no universally applicable conditions that guarantee successful fruiting, and the underlying mechanisms of fruit–body production remain inadequately understood. Nevertheless, recent years have witnessed considerable progress in elucidating the genes, enzymes, and proteins that play a crucial role in the initiation of fruit–body formation (Kues and Liu 2000). However, these breakthroughs are largely confined to specific model organisms where molecular genetics can be readily applied, making it challenging to generalize findings to wild mushroom species. In their natural habitats, a multitude of factors interact to influence mushroom growth. Martinez de Aragon et al. (2007) highlight various environmental, ecological, anthropogenic, and silvicultural factors that significantly impact the fruiting period of mushrooms, particularly during the autumn season.

The production of mycorrhizal mushrooms is still limited to the climatic regions where they naturally occur, and their availability remains constrained by seasonal factors (Peksen et al. 2013). One of the key challenges in cultivating mycorrhizal species on organic waste substrates is the deficiency of essential enzymes, including cellobiohydrolase, endoglucanase, and invertase. These enzymes are critical for the metabolism of complex carbohydrates, which are necessary for the growth and development of these fungi (Mei et al. 2021). The dependence of mycorrhizal fungi on glucose or fructose as a carbon source for the formation and development of sporocarps has fostered the evolution of mutualistic relationships with their woody hosts. In these associations, the host plant supplies monosaccharides both prior to and during sporocarp development (Treseder et al. 2006). A significant challenge in cultivating mycorrhizal mushrooms on organic substrates is the mycelium's inability to hydrolyze lignocellulosic materials. Consequently, it is essential to incorporate glucose or fructose into the basal substrate to provide the necessary carbon sources, along with adequate levels of metabolic nitrogen (Courty et al. 2010). However, maintaining these nutritional conditions can also pose the challenge of contamination, which complicates the cultivation process (Khalid and Keller 2021). Moreover, beyond the technological challenges, several additional hurdles exist.

3.5 Some Nonfarmable Mushrooms

A few of the more cherished cultivated mushrooms originate from mycorrhizal fungi living in association with the trees of its genera Quercus, Pinus, and Corylus. The genera involved include Tuber and Morchella together with Boletus, Tricholoma, and Cantharellus. In contrast to saprophytic fungi, the mentioned fungi have a much more complicated life cycle, which provokes the complexity and nontriviality of the cultivation process. Consequently, their availability was limited in the past by natural production and seasonal distribution, where the product depends on the season of the year and the type of species, depending on the type of area (Hall et al. 2003).

3.5.1 Morels

The Morchella genus produces the highly prized morel mushrooms, with *Morchella esculenta* being particularly notable. Initially classified as saprophytic, this species can also form mycorrhizal associations. Morels are cosmopolitan, occurring across various areas, but they are mostly associated with North America. While *M. esculenta* is not a mycorrhiza-only fungus and the organism is capable of growing saprotrophically (Tedersoo et al. 2013), its growth is very difficult. Low-intensity fire has a vital role in increasing yields of wild mushroom crops (Wurtz et al. 2005). The effects of fire are normally most apparent 1 or 2 years after the fire has occurred, during the summer season. Another peculiar characteristic of the morel fruiting body is the formation of sclerotia, which is formed chiefly by the secondary mycelium. These sclerotia actually act as survival structures, which can remain viable for a long time in stable environmental conditions and germinate only when favorable stimulants are provided. When a sclerotium is stimulated, for instance by a fire, it germinates to form the ascogenic mycelium and in the process gives rise to the morel fruiting body. Besides the environmental conditions include temperature stress, especially the highly stressful temperatures, heavy rainfall, low humidity, or continuous cold stress break the sclerotial dormancy, but fire remains the most effective. This is because fire interferes not only with the biological state of the soil but also the chemical conditions that are favorable to morel growth. Sclerotium mass formation is therefore a very important event in the life cycles of the fungus, because it determines the formation and the number of fructification structures. Such a connection of fire with the growth of morels is crucial for understanding because of the part played by the germination of the sclerotia in the morel life cycle and the potential for cultivation.

3.5.2 Chanterelle

Chanterelles are known as cantharel mushrooms, such as *Cantharellus cibarius* and *Cantharellus formosus*, being popular in North America (Pilz and Molina 2001) and consumed in Europe. As with many other species, these mushrooms have reduced populations in the wild. There are many different types of fungi with chanterelle in their name. Still, all true species of chanterelles belong to the Cantharellus genus, and do not include the Craterellus, Gomphus, or Polyozellus in the true chanterelles category. Chanterelles are ectomycorrhizal with the spruce pine, pine, oak, and Douglas fir. Cultivation of both of them is connected with additional complexities, and their ecological specificity is emphasized by this symbiosis. As with other species of temperate edible mycorrhizal fungi, the genera of chanterelles spread as far as tropical Africa, Asia, and Latin America. Pest is able to produce fruits at warm periods with adequate humidity in temperate areas and broad ecological generalism, thereby guaranteeing the pest a wide geographic distribution (Yamada 2022). The main issue in cultivating chanterelles has thus been the shortage of pure cultures in terms of growth. The replanting of pure cultures of chanterelles is quite challenging when compared to other ectomycorrhizal edible mushrooms such as Tricholoma, Boletus, and Lactarius. Mycelia of *Cantharellus cibarius* can develop on nutrient media, but strains isolated from basidiomata are pure cultures, which are the problem (Danell 2000). The challenge is that basidioma tissues of *C. cibarius* host various bacteria and fungi, which interfere with mycelial growth on nutrient agar media

(Rangel-Castro et al. 2002). Other members of the Chanterelle and Craterellus genera have been reported to exhibit similar problems. Isolation of pure culture from *Cantharellus anzutake* basidiomata is particularly rare, as described by Ogawa et al. (2019). These samples will always be contaminated regardless of the fact that they are cultured in a laboratory environment.

3.5.3 The Matsutake

Matsutake (*Tricholoma matsutake*) is a mushroom that is deeply rooted in Japanese culture and is often considered the equivalent of truffles in Europe. Like the European truffle, matsutake has suffered a dramatic decline in natural yields and has not yet been successfully cultivated on a commercial scale. In contrast, the American matsutake, *Tricholoma magnivelare*, is still abundant in North America. The name "matsutake" comes from the fact that the primary host tree species is a pine tree, and the word "take" means mushroom. Matsutake is found in association with various trees, including pines (Pinus), oaks (Quercus), chestnuts (Castanopsis), and spruces (Picea), depending on the region and specific Tricholoma species. Matsutake is native to the Northern Hemisphere, and like many other mycorrhizal mushrooms discussed here, its natural fruiting season is in late summer to autumn, when the temperature is cool. Production of *Tricholoma matsutake* mushrooms is associated with a structure called the "shiro" (Vaario et al. 2011), a dense mass of mycelium that surrounds roots and is found within soil particles. These structures are dominated by *Tricholoma hyphae*, but they are not solitary, and a diverse array of prokaryotes and other fungi coexist within the shiro. Tricholoma switches to an alternative energy strategy during fruiting, utilizing organic compounds such as hemicelluloses as a substrate. It is likely that the microbial community within the shiro plays a major role in this process by producing the enzymes needed to break down organic matter. These interactions illustrate the complex life cycle of *Tricholoma matsutake* and the difficulties in cultivating it.

3.6 Characteristics/Specification of Nonfarmable Mushroom

A diverse range of fungal species, which are not suitable for farming, present challenges in cultivation within controlled environments because of their intricate ecological needs and distinctive symbiotic relationships. Obligate mycorrhizal associations are characteristic of these fungi, which engage in obligatory symbiotic relationships with plant roots that support the growth and persistence of their own folklore. Unlike their sister, aligned fungi fared for their role in organic matter decomposition, the nonfarmed mushrooms are reliant upon specific, environmentally sensitive signals, as well as interaction with host plants, processes that are difficult to replicate under artificial conditions (Gupta and Bansal 2021).

Tuber spp. (truffles), as other mycorrhizal fungi, are prime examples of these complex relationships. The latter species depend on a symbiotic relationship with the roots of particular tree species, as well as specific soil conditions, for proper growth. However, attempting to grow them elsewhere has been quite difficult because the ecological conditions necessary to minimize these factors have proven to be extremely challenging. They are also similar to Morchella spp. (morels) in that there are particular saprophytic fungi, which are even more difficult. Although morels are saprotrophic organisms, their requirements for a specific habitat and slow growth rates make cultivation of morels unlikely, and species are reliant on wild harvesting (Siwulski et al. 2019). For example, parasitic fungi like *Cordyceps sinensis* are mushrooms that will not grow on farms. It requires a specific insect host in its life cycle and is, therefore, nearly impossible to cultivate in a controlled environment (Abdullah and Kumar 2023). Identification morphology of nonfarmable mushrooms represents the ecological dependencies of those mushrooms that are a crucial part. Various distinguishing traits between edible and toxic species are based on the shape of the cap, gill attachment, and the color of the spore print. The Amanita genus is of great concern to foragers, containing numerous highly toxic species that closely resemble edible varieties.

3.6.1 Ecological Requirements

The growth of wild edible fungus (WEF) is encouraged and reproduced in certain ecological environments. Substrate, climatic conditions, and relationships with surrounding flora and fauna are extremely important in determining particular points of ecological requirements. The necessity to be familiar with all assigned needs to achieve a sustainable harvest, just as investigating potential farming techniques, is as important as ever (Zhang et al. 2020). The WEF displays the different trophic strategies, including saprotrophic, ectomycorrhizal, and parasitic lifestyles. Such fungi can live only on dead organic matters, such as dead plant materials, composted leaves, and decaying wood. In most environments, the quantity and quality of organic matter present have influenced the distribution of organisms. Both animals and plants are commonly grown on substrates rich in lignin and cellulose. On the other side, ectomycorrhizal fungi that are cultivated in association with certain tree species in exchange for extra chemicals that contain additional carbon (Atri 2024).

The proliferation and establishment of these species rely on the availability of appropriate host plants from the families Pinaceae, Fagaceae, and Betulaceae. To ensure the climatic conditions essential for the growth of Boletus and Cantharellus, the presence of established old woods with robust root systems is important. Findings indicate that the WEF plays an important role as a climatic factor, especially concerning temperature and moisture levels (Yun and Hall 2004). The mycelium of these fungal species widely grows in conditions where temperatures are between 15 °C and 25 °C, and humidity levels from moderate to high. These conditions are optimal for the proliferation and spore formation of the fungal species.

Diverse rainfall patterns are critically important for both the decomposition of organic waste and the hydration of fungal structures. Mushrooms mainly grow during periods when conditions are most favorable for spore dispersal and growth, particularly following rainfall, which provides an ideal environment for their development. Other important factor, including pH, texture, and nutrient availability, can influence the pore structure of soil, which in turn can significantly affect fungal proliferation. Ectomycorrhizal fungi can flourish on diverse substrates; however, they exhibit a notable preference for nutrient-poor soils, as these environments allow them to obtain essential nutrients from their associated plants. Some types of WEF show greater tolerance to acidity and even alkalinity, yet most species proliferate in soils that range from slightly acidic to neutral (Muthukumar et al. 2014).

The conditions regarding light and air circulation appear to be optimal for the proliferation of wild fungus. Mushrooms, similar to other heterotrophs, do not rely on sunlight for energy; however, additional light can enhance the growth of their fruiting bodies. Some wild edible mushroom species show such a high level of shade tolerance that they prefer habitats characterized by shadow or partial shade (Győrfi et al. 2011). For example, they selected trees situated in the forest due to their providing of etiolation and their ability to shield against extreme heat and dryness. To promote the proper growth of fungal mycelium, it is essential to prevent the accumulation of substances such as carbon dioxide by ensuring adequate aeration. Furthermore, it has been demonstrated that biotic interactions significantly influence the development of WEF systems. The ecological factors of insects and animals are influenced by their predators, their relationships with plants, interactions with fungi and bacteria, as well as their connections with other fungi. Fungal populations were influenced by numerous factors, such as the transfer of spores by animals, the breakdown of deceased animals, along with various disturbances and variations in forest composition (Tomao et al. 2020). The influence of these factors extends to the availability of suitable habitats that can be accessed.

Tuber melanosporum truffles are able to withstand elevated temperature diurnal changes as well as variations in soil moisture, especially in areas without a canopy (Hall and Haslam 2013). Similarly, *T. uncinatum* and *Lactarius deliciosus* are able to produce spore in similar conditions, with complete canopy closure (Guerin-Laguette 2021). *Tricholomas matsutake* occurs in *Pinus densiflora* stands greater than 20 years old with a canopy density of 75%. Less abundant than this fungus is found when the initial host is replaced by other tree species later on in the cycle (Lian et al. 2006). Production of the mycorrhizal edible mushrooms is closely related to the soil and climatic conditions that have not yet been fully explained. Much has also been learned in regard to the ecological needs of *Tuber melanosporum* and *Tuber uncinatum* mushrooms (Wang and Chen 2014), but the requirements of other edible ectomycorrhizal species are still unknown. Information about *T. melanosporum* is mainly based on growers' or scientists' observations, experimental data are rare and not very specific, and more research is needed to see whether *T. melanosporum* symbiosis is formed with younger or older host trees, known as early or late-stage mushrooms (Karwa et al. 2010). Finally, there is little understanding of what specific soil and climate conditions are required for growth. Less is known, for example, of the effects of environmental changes on mycorrhizal community, of the relationships between these fungi and other soil organisms, and of those factors inducing fruiting of these fungi. It is anticipated that biochemical and molecular techniques will fill the gaps and be valuable sources of information on the ecological limits and possibilities of mushroom development (Zambonelli et al. 2012).

3.6.2 Environmental and Ecological Factors of Selected Mushrooms

Cordyceps sinensis are known to thrive in cold (10–15 °C) and high altitude (the Himalayan Plateau) environments. High humidity of 70–90% is needed for fungal invasion and fruiting body development. The fungus requires oxygen for its respiration only during its growth stages and does well with low oxygen environments common to insect subterranean habitats where it is grown. Natural habitats of *Agaricus campestris* include meadows, grasslands, lawns, and parks, usually in open, or partly sunsets (Krishna et al. 2024). As this species is extremely specific in its need for special soil conditions rich in organic matter, it is harder to cultivate in artificial growth chambers than *Agaricus bisporus*. Being able to adopt these natural conditions accurately is essentially a requirement of its ecological needs and severely limits its potential for large-scale commercial cultivation.

The species is *Agaricus asiaticus*, a related species to *A. campestris*, which is native to dry lowland Dipterocarpaceae forest. It occurs in sandy or lateritic soils, frequently in near association with tree roots, indicating a mycorrhizal association. It grows in the warm, humid climate of these forested areas, so that its success depends on particular ecological conditions for growth (Wang and Bau 2024). Some mushroom needs some kind of specific ecological and nutritional requirements, such as *Termitomyces heimii* is difficult to culture because it has a symbiotic relationship with termites. This species is a Chemosymbiosis and depends on a specialized substrate, the fungus comb, within termite nests. The unique environment of termite nests is extremely challenging to provide artificially in controlled environments, and thereby, it is limited what we know about how it grows and develops (Paloi et al. 2023).

Hygrocybe conica is common in wide open grasslands, meadows, and grassy woodlands, especially in nutrient-rich, acidic soils with a pH range from 5.0 to 6.0. Well drained yet maintain enough moisture to sustain fungal growth during the fruiting season is what these soils need to be. As with other members of its genus, *H. conica* forms mycorrhizal relationships (symbiotic relationships where the fungus helps the plant in exchange for food and other nutrients) with grasses, herbaceous plants, and some trees, and so is strongly dependent on the plant for its survival and growth (Silva-Filho et al. 2019). All these mushrooms show how involved they are with particular environmental and ecological factors, and how difficult it is to replicate those conditions for cultivation.

3.7 Nutritional Needs

Mushrooms are living organisms, and like all living organisms, they have nutritional needs if they are to grow and mature. Whether a given organism does well as a decomposer, symbiotic, or pathogen depends on how it obtains and metabolizes the various nutrients it can get in its surroundings. Carbon is an important nutritional component of mushrooms and is the primary energy source. However, mushrooms use carbon from complex organic compounds as cellulose, hemicellulose, and lignin, which constitute plant material (Corbu et al. 2023). *Agaricus bisporus* and *Pleurotus ostreatus* are great decomposers of matter through sugars as simple as glucose and fructose when one of its components is part of its substrate. While they are not typical mycorrhizal mushrooms, however, Boletus and Amanita species associate with plants, which in turn provide sugars from photosynthesis, a strategy showing carbon acquisition diversity among diverse types of species (Patyshakuliyeva et al. 2013).

Nevertheless, nitrogen is an incredibly vital nutrient in mushroom growth, where it is involved in the synthesis of protein, the production of enzyme, and DNA replication. Conditions for mycelial development and fruiting body formation would be dependent on nitrogen availability. Saprotrophic mushrooms grow on compost, manure, or agricultural residues, and fungus feeds on organic nitrogen source (Nunes et al. 2012). On the other hand, the mycorrhizal mushrooms do depend upon their plant hosts for inorganic sources of nitrogen, such as ammonium or nitrate, for this. Achieving an appropriate carbon-to-nitrogen (C: N) ratio—if it is too out Wack you can get “out Wack” plant growth. Many cultivated mushrooms perform best with C:N ratios 25:1, or larger, depending on the species and growth conditions (Tian et al. 2010).

So saluted are mushrooms as a moisture requirement; they help enzymatic reactions, absorption of nutrients, and cellular expansion. Mushrooms do not need light for photosynthesis, though they do grow in a humid and moist area. Securing establishment for the mycelium in the first stages of colonization of the substrate requires proper moisture in the substrate. The perfect humidity level for the development of a healthy fruiting body is 85–95% while the fruiting body is approaching a fruiting stage. Despite being grown on other substrates, such as *A. bisporus* and *P. ostreatus*, these even species can be grown under controlled moisture conditions in order to prevent too much water, which results in anaerobic conditions and prevents fungal metabolism (Manan et al. 2021).

All mushroom species require carbon and nitrogen, but each species has other nutritional requirements as well. In which case, they may need to get a couple of vitamins, minerals, and other micronutrients essential for enzymatic and metabolic functions. For example, the ability of substrates with minerals such as magnesium, zinc, and iron may be good for *Ganoderma*, as it can aid in the metabolic process and create more medicinal substances. For commercial cultivation or research, an important part is understanding and delivering these unique needs.

Unlike cultivated varieties, wild edible mushrooms depend on natural ecosystems for food, and as a result have special nutrient requirements. Decaying fallen leaves, dead wood, forest litter, and other forest debris give these mushrooms nutrients. The decomposition of these materials, however, produces an extraordinarily complex collection of carbon compounds and trace elements essential to their growth, but which vary with individual hosts. WEF are not like cultivated mushrooms because cultivated mushrooms only do well in rich nutrients—natural decomposition only exists, but WEF can be grown in simplified, controlled substrates (Hussain et al. 2023). Further, many wild edible mushrooms, especially those mycorrhizal, depend upon symbiotic associations with specific host plant species, and they tend to associate preferentially with specific host plant species or host forest tree species. The mushrooms can assimilate carbohydrates by obtaining them from themselves and the plant via photosynthetic conversion (these associations), and in return, the host is supplied with minerals and water taken up through the fungal mycelial network. In addition, the ecological needs, the specific host species on which they depend (such as oaks, pines, or birches), further narrow things down and make it difficult for growing them outside of their native habitats (Niego et al. 2023). Simply put, we all need the stuff that decays (on plants and the like), without which we won't make it to cultivation.

An attempt is made to describe together in carbon and nitrogen sources, moisture, and small elements, the nutritional requirements for mushrooms. The balance of these factors is necessary for the growth and development of these factors in either a natural ecosystem or for controlled cultivation. Advances in microbiology, biotechnology, and substrate (substrate management) enable cultivators to provide the needed specific nutrition for the distinct species of mushrooms to encourage their growth and production (Klaus and Wan 2022).

3.7.1 Nutritional Needs of Selected Fungi

Cordyceps sinensis gets its nutrients from the insect host (ghost moths), with specialized enzymes that break down the body tissues of its insect host. Nitrogen, as protein, in the muscles and hemolymph of the insect, is a key nutrient needed by fungi for reproduction and growth. *C. sinensis* utilizes amino acids lysine and arginine may be useful for fruiting body development of the mushroom. Stored carbohydrates (glycogen) and fats are the primary energy sources that provide carbon and so the electrons making up carbon chain components (Abdullah and Kumar 2023). Also, potassium and phosphorus and magnesium fibers of the host's body help the fungus to metabolize the cell. To access these nutrients, Cordyceps secretes a range of enzymes: chitinases break chitin, a component of the insect exoskeleton;

proteases break down proteins into amino acids, lipases dissolve lipids from the insect's fat body; amylases break down glycogen and other carbohydrates (Manoharachary et al. 2014). In addition, the chemical makeup of the insect exoskeleton helps promote fungal germination. Fungal chitinases is a chitin-degrading enzyme for the polymer of N-acetyl glucosamine (chitin), which is degraded to simpler sugars, thus facilitating depth penetration into the internal tissues of the host (Krishna et al. 2024). Chemical signals and recognition among cuticular lipids, including hydrocarbons, wax esters, free fatty acids, and sterol esters, may trigger fungal spore germination, fungal pathogen germination, and the initiation of parasitic growth.

Termitomyces heimii has a highly specialized nutritional profile that involves a symbiotic relationship with termites and the surrounding soil environment. Scientific observations at different developmental stages show that protein synthesis and nutrient acquisition are complexly regulated, thus artificial cultivation is quite difficult (Rahmad et al. 2014). *T. heimii*, like wild mushrooms, has developed intricate interactions with their environment (soil and vegetation) in order to take up essential minerals like potassium, magnesium, and phosphorus, and trace elements too. The interactions both contribute to the biochemical composition of the fungus and increase the production of bioactive compounds, volatile organic compounds (VOCs), and secondary metabolites. The addition of such compounds is what makes wild mushrooms so unique in their aroma and flavor (Morath et al. 2012).

The flavor complexity of black truffles (*Tuber melanosporum*), so characteristic of this fungus, is dependent on symbiotic associations with the roots of oak and hazelnut trees. By using this partnership, truffles can absorb specific nutrients from nutrient-rich soils while also producing sulfur-containing VOCs, such as dimethyl sulfide, which is responsible for their distinctive pungent aroma (Epping et al. 2024). Likewise, chanterelles (*Cantharellus cibarius*) associated with coniferous and broadleaf forests present enriched minerals making their distinct, fruity, and peppery flavors. In addition to terpenes and esters, the production of this flavor profile is further enriched. We conclude that the nutritional strategies of these fungi are intimately linked to their ecological niches and associated symbioses, relying on environmental and host-derived resource inputs to drive growth, reproduction, and biochemical diversity (Zhang et al. 2022).

3.7.2 Life Cycle of Mushrooms and Its Complexity

The life cycle of mushrooms is a multifaceted process comprising several distinct stages, each influenced by the organism's adaptation to its habitat and its ecological role. The environmental conditions and availability of resources significantly impact the progression through these stages, which include spore germination, hyphal growth, mycelium development, sporocarp formation, and spore dispersal.

Cordyceps sinensis exhibits a highly specialized life cycle intricately tied to its insect host, specifically caterpillars. The host provides critical biochemical and hormonal triggers necessary for fungal germination and development. Cuticle lipids on the insect exoskeleton serve as chemical signals for spore germination, while insect hormones, such as juvenile hormones, promote the fungus's invasive growth and subsequent life cycle stages. This specialized interaction highlights the complexity of the organism's ecological and biological adaptations (Baral 2017).

Termitomyces heimii undergoes a similarly intricate life cycle with three main developmental stages: mycelium formation, primordium development, and fruiting body maturation. The successful progression through these stages requires precise environmental conditions that are challenging to replicate artificially. Some important essential fatty acids, like linoleic and oleic acid that help spore germination. After the mummification of an insect by *T. heimii* dead larva helps the formation of the fruiting body. Despite its ecological significance and nutritional value, the life cycle and growth requirements of *T. heimii* remain poorly understood (Rahmad et al. 2014). Limited research on its development hinders the creation of effective cultivation strategies, emphasizing the need for further investigation into its biology and ecological interactions.

The life cycles of mushrooms like *Cordyceps sinensis* and *Termitomyces heimii* demonstrate the complexity of fungal development and its dependence on specific environmental and biological cues. Understanding these intricate processes is essential for unlocking the potential of these fungi for ecological, medicinal, and nutritional applications.

3.7.3 Spore Germination

The spores, microscopic reproductive units thousand times larger in number than gametes, are extremely resistant to adverse environmental conditions, and this life begins with their dispersal. Wind, water, or animal transport these spores to new locations. The spores land in a good place—moist, rich in nutrients, and at the right temperature—the only thing that allows it to germinate (Money 2016). The stage in which the spores absorb water, swell, and with metabolic activity begin. As energy for the emerging hyphae, a readily available carbon source, such as glucose or other simple sugars, is needed for the germination process. The factors that might hinder germination include either inadequate humidity or nutrients, or competition from other microorganisms (Gottlieb 1950). It documents this first stage as the mushroom's response to its environment, being sensitive to its surroundings, as a prerequisite for the start of its life cycle.

3.7.4 Hyphal Growth

Germinated spores develop into hyphae, which elongate and branch into a network. The vegetative body of the fungus is this, the mycelium, or network. Tip formation and substrate nutrient absorption are by hyphal growth. To be compatible with other differences of mating types, the hyphae of sexually reproducing mushrooms may fuse, becoming dikaryotic hyphae with two nuclei per cell (Fricker et al. 2017). This fusion is necessary to increase the diversity of its population, as genetic recombination does, in order to achieve the fertility of its population (or adaptability, or survival of the population). Determining the success of this stage depends on environmental factors: substrate quality, pH, temperature, and competition with other fungi or bacteria (Boddy and Hiscox 2016).

3.7.5 Mycelium Development

It extends the substrate as well as to adjacent soils and forms a firm matrix that attracts nutrients. It forms a mycorrhizal fungus, a mutualism in the roots of plants, which is the stage. The fungus, in return, helps plants become better at taking in water and nutrients, specifically phosphorus and nitrogen. Mutualism is indispensable in forestry production, not only to the nutrient cycling processes but also to the condition of plants. These are saprotrophic mushrooms, those which eat dead organic substances, and then make them available to the environment as nutrients (Nielsen 2005). Mycelium growth is an extremely complicated thing with varying influences besides the direct stage of environment—that is, for example, the influence of soil moisture, of organic matter, or of the plant host.

3.7.6 Sporocarp Formation

In ideal circumstances, though—temperature is high enough, moisture is enough, and there is nutrient enough—the mycelium goes into the reproductive phase, producing fruiting bodies, or sporocarps. This is the stage that the fungus is in now, as the fungus starts to expend its energy in making its reproductive structures. The mushrooms are the only part of the fungus that is seen commonly, and they are specialized for spore production: sporocarps (Fernández-Ruiz et al. 2023). The transition to sporocarp formation can be induced by behavior, causing the behavior to induce transition to sporocarp formation, and for example, by seasonal changes, by rainfall, rather than just a given temperature threshold. If delay or inhibition at this stage stops development, this perennial becomes dependent on environmental conditions or nutrient availability.

3.7.7 *Spore Dispersion and Spore Production*

The final stage of the mushroom life cycle is spore production and dispersal. The spores form in distinct types of mushroom specialized structures (known as gills, pores, or spines). Afterward, they release spores into the environment by wind currents, water flow, or by animal activity. To ensure this continuity and occurrence of other habitats, the spores need to be dispersed to reach (Spatafora et al. 2017). Colonization of dispensing bodies of dispensers is very dependent on environmental parameters: humidity, airflow, disposition of dispersive organs, and the release of spores.

3.7.8 *Environmental Influence and Complexity*

Each stage of the mushroom life cycle is affected by environmental conditions for survival, proper growth, and reproduction. The availability of moisture, nutrients, and suitable substrates determines the success of spore germination, hyphal growth, and mycelium development. Additionally, interactions with other organisms, including plants, animals, and competing microbes, further shape the life cycle's complexity (Naranjo-Ortiz and Gabaldón 2020). The life cycle of mushrooms not only highlights their adaptability and ecological significance but also emphasizes the intricate balance required for their existence. From nutrient recycling to symbiotic relationships, mushrooms play a vital role in maintaining ecosystem health and biodiversity.

3.8 Challenges in Cultivation

The global demand for nutrients and healthy food sources is largely the reason for diversification in cultivating WEF through edible mushroom cultivation. Meeting this demand in an ecologically sustainable fashion presents a difficult balance to strike: At the same time, we require protecting natural ecosystems while sustaining human livelihoods. This dual objective aligns with global actions to address climate change, safeguard biodiversity, and support equitable socioeconomic development indicated by the 17 United Nations (UN) sustainable development goals (SDGs).

It is necessary to target interventions to reach these goals. Cultivation of ectomycorrhizal fungi is a critical area for their growth, as they lack rigorous and effective cultivation methods, diversification of saprobic fungi culture, and advanced bioinoculants for reforestation projects. Sustainable harvesting for WEF natural populations must be promoted, and the economic value of these fungi promoted by innovation, like the stimulation of mycotourism and value in high-end culinary applications stemming from WEF (Yamada 2022). The changing of long-term sustainability of the project, as well as the revitalization of traditional knowledge with kids educated about mycological resources.

Implementation of these measures will be effective if rural and urban communities, policymakers, researchers, and entrepreneurs are involved themselves actively. Depending on the appropriate cultural, ecological, and geographical conditions, context specific strategies are called for. The COVID-19 pandemic has worsened financial and sociocultural vulnerabilities that necessitate the need to develop sustainable management practices for the marginalized layers and build resilience. Sustainable WEF management and utilization is a great solution to food security, improved rural livelihoods, and ecosystem services. However, doing so requires its own hurdles, such as resource constraints, knowledge deficits, and technological boundaries. WEF has a unique opportunity to help advance global sustainability objectives by enabling innovation and taking an inclusive and science-based approach to tackling the multiple challenges of today (Çakmakçı et al. 2024).

3.8.1 Lack of Suitable Mushroom Spawn

Mushroom cultivation depends on mass-produced spawn of selected fungal strains that have been optimized for controlled growth, this approach is not viable with wild mushrooms. Wild species have greater genetic diversity and a unique genetic makeup that makes it difficult to reproduce in farm condition. Because this genetic variability is present, growth patterns are often inconsistent, and in some cases, the mycelium may be slow growing and therefore less vigorous compared to cultivated mushroom spp. For instance, Porcini and Matsutake are highly genetically distinct, so it is extremely difficult to grow in farmed cultivation (Pipaliya and Ansari 2023). High-quality mushroom spawn is clearly a costly and still a key critical requirement for successful cultivation when many regions might lack capacity for production or might even have little or no production capacity at all. Low-quality spawn produces low yields and is very susceptible to contamination /poison. Investments in spawn production infrastructure, quality control standards, and spawn distribution are required to solve the problem. Training programs that teach producers how to produce their own spawn increase self-sufficiency and reduce reliance on external suppliers (Borah et al. 2019).

3.8.2 Dependency on Complex Natural Substrates

Nonfarmed mushrooms depend on specific natural substrates and environmental conditions and thus face many challenges in farmed cultivation. Wild mushrooms require unique substrates like decaying wood, certain soils, tree roots, organic compost, and living host substrates are difficult to reproduce in farmed cultivation. The cultivation becomes more difficult because some of the species even require symbiotic relationships with trees or microorganisms. Wild mushrooms grow in complex ecosystems where humidity, light cycles, as well as nutrient interactions are very important and impossible to manage in controlled conditions. Nutrient requirements are going imbalanced when alternative substrates such as sawdust or straw are used. Wild mushrooms are very sensitive to their carbon, nitrogen, and micronutrient balance. Morchella mushroom spp. cultivation is extremely difficult because that depends on specific soil microbiota. Matsutake mushroom also depends on mycorrhizal associations with trees, that's why, on a large scale, commercially cultivation is nearly impossible (Kaliyaperumal et al. 2018).

Commercial cultivation of the wild edible mushrooms (WEF) cannot have a stable nutrient source and often relies on decaying organic material to support them, which is a major problem. WEF are grown on substrates of similar, rather than standardized, artificial substrates as cultivated varieties. In fact, it includes all the decomposed wood, the leaf litter, all plant debris that is unusable through decomposition by microorganisms that release the carbon, the nitrogen, and phosphorus in a form to make it usable again for vegetation. Anyway, they hold microbial communities to cycle nutrients and also help mushroom growth, and they are good because they are natural substrates. This is, unfortunately, something that cannot be reproduced in a controlled environment without an exceptionally fine modification of the microbial activity and substrate composition of the ecosystem. Wild edible mushroom cultivation is also further realized by scientific studies reporting the selection of substrate utilization amongst. Species requiring specifically some special residues from trees, which are not easy to purchase and process on an industrial scale (Corbu et al. 2023). These challenges demonstrate that it's difficult to grow wild mushrooms on farmed for commercial production. Dependence upon complex substrates severely restricts the conditions under which cultivation can occur, including such areas of natural ecosystems that support wild mushroom populations.

3.8.3 Obligate Symbiotic Relationships

The growth and development of many wild edible fungi (WEF) depend on complex obligate symbiotic relationships with other organisms essential for survival and ecological roles. Those are the fungi that make a mutualistic association with the roots of some particular trees, such as pine Pinus spp. and oak Quercus spp., by helping exchange nutrients between the trees and the fungi. These fungi provide their host plants with water and minerals, and carbohydrates derived from photosynthesis. Because of this strict dependence, these fungi cannot be cultivated on artificial substrates such as compost or sawdust used by saprotrophic species. A major challenge to replicating the natural conditions necessary for mycorrhizal fungi. The composition of soil microbiome, pH, moisture, temperature, and the presence of specific microbial partners are key to the establishment and formation of fungal fruiting bodies. However, these factors are not achieved in controlled environments, so cultivation is extremely unsuccessful. In addition, it takes multiple years to establish a stable symbiotic relationship, which makes the cultivation process time-consuming and expensive. Additionally, wild edible mushrooms have complex genetic and biochemical interactions in the life cycle. Many wild mushroom species require specific bacterial or fungal partners to complete their reproductive cycle (Albuquerque-Martins et al. 2019). In *T. heimii* cellulolytic bacteria in the termite nest break down cellulose in plant material eaten by termites. For their part, the termites feed them, breaking down fibers of complex plants into simpler compounds that supply both themselves and the fungus that makes these mounds part of its habitat (Albuquerque-Martins et al. 2019). Furthermore, because atmospheric nitrogen does not occur in ways usable by termites and the fungus, nitrogen-fixing bacteria inside the nest convert atmospheric nitrogen into forms that the termites and fungus can use. *The heimii* also survives with fungal symbionts. In the termite nest, the actinobacteria and ascomycetes help degrade plant material and create the right substrate for fungal growth. *T. heimii* may coexist with other basidiomycetes in the nest and contribute to the breakdown of lignocelluloses, which support the development of *T. heimii*. Like many WEF species, these species also form a mutualistic association with the roots of specific trees or plants and rely on a mycorrhizal relationship with those trees, and are vital for the survival of such species (Paloi et al. 2023). Controlled cultivation of the fungi is highly challenging due to these processes and the fungi's sensitivity to environmental conditions such as soil composition, moisture, and temperature. The symbiotic relationships with specific host plants and the associated ecological dynamics impose barriers to the commercial production of WEF, and they need to be studied in greater detail before further attempts at producing these in commercial quantities. Truffles and chanterelles, such as these fungi, feed on the host plants and receive nutrients from and water for themselves. For instance, truffles prefer oaks, pines, or birches, while chanterelles are associated with coniferous or broadleaf forests. These associations are essential for the fungi's growth, but they present a major challenge to commercial cultivation because these interactions are not easily replicated in artificial settings. The complexity of signaling mechanisms and nutrient exchange processes described in scientific studies reviews mycorrhizal relationships (Hall and Haslam 2013).

3.8.4 Biochemical and Genetic Variability

Wild edible mushrooms show a remarkable biochemical and genetic diversity that complicates the standardization of their cultivation. The cultivation of wild mushrooms faces significant difficulties because of their high biochemical and genetic diversity, which impacts their development. Commercially cultivated mushrooms show varying biochemical variability because their genetics and environmental conditions affect them. The biochemical variability among mushrooms produces unique protein and lipid levels, building precious bioactive compounds. Wild mushrooms enhanced concentrations of antioxidant molecules, ergothioneine, and immune-boosting beta-glucans. Cultivated mushrooms lack certain medicinal compounds, including terpenoids, indole derivatives, psilocybin, and norbaeocystin, that wild mushrooms naturally produce. Wild mushrooms grow a higher amount of vitamin D2 when exposed to sunlight. The biochemical makeup of wild mushrooms demands specific nutrient-rich substrates such as wood in combination with compost, straw, and minerals that promote their maximum growth potential. Although these variations in biochemical composition result in inconsistent yields, unpredictable bioactivity and varying nutritional content, cultivation on commercial scales is challenging. Domestication processes have become more complex due to the need for specific environmental factors and specialized growth substrates. Secondary metabolites that many WEF produce include toxins and antimicrobial compounds that aid in successful competition in their natural habitat. Other organisms grow well because these compounds may inhibit the growth of other organisms, but do not create favorable conditions for artificial cultivation (Kissanga et al. 2022).

Additionally, wild mushroom species exhibit genetic diversity, and their growth requirements, environmental tolerances, and substrate response are affected by genetic variation between species. Wild mushrooms have genetic variability, which makes it more difficult to cultivate because of reproduction by spores, resulting in great genetic variability. The variability in growth rate, yield, and morphology makes the commercial production of wild mushrooms more complicated. More diversity is introduced by genetic recombination and sexual reproduction, as well as spore dispersal, to increase the genetic exchange between populations. Fruiting time, growth rate, and biochemical composition of fruit differ for different genotypes, and yield predictability and commercial viability are affected. In addition, size, shape, color, and bioactive metabolite production vary in their nutritional and medicinal properties. However, due to this variability, developing uniform cultivation protocols suited for large-scale production is difficult. However, research has shown that genetic strains can have different growth behaviors, making cultivation efforts highly unpredictable even within a species. Further, the triggering of the production of reproductive structures, including fruit bodies, is typically promoted by particular environmental cues that are not easily reproduced repeatedly (Alain and Querellou 2009). Combining the genetic and biochemical complexity of WEF underscores the difficulty of transferring WEF from natural ecosystems to controlled commercial systems.

3.8.5 Challenges in Reproduction and Fruiting

In most cases, wild mushroom reproduction depends on complex ecological interactions like the presence of insects or environmental factors. While fungus fruiting in an arterial environment is a very challenging task, we do not know what factors influence sporocarp formation. An example is that some species functionally require certain light cycles or must develop symbiotic interactions for fruiting, which makes it more challenging under controlled conditions (Suwannarach et al. 2022).

Cultivation of nonfarmed mushrooms in controlled environments is particularly challenging because of their complex ecology and the requirement of specialized environmental conditions. Both their reproduction and their fruiting are primarily limited by these challenges, making them commercially impractical. Wild mushrooms have many spores requiring extremely precise conditions for spore germination, such as temperature, humidity, and microbial interactions. Some species have developed complex mating systems that necessitate a series of mating types for successful reproduction. *Lentinula edodes* follow a tetrapolar mating system with outdoor mating (22–25 °C, high humidity) taking roughly 1–2 weeks, but complete colonization and fruiting in 3 months. Some other species have slow mycelial development and delayed fruiting. For example, *Tricholoma matsutake* and *Cantharellus cibarius* have been determined to require several years to establish and fruit as a mycorrhizal species. The absence of natural microbial partners may limit growth, and the presence of competing bacteria and fungi may block mycelial development and flowering (Carrasco and Preston 2020).

3.8.6 Complex Biotic Interactions

Pseudomonas spp., Bacillus spp., Bradyrhizobium, Rhizobium, and Streptomyces spp. are important for the ecology of ectomycorrhizae. These bacteria are commonly known as mycorrhizae-associated bacteria. Some mushrooms need complex interaction with specific bacteria and fungi; they need soil bacteria to break dormancy for spore germination. These bacteria enhance germination and mycorrhizal formation. Some microbes, such as Streptomyces spp. play a beneficial role by developing early-stage mycelial colonization and supporting spore germination in matsutake mushrooms. In Chanterelle mushroom, spore germination needs specific tree roots and associated fungi, Sebacina spp. that help break the dormancy and benefit mycelial growth and development. Literature shows some bacteria that have an antimycotic action of a specific mushroom. Like Pseudomonas spp. against the fungal complement of *Tuber borchii* ascocarps (Rodríguez-Gutiérrez et al. 2019). Several of these bacteria produce metabolites under controlled conditions that stimulate the growth of *T. borchii* mycelia. The observations presented here suggest that Pseudomonas bacteria may be involved in forming *Sigillus ascocarps* and *Sigillus mycorrhizae* (Ma et al. 2019). In addition, *T. borchii* mycelia harbored a Cytophaga, Flexibacter, and Bacteroides bacterium, as in Barbieri et al. (2000); however, its use has not yet been established. The bacterial and fungal interactions showed proof that the previously mentioned type of interaction controls fungal growth and sporing. Part of the ectomycorrhizal fungi has been occasionally associated with saprobic fungi or other ectomycorrhizal forms. For instance, Rhizopogon with *Suillus bovinus, Gomphidiu sroseus* with Chroogomphus species with Suillus spp. and *Scleroderma citrinum* with *Boletus parasiticus* (Mello and Balestrini 2018). A *Boletus edulis, Amanita excelsa,* or *Amanita muscaria* can be found in Austria, Italy, England, Sweden, New Zealand, the United States, etc.

3.8.7 Lack of Technical Supportive Policies

Production and conservation of wild edible mushrooms are major constraining factors to the sector's growth due to the absence of comprehensive policies to promote the production and conservation of wild edible fungi. Producers have no tools for success if producer policies do not tackle resource access, training, developmental marketing, and research funding (Guerin-Laguette 2021). Additionally, unsustainability in harvesting practices and market inefficiency may be exacerbated in the absence of adequate regulatory oversight typical of such markets. Therefore, edible mushrooms can be a necessary target for today's governments and international organizations to establish supportive frameworks for WEF in agricultural and forestry policies. These should include incentives for sustainable practices, investments in research and development, and programs in capacity building for local communities (Pipaliya and Ansari 2023). Collaborative work between policymakers and other stakeholders of the mushroom sector, in partnership with researchers and industry stakeholders.

3.8.8 Undeveloped Technology

The technological gap is a major barrier to WEF cultivation and processing at a large scale and with high efficiency. However, most producers rely on conventional means to maximize yields and ensure stable quality. Advanced technologies, such as those systems suitable for controlled environment, automated harvesting, precision agriculture, and WEF production, will be developed and disseminated (Khan et al. 2024). Shared responsibility is required between research institutions and private sector actors to transfer technology from research institutions and private sector actors into producers. Furthermore, public–private partnerships can avoid funding gaps and expedite technological breakthroughs (Sangeeta et al. 2024).

3.9 Conclusion

Nonfarmed mushrooms have immense nutritional, ecological, and medicinal value, but their commercial cultivation remains challenging due to intricate ecological dependencies. This chapter explored the biochemical composition of nonfarmable mushrooms rich in protein, carbohydrate, unique lipids, vitamins profiling, and various bioactive compounds. Their nutritional value highlights the significance of these mushrooms for human health. Wild mushrooms cannot be cultivated commercially due to their obligate symbiotic relationships with plants, insects, and microbial communities and their dependency on specific environmental conditions such as soil composition, moisture levels, and microbial interactions. This chapter concluded that wild mushrooms are important in promoting ecological sustainability, nutrient cycling, biodiversity preservation, and addressing and improving food security and ecosystem health. Due to their dependency on specific host organisms and environmental factors, they are almost impossible to cultivate on a large scale. Future research should focus on the development of biotechnological approaches and genetic engineering advancements that minimize their natural habitats and symbiotic associations, toward sustainable cultivation while preserving their nutritional and medicinal properties. Research and genetic engineering advancements could provide insights into their growth mechanisms, allowing for sustainable cultivation while preserving their nutritional and medicinal properties.

References

Abdullah S, Kumar A. A brief review on the medicinal uses of Cordyceps militaris. Pharmacol Res-Mod Chinese Med. 2023;7:100228.

Adamska I, Felisiak K. Black trumpet [Craterellus cornucopioides (L.) Pers.]—bioactive properties and prospects for application in medicine and production of health-promoting food. Nutrients. 2024;16(9):1325.

Alain K, Querellou J. Cultivating the uncultured: limits, advances and future challenges. Extremophiles. 2009;13:583–94.

Albuquerque-Martins R, Carvalho P, Miranda D, Gonçalves MT, Portugal A. Edible ectomycorrhizal fungi and Cistaceae. A study on compatibility and fungal ecological strategies. PLoS One. 2019;14:e0226849.

Alzand KI, Bofaris M, Ugis A. Chemical composition and nutritional value of edible wild growing mushrooms: a review. World Journal of Pharmaceutical Research 2019;8:31–46.

Anusiya G, Gowthama Prabu U, Yamini NV, Sivarajasekar N, Rambabu K, Bharath G, Banat F. A review of the therapeutic and biological effects of edible and wild mushrooms. Bioengineered. 2021;12(2):11239–68.

Assemie A, Abaya G. The effect of edible mushroom on health and their biochemistry. Int J Microbiol. 2022a;2022:1–7.

Assemie A, Abaya G. The effect of edible mushroom on health and their biochemistry. Int J Microbiol. 2022b;2022(1):8744788.

Atri N. Mushrooms in service of the society and the environment. KAVAKA. 2024;60:54–76.

Badalyan SM. Antiprotozoal activity and mitogenic effect of mycelium of culinary medicinal shiitake mushroom Lentinusedodes (Berk.) Singer (Agaricomycetidae). Int J Med Mushrooms. 2004;6:131–8.

Baral B. Entomopathogenicity and biological attributes of Himalayan treasured fungus Ophio*Cordyceps sinensis* (Yarsagumba). J Fungi. 2017;3:4.

Barbieri E, Potenza L, Rossi I, Sisti D, Giomaro G, Rossetti S, Beimfohr C, Stocchi V. Phylogenetic characterization and in situ detection of a Cytophaga-Flexibacter-Bacteroides phylogroup bacterium in tuber borchii Vittad. Ectomycorrhizal mycelium. Appl Environ Microbiol. 2000;66:5035–42.

Barros L, Baptista P, Correia DM, Morais JS, Ferreira I. Fatty acid and sugar compositions, and nutritional value of five wild edible mushrooms from Northeast Portugal. Food Chem. 2008;105(1):140–5.

Bell V, Silva CRPG, Guina J, Fernandes TH. Mushrooms as future generation healthy foods. Front Nutr. 2022;9:1050099.

Biswas D, Biswas J. Major deteriorative, pathogenic and beneficial fungi reported from various subterranean caves of the world: a mini review. Int J Ecosyst. 2017;7:11–6. https://doi.org/10.5923/j.ije.20170701.02.

Boa E. Wild edible fungi: a global overview of their use and importance to people. Non-wood Forest products 17. Rome: FAO; 2004a.

Boa ER. Wild edible fungi: a global overview of their use and importance to people. Rome: Food and Agriculture Organization of the United Nations; 2004b.

Boddy L, Hiscox J. Fungal ecology: principles and mechanisms of colonization and competition by saprotrophic fungi. Microbiol Spectr. 2016;4 https://doi.org/10.1128/microbiolspec.funk-0019-2016.

Borah TR, Singh AR, Paul P, Talang H, Kumar B, Hazarika S. Spawn production and mushroom cultivation technology, vol. 46. ICAR research complex for NEH region; 2019.

Çakmakçı S, Polatoğlu B, Çakmakçı R. Foods of the future: challenges, opportunities, trends, and expectations. Foods. 2024;13:2663.

Cardwell G, Bornman JF, James AP, Black LJ. A review of mushrooms as a potential source of dietary vitamin D. Nutrients. 2018;10(10):1498.

Carrasco J, Preston GM. Growing edible mushrooms: a conversation between bacteria and fungi. Environ Microbiol. 2020;22:858–72.

Cerletti C, Esposito S, Iacoviello L. Edible mushrooms and beta-glucans: impact on human health. Nutrients. 2021;13(7):2195.

Chang ST. The Chinese mushroom (Volvariellavolvacea): morphology, cytology, genetics, nutrition and cultivation. Chinese University Press; 1972.

Chang ST. The origin and early development of straw mushroom cultivation. Econ Bot. 1977;31(3):374–6. https://doi.org/10.1007/BF02866890.

Chang ST. Mushroom research and development-equality and mutual benefit. In: Proceedings of the 2nd international conference on mushroom biology and mushroom products. Pennsylvannia: Pennsylvania State University; 1996. p. 1–10.

Chang ST. The world mushroom industry: trends and technological developments. Int J Med Mushrooms. 2006;8:297–314.

Cheung PCK. Nutritional value and health benefits of mushrooms. In: Mushrooms as functional foods. Wiley; 2008. p. 71–109. https://doi.org/10.1002/9781118944653.ch3.

Chong PS, Fung ML, Wong KH, Lim LW. Therapeutic potential of Hericium erinaceus for depressive disorder. Int J Mol Sci. 2020;21(1):163.

Corbu VM, Gheorghe-Barbu I, Dumbravă AŞ, Vrâncianu CO, Şesan TE. Current insights in fungal importance—a comprehensive review. Microorganisms. 2023;11:1384.

Courty P, Buée M, Diedhiou AG, Frey-Klett P, Tacon FL, Rineau F, Turpault M, Uroz S, Garbaye J. The role of ectomycorrhizal communities in forest ecosystem processes: new perspectives and emerging concepts. Soil Biol Biochem. 2010;42:679–98. https://doi.org/10.1016/j.soilbio.2009.12.006.

Crescenzi A, Latronico T, Petraglia T, Rosano R, Liuzzi GM. The edible mushrooms as potential candidate of antioxidants as a treatment for age related neurodegenerative diseases. Nutrients. 2023;15(8):1913.

Danell E. Cantharellus. In: Cairney JWG, Chambers SM, editors. Ectomycorrhizal fungi: key genera in profile. Springer-Verlag; 2000. p. 253–67.

De Silva DD, Rapior S, Hyde KD, Bahkali AH. Medicinal mushrooms in prevention and control of diabetes mellitus. Fungal Divers. 2012;56:1–29.

De-Miguel S, Bonet JA, Pukkala T, de Aragón JM. Impact of forest management intensity on landscape-level mushroom productivity: a regional model-based scenario analysis. For Ecol Manag. 2014;330:218–27. https://doi.org/10.1016/j.foreco.2014.07.014.

El-Ramady H, Abdalla N, Badgar K, Llanaj X, Törős G, Hajdú P, Eid Y, Prokisch J. Edible mushrooms for sustainable and healthy human food: nutritional and medicinal attributes. Sustainability. 2022;14:4941.

Epping R, Lisec J, Koch M. Changes in Black truffle (*Tuber melanosporum*) aroma during storage under different conditions. J Fungi. 2024;10:354.

Fernandes T, Garrine C, Ferrão J, Bell V, Varzakas T. Mushroom nutrition as preventative health care in Sub-Saharan Africa. Ann Res Rev Biol. 2021;34(6):1–13. https://doi.org/10.3390/app11094221.

Fernández-Ruiz A, Vicente-Villardón JL, Sánchez-Sánchez J, García-Jiménez P, Sánchez-Durán S, La Cruz DR-D. A statistical approach to macrofungal diversity in a mediterranean ecosystem of the Iberian Peninsula Dominated by the Holm Oak (Quercus ilex L. subsp. ballota (Desf.) Samp.). Forests. 2023;14:1662.

Fogarasi M, Fărcas AC, Socaci SA, Socaciu MI, Semeniuc CA. Health benefits of edible wild mushrooms. In: Wild mushrooms. CRC Press; 2022. p. 193–210.

Fricker MD, Heaton LL, Jones NS, Boddy L. The mycelium as a network. In: The fungal kingdom. Wiley; 2017. p. 335–67.

Ganesan K, Xu B. Anti-obesity effects of medicinal and edible mushrooms. Molecules. 2018;23(11):2880.

García-Melo KS, García-Montalvo IA, Pérez-Santiago AD, Sánchez-Medina MA, Matías-Pérez D. Estimation of chemical comprise and antioxidants in edible wild mushrooms. J Adv Food Sci Technol. 2022;9(1):45–52.

Gayley D. Experimental spawn and mushroom culture II artificial composts. Ann Appl Biol. 1938;25(2):322–40. https://doi.org/10.1111/j.1744-7348.1938.tb02337.x.

Gottlieb D. The physiology of spore germination in fungi. Bot Rev. 1950;16:229–57.

Gravina AG, Pellegrino R, Auletta S, Palladino G, Brandimarte G, D'Onofrio R, et al. Hericium erinaceus, a medicinal fungus with a centuries-old history: evidence in gastrointestinal diseases. World J Gastroenterol. 2023;29(20):3048.

Guerin-Laguette A. Successes and challenges in the sustainable cultivation of edible mycorrhizal fungi–furthering the dream. Mycoscience. 2021;62:10–28.

Guillamón E, García-Lafuente A, Lozano M, Rostagno MA, Villares A, Martínez JA. Edible mushrooms: role in the prevention of cardiovascular diseases. Fitoterapia. 2010;81(7):715–23.

Gupta A, Bansal V. Classification and species: mushroom. Webology. 2021;18:1653–60.

Győrfi J, Kovács A, Szabó A. Increasing the vitamin D level of oyster mushrooms by UV light. Int J Horticultural Sci. 2011;17:119–23.

Hall IR, Haslam W. Truffle cultivation in the Southern Hemisphere. In: Edible Ectomycorrhizal mushrooms: current knowledge and future prospects. Springer; 2013.

Hall IR, Stephenson SL, Buchanan PK, Cole AL, Yun W. Edible and poisonous mushrooms of the world, vol. 372. Portland: Timber Press; 2003.

Hanuš LO, Shkrob I, Dembitsky VM. Lipids and fatty acids of wild edible mushrooms of the genus boletus. J Food Lipids. 2008;15(3):370–83.

Hawksworth DL, Lücking R. Fungal diversity revisited: 2.2 to 3.8 million species. Microbiol Spectr. 2017;5(4):1–17. https://doi.org/10.1128/microbiolspec.FUNK-0052-2016.

Heleno SA, Barros L, Sousa MJ, Martins A, Santos-Buelga C, Ferreira IC. Targeted metabolites analysis in wild boletus species. LWT-Food Sci Technol. 2011;44(6):1343–8.

Hess J, Wang Q, Gould T, Slavin J. Impact of Agaricusbisporus mushroom consumption on gut health markers in healthy adults. Nutrients. 2018;10(10):1402.

Hong JS, Kim YH, Kim MK, Kim TY, Kim KJ. Studies on the lipids composition of Korean edible mushrooms. J Korean Soc Food Culture. 1990;5(4):437–42.

Hussain S, Sher H, Ullah Z, Elshikh MS, Al Farraj DA, Ali A, Abbasi AM. Traditional uses of wild edible mushrooms among the local communities of Swat, Pakistan. Foods. 2023;12:1705.

Ichimura T, Watanabe O, Muruyama S. Inhibition of HIV1 protease by water-soluble lignin-like substance from an edible mushroom, Fuscoporia oblique. Biosci Biotechnol Biochem. 1998;62:575–7. https://doi.org/10.1271/bbb.62.575.

Inoue A, Kodama N, Nanba H. Effect of maitake (Grifolafrondosa) D-fraction on the control of the T lymph node Th-1/Th-2 proportion. Biol Pharm Bull. 2002;25:536–40.

Jaiswal A, Heera LS, Gugalia G. Mushroom-a nutritious and environment friendly crop: a review. Bhartiya Krishi Anusandhan Patrika. 2024; https://doi.org/10.18805/bkap694.

Kabuyi MK, Kapepula PM, Kabengele JK, Moni B, Makengo G, Mungitshi PM, Fundu TM. Selenium content and antioxidant potential of some edible wild mushrooms from Bandundu area, DR Congo. Nat Resour. 2017;8(2):103–13.

Kalač P. A review of chemical composition and nutritional value of wild-growing and cultivated mushrooms. J Sci Food Agric. 2013;93(2):209–18.

Kaliyaperumal M, Kezo K, Gunaseelan S. A global overview of edible mushrooms. Biol Macrofungi. 2018:15–56.

Karwa A, Varma A, Rai M. Edible ectomycorrhizal fungi: cultivation, conservation and challenges. In: Diversity and biotechnology of Ectomycorrhizae. Springer; 2010.

Kewessa G, Dejene T, Alem D, Tolera M, Martín-Pinto P. Forest type and site conditions influence the diversity and biomass of edible macrofungal species in Ethiopia. J Fungi. 2022;8:1023.

Khalid S, Keller NP. Chemical signals driving bacterial–fungal interactions. Environ Microbiol. 2021;23(3):1334–47. https://doi.org/10.1111/1462-2920.15410.

Khan A, Murad W, Ali S, Shah SS, Halim SA, Khalid A, Kashtoh H, Khan A, Al-Harrasi A. Contribution of mushroom farming to mitigating food scarcity: current status, challenges and potential prospects in Pakistan. Heliyon. 2024;10:e40362.

Khatua S, Paul S, Acharya K. Mushroom as the potential source of new generation of antioxidant: a review. Res J Pharm Technol. 2013;6(5):3.

Kissanga R, Liberal Â, Diniz I, Rodrigues AS, Baptista-Ferreira JL, Batista D, Ivanov M, Soković M, Ferreira IC, Fernandes Â. Biochemical and molecular profiling of wild edible mushrooms from Huila, Angola. Foods. 2022;11:3240.

Klaus A, Wan WAAQI. Cultivation strategies of edible and medicinal mushrooms. In: Wild mushrooms. CRC Press; 2022.

Kosanić M, Ranković B, Stanojković T, Radović-Jakovljević M, Ćirić A, Grujičić D, Milošević-Djordjević O. Craterellus cornucopioides edible mushroom as source of biologically active compounds. Nat Prod Commun. 2019;14(5):1934578X19843610.

Krishna KV, Balasubramanian B, Park S, Bhattacharya S, Kadanthottu Sebastian J, Liu W-C, Pappuswamy M, Meyyazhagan A, Kamyab H, Chelliapan S. Conservation of endangered *Cordyceps sinensis* through artificial cultivation strategies of C. militaris, an alternate. Mol Biotechnol. 2024;67:1–16.

Krittanawong C, Isath A, Hahn J, Wang Z, Fogg SE, Bandyopadhyay D, Tang WW. Mushroom consumption and cardiovascular health: a systematic review. Am J Med. 2021;134(5):637–42.

Kues U, Liu Y. Fruiting body production in basidiomycetes. Appl Microbiol Biotechnol. 2000;54:141–52.

Kumar S, Mishra AK, Mishra S, Marndi S. Economic importance of wild mushrooms in Mayurbhanj District, Odisha, India. Asian J Biol. 2022a;15:20–5.

Kumar S, Mishra AK, Sanath KN, Mishra S. Economically important wild edible mushrooms of Bonai Forest division, Odisha, India. Asian J Biol. 2022b;16:31–40.

Kumar A, Devi R, Dhalaria R, Tapwal A, Verma R, Rashid S, Verma T. Nutritional, nutraceutical, and medicinal potential of *Cantharellus cibarius* Fr.: a comprehensive review. Food Sci Nutr. 2024;13:e4641.

Kuznetsov OI, Milkova EV, Sosnia AE, Sotnikova NI. Antimicrobial action of Lentinus edodes juice on human microflora. Mikrobiol Epidemiol Immunobiol. 2005;1:80–2.

Li J, Wen S, Zhang B, Wang F. Selenium enrichment of the edible medicinal mushroom Antrodia camphorata by submerged fermentation. Molecules. 2023;28(7):3036.

Lian C, Narimatsu M, Nara K, Hogetsu T. Tricholoma matsutake in a natural Pinus densiflora forest: correspondence between above-and below-ground genets, association with multiple host trees and alteration of existing ectomycorrhizal communities. New Phytol. 2006;171:825–36.

Liuzzi GM, Petraglia T, Latronico T, Crescenzi A, Rossano R. Antioxidant compounds from edible mushrooms as potential candidates for treating age-related neurodegenerative diseases. Nutrients. 2023;15(8):1913.

Lu Z, Shi P, He Y, Zhang D, He Z, Chen S, Zeng W. Review on natural enemies and diseases in the artificial cultivation of Chinese caterpillar mushroom, Ophiocordyceps sinensis (Ascomycetes). International Journal of Medicinal Mushrooms. 2015;17: 7.

Ma YJ, Zheng LP, Wang JW. Bacteria associated with Shiraia fruiting bodies influence fungal production of hypocrellin A. Front Microbiol. 2019;10:2023.

Manan S, Ullah MW, Ul-Islam M, Atta OM, Yang G. Synthesis and applications of fungal mycelium-based advanced functional materials. J Bioresour Bioprod. 2021;6:1–10.

Manikandan K. Nutritional and medicinal values of mushrooms. In: Mushrooms cultivation, marketing and consumption. Directorate of Mushroom Research, Indian Council of Agricultural Research; 2011. p. 11–4.

Manjula BL, Kumar S. Wild mushrooms: a source of livelihood and future urban food. Ecol Environ Conserv. 2024;30:187–91.

Manoharachary C, Kunwar I, Rajithasri A. Advances in applied mycology and fungal biotechnology. KAVAKA. 2014;43:79–92.

Manzi P, Aguzzi A, Pizzoferrato L. Nutritional value of mushrooms widely consumed in Italy. Food Chem. 2001;73(3):321–5.

Martinez de Aragon J, Bonet JA, Fischer CR, Colinas C. Productivity of ectomycorrhizal and selected edible saprotrophic fungi in pine forests of the pre-Pyrenees mountains, Spain: predictive equations for forest management of mycological resources. For Ecol Manag. 2007;252:239–56.

Mei Y, Liu C-Y, Li S-H, Guerin-Laguette A, Xiao Y-J, Tang P, Wan S-P, Bonito G, Wang Y. Phlebopusroseus, a new edible bolete from China, is associated with insects and plants. Mycologia. 2021;113(1):33–42. https://doi.org/10.1080/00275514.2020.1816781.

Mello A, Balestrini R. Recent insights on biological and ecological aspects of ectomycorrhizal fungi and their interactions. Front Microbiol. 2018;9:216.

Mishra M, Srivastava D. Harnessing the power of medicinal mushrooms into cosmetics: a fountain of youth from nature. J Indian Botanical Soc. 2024; https://doi.org/10.61289/jibs2024.06.01.1214.

Mishra S, Sangeeta SK, Gupta S. Culinary techniques for maximizing nutritional value of mushrooms. Mushroom Magic. CRC Press; 2024. p. 191–209.

Money NP. Spore production, discharge, and dispersal. The fungi: Elsevier; 2016.

Morath SU, Hung R, Bennett JW. Fungal volatile organic compounds: a review with emphasis on their biotechnological potential. Fungal Biol Rev. 2012;26:73–83.

Mothana RAA, Jansen R, Julich WD, Lindequist U. Ganomycin A and B, new antimicrobial farnesyl hydroquinones from the basidiomycete Ganoderma pfiefferi. J Nat Prod. 2000;63:416–8.

Muthukumar T, Priyadharsini P, Uma E, Jaison S, Pandey RR. Role of arbuscular mycorrhizal fungi in alleviation of acidity stress on plant growth. In: Use of microbes for the alleviation of soil stresses, vol. 1. Springer; 2014. p. 43–71.

Naranjo-Ortiz MA, Gabaldón T. Fungal evolution: diversity, taxonomy and phylogeny of the fungi. Biol Rev. 2019;94(6):2101–37. https://doi.org/10.1111/brv.12550.

Naranjo-Ortiz MA, Gabaldón T. Fungal evolution: cellular, genomic and metabolic complexity. Biol Rev. 2020;95:1198–232.

Nasiry D, Khalatbary AR, Ebrahimzadeh MA. Anti-inflammatory and wound-healing potential of golden chanterelle mushroom, *Cantharellus cibarius* (Agaricomycetes). Int J Med Mushrooms. 2017;19(10):893.

Niego AGT, Rapior S, Thongklang N, Raspé O, Hyde KD, Mortimer P. Reviewing the contributions of macrofungi to forest ecosystem processes and services. Fungal Biol Rev. 2023;44:100294.

Nielsen J. Modelling the growth of filamentous fungi. In: Modern biochemical engineering. Springer; 2005. p. 187–223.

Nunes MD, Da Luz JMR, Paes SA, Ribeiro JJO, Da Silva MDCS, Kasuya MCM. Nitrogen supplementation on the productivity and the chemical composition of oyster mushroom. J Food Res. 2012;1:113.

Ogawa W, Endo N, Takeda Y, Kodaira M, Fukuda M, Yamada A. Efficient establishment of pure cultures of yellow chanterelle *Cantharellusanzutake* from ectomycorrhizal root tips, and morphological characteristics of ectomycorrhizae and cultured mycelium. Mycoscience. 2019;60:45–53.

Pal A, Ray R, Acharya K, Paul S. Assessment of the anti-leukemic and antioxidant potential of the methanol extract of a wild, edible, and novel mushroom, Astraeus hygrometricus, and unraveling its metabolomic profile. J Adv Biotechnol Exp Ther. 2021;4(3):388–404.

Paloi S, Kumla J, Paloi BP, Srinuanpan S, Hoijang S, Karunarathna SC, Acharya K, Suwannarach N, Lumyong S. Termite mushrooms (termitomyces), a potential source of nutrients and bioactive compounds exhibiting human health benefits: a review. J Fungi. 2023;9:112.

Panda SK, Sahoo G, Swain SS, Luyten W. Anticancer activities of mushrooms: a neglected source for drug discovery. Pharmaceuticals (Basel). 2022;15(2):176. https://doi.org/10.3390/ph15020176. PMID: 35215289; PMCID: PMC8876642.

Patyshakuliyeva A, Jurak E, Kohler A, Baker A, Battaglia E, De Bruijn W, Burton KS, Challen MP, Coutinho PM, Eastwood DC. Carbohydrate utilization and metabolism is highly differentiated in *Agaricus bisporus*. BMC Genomics. 2013;14:1–14.

Peksen A, Kibar B, Yakupoglu G. Favourable culture conditions for mycelial growth of Hydnum repandum, a medicinal mushroom. Afr J Trad Comp Alt Med. 2013;10(6):431–4. https://doi.org/10.4314/ajtcam.v10i6.4.

Pena R, Tibbett M. Mycorrhizal symbiosis and the nitrogen nutrition of forest trees. Appl Microbiol Biotechnol. 2024;108:461.

Pilz D, Molina R. Commercial harvests of edible mushrooms from the forests of the Pacific Northwest United States: issues, management, and monitoring for sustainability. For Ecol Manag. 2001;5593:1–14. https://doi.org/10.1016/S0378-1127(01)00543-6.

Pipaliya G, Ansari M. Exploring the challenges and constraints encountered by mushroom growers in Uttarakhand: a comprehensive study. Asian J Agric Ext Econ Sociol. 2023;41:1–8.

Podkowa A, Kryczyk-Poprawa A, Opoka W, Muszyńska B. Culinary–medicinal mushrooms: a review of organic compounds and bioelements with antioxidant activity. Eur Food Res Technol. 2021;247:513–33.

Popovic V, Zivkovic J, Davidovic S, Stevanovic M, Stojkovic D. Mycotherapy of cancer: an update on cytotoxic and antitumor activities of mushrooms, bioactive principles and molecular mechanisms of their action. Curr Top Med Chem. 2013;13(21):2791–806.

Radović J, Leković A, Tačić A, Dodevska M, Stanojković T, Marinković T, Kundaković-Vasović T. Black trumpet, Craterellus cornucopioides (L.) Pers.: culinary mushroom with angiotensin converting enzyme inhibitory and cytotoxic activity. Polish J Food Nutr Sci. 2022;72(2):17.

Rahmad N, Al-Obaidi JR, Rashid NMN, Zean NB, Yusoff MHYM, Shaharuddin NS, Jamil NAM, Saleh NM. Comparative proteomic analysis of different developmental stages of the edible mushroom *Termitomyces heimii*. Biol Res. 2014;47:1–8.

Rangel-Castro JI, Danell E, Pfeffer PE. A C-NMR study of exudation and storage of carbohydrates and amino acids in the ectomycorrhizal edible mushroom *Cantharelluscibarius*. Mycologia. 2002;94:190–9.

Rathod MG, Gadade RB, Thakur GM, Pathak AP. Oyster mushroom: cultivation, bioactive significance and commercial status. Front Life Sci. 2021;2. https://www.bhumipublishing.com/book/.

Ribeiro B, de Pinho PG, Andrade PB, Baptista P, Valentão P. Fatty acid composition of wild edible mushrooms species: a comparative study. Microchem J. 2009;93(1):29–35.

Rodríguez-Gutiérrez I, Ramírez-Martínez D, Garibay-Orijel R, Jacob-Cervantes V, Pérez-Moreno J, Ortega-Larrocea MDP, Arellano-Torres E. Sympatric species develop more efficient ectomycorrhizae in the Pinus-Laccaria symbiosis. Revista mexicana de biodiversidad. 2019;90 https://doi.org/10.22201/ib.20078706e.2019.90.2868.

Roszczenko P, Szewczyk-Roszczenko OK, Gornowicz A, Iwańska IA, Bielawski K, Wujec M, Bielawska A. The anticancer potential of edible mushrooms: a review of selected species from Roztocze, Poland. Nutrients. 2024;16(17):2849.

Rout Y, Behera F, Kumar S, Sahoo MP, Devi RS. Mushroom diversity of Dhenkanal District, Odisha, India: source of alternative foods and medicines. Eur J Med Plants. 2020;31:33–41.
Sangeeta S, Sharma D, Ramniwas S, Mugabi R, Uddin J, Nayik GA. Revolutionizing mushroom processing: innovative techniques and technologies. Food Chem: X. 2024;23:101774.
Sanodiya BS, Thakur GS, Baghel RK, Prasad GBKS, Bisen PS. Ganodermalucidum: a potent pharmacological macrofungus. Curr Pharm Biotechnol. 2009;10(8):717–42.
Schunko C, Li X, Klappoth B, Lesi F, Porcher V, Porcuna-Ferrer A, Reyes-García V. Local communities' perceptions of wild edible plant and mushroom change: a systematic review. Glob Food Secur. 2022;32:100601.
Sethi J, Mishra S, Kumar S. Wild edible mushrooms of Rourkela Forest division, Odisha. India Asian J Environ Ecol. 2024;23:83–7.
Shen XJ, Wang Q, Liu KY, Cai J, Wang H, Zhang Q, et al. Main functional ingredients, nutritional, and medicinal values of common wild edible fungi: a review. Int Food Res J. 2022;29(1):1–9.
Shirur M, Barh A, Annepu SK. Sustainable production of edible and medicinal mushrooms: implications on mushroom consumption. In: HebsaleMallappa VK, Shirur M, editors. Climate change and resilient food systems. Singapore: Springer Singapore; 2021. p. 315–46.
Shungchang J, Mark M, Qi T. The 86th anniversary of the invention of grain spawn: witnessing the modernization of commercial cultivation of the mushroom Agaricusbisporus. ActaEdulis Fungi. 2019;26(1):77–98. https://www.cabdirect.org/cabdirect/abstract/20193278412.
Silva-Filho AG, Bottke CC, Baseia IG, Cortez VG, Wartchow F. Morphological description and new records of Hygrocybe conica var. conica and H. nigrescens var. brevispora (Hygrophoraceae) in Brazil. Hoehnea. 2019;46:e012019.
Siwulski M, Rzymski P, Budka A, Kalač P, Budzyńska S, Dawidowicz L, Hajduk E, Kozak L, Budzulak J, Sobieralski K. The effect of different substrates on the growth of six cultivated mushroom species and composition of macro and trace elements in their fruiting bodies. Eur Food Res Technol. 2019;245:419–31.
Smania A, Delle MF, Smania EFA, Cuneo RS. Antibacterial activity of steroidal compounds isolated from Ganoderma applanatum (Pers.) Pat. (Aphyllophoromycetideae) fruit body. Int J Med Mushrooms. 1999;1:325–30.
Spatafora JW, Aime MC, Grigoriev IV, Martin F, Stajich JE, Blackwell M. The fungal tree of life: from molecular systematics to genome-scale phylogenies. In: The fungal kingdom. Wiley; 2017. p. 1–34.
Su Z, Xu B. Chemical compositions and health promoting effects of wild edible mushroom milk-cap (Lactarius deliciosus): a review. Food Biosci. 2024;62:105118.
Sundar SM. Edible & Medicinal Mushrooms of India, Vol I; 2024. https://doi.org/10.5281/zenodo.10947246.
Suwannarach N, Kumla J, Zhao Y, Kakumyan P. Impact of cultivation substrate and microbial community on improving mushroom productivity: a review. Biology. 2022;11:569.
Szućko-Kociuba I, Trzeciak-Ryczek A, Kupnicka P, Chlubek D. Neurotrophic and neuroprotective effects of Hericium erinaceus. Int J Mol Sci. 2023;24(21):15960.
Tan X, Chen W, Jiao C, Liang H, Yun H, He C, Xie Y. Anti-tumor and immunomodulatory activity of the aqueous extract of Sarcodon imbricatus in vitro and in vivo. Food Funct. 2020;11(1):1110–21.
Tellez-Tellez M. Wild edible mushrooms as an alternative for the consumption of antioxidants and phenolic compounds: an overview. Bioresources. 2024;19(2):3945.
Thatoi H, Singdevsachan SK. Diversity, nutritional composition and medicinal potential of Indian mushrooms: a review. Afr J Biotechnol. 2014;13(4):523–45.
Tian C, Kasiborski B, Koul R, Lammers PJ, Bücking H, Shachar-Hill Y. Regulation of the nitrogen transfer pathway in the arbuscular mycorrhizal symbiosis: gene characterization and the coordination of expression with nitrogen flux. Plant Physiol. 2010;153:1175–87.
Tibuhwa DD. Folk taxonomy and use of mushrooms in communities around Ngorongoro and Serengeti National Park. Tanzania J Ethnobiol Ethnomed. 2012;8:36.

Tomao A, Bonet JA, Castano C, De-Miguel S. How does forest management affect fungal diversity and community composition? Current knowledge and future perspectives for the conservation of forest fungi. For Ecol Manag. 2020;457:117678.

Treseder K, Torn M, Masiello C. An ecosystem-scale radiocarbon tracer to test use of litter carbon by ectomycorrhizal fungi. Soil Biol Biochem. 2006;38(5):1077–82. https://doi.org/10.1016/j.soilbio.2005.09.006.

Vaario L-M, Fritze H, Spetz P, Heinonsalo J, Hanajík P, Pennanen T. *Tricholomamatsutake* dominates diverse microbial communities in different forest soils. Appl Environ Microbiol. 2011;77:8523–31. https://doi.org/10.1128/AEM.05839-11.

Valverde ME, Hernández-Pérez T, Paredes-López O. Edible mushrooms: improving human health and promoting quality life. Int J Microbiol. 2015;2015(1):376387.

Virágh M, Merényi Z, Csernetics Á, Földi C, Sahu N, Liu X-B, Hibbett DS, Nagy LG. Evolutionary morphogenesis of sexual fruiting bodies in basidiomycota: toward a new evo-devo synthesis. Microbiol Mol Biol Rev MMBR. 2021;86:e00019–21.

Voynova M, Shkondrov A, Kondeva-Burdina M, Krasteva I. Toxicological and pharmacological profile of amanita muscaria (L.) Lam.–a new rising opportunity for biomedicine. Pharmacia. 2020;67:317–23.

Wang Y. Some ectomycorrhizal mushrooms. In: Hall IR, et al., editors. Edible Mycorrhizal mushrooms and their cultivation, proceedings of the second international conference on edible Mycorrhizal mushrooms. CD-ROM, New Zealand Institute for Crop & Food Research Limited; 2002.

Wang S-E, Bau T. Six new species of Agaricus (Agaricaceae, Agaricales) from Northeast China. J Fungi. 2024;10:59.

Wang Y, Chen YL. Recent advances in cultivation of edible mycorrhizal mushrooms. In: Mycorrhizal fungi: use in sustainable agriculture and land restoration. Springer; 2014. p. 375–97.

Wong JY, Chye FY. Antioxidant properties of selected tropical wild edible mushrooms. J Food Compos Anal. 2009;22(4):269–77.

Wurtz TL, Wiita AL, Weber NS, Pilz D. Harvesting morels after wildfire in Alaska. U.S. Department of Agriculture, Forest Service, Pacific Northwest Research Station, Research Note PNW-RN-546, February; 2005.

Xie H, Zhao JW, Zhou HW, Ren SH, Zhang RX. Secondary utilizations and perspectives of mined underground space. Tunn Undergr Space Technol. 2020;96(8):1–14. https://doi.org/10.1016/j.tust.2019.103129.

Yamada A. Cultivation studies of edible ectomycorrhizal mushrooms: successful establishment of ectomycorrhizal associations in vitro and efficient production of fruiting bodies. Mycoscience. 2022;63:235–46.

Yun W, Hall IR. Edible ectomycorrhizal mushrooms: challenges and achievements. Can J Bot. 2004;82:1063–73.

Zambonelli A, Iotti M, Boutahir S, Lancellotti E, Perini C, Pacioni G. Ectomycorrhizal fungal communities of edible ectomycorrhizal mushrooms. In: Edible ectomycorrhizal mushrooms: current knowledge and future prospects. Springer; 2012. p. 105–24.

Zeb M, Lee CH. Medicinal properties and bioactive compounds from wild mushrooms native to North America. Molecules. 2021;26(2):251.

Zhang Y, Li S, Li H, Wang R, Zhang K-Q, Xu J. Fungi–nematode interactions: diversity, ecology, and biocontrol prospects in agriculture. J Fungi. 2020;6:206.

Zhang M, Wang C-Q, Gan M-S, Li Y, Shao S-C, Qin W-Q, Deng W-Q, Li T-H. Diversity of Cantharellus (Cantharellales, Basidiomycota) in China with description of some new species and new records. J Fungi. 2022;8:483.

Chapter 4
Mushrooms: The Miraculous Medicine

Saifullah, Imran Ul Haq, Siddra Ijaz, Nabeeha Aslam Khan, Muhammad Aslam, Jamila Fatima, Ahmad Nisar, and Abdullah

Abstract The worldwide human health system is threatened by the spread of numerous diseases and microbial contamination, further increased by climate change and its environmental consequences. Scientists are actively researching innovative medication sources, such as plants, animals, microorganisms, and fungi. Fungi, particularly medicinal mushrooms, are gaining popularity because of their high potential to produce different bioactive metabolites for drug discovery. Different medicinal mushrooms produce compounds that are both medicinal and beneficial to human nutrition. These metabolites can potentially treat various health issues, including antibacterial, immunomodulatory, anticancer, and anti-inflammatory properties. This chapter thoroughly examines medicinal mushrooms, their beneficial bioactive metabolites, and their applications in the treatment of human health disorders. It also discusses the mechanics and applications of these metabolites, highlighting their significance in establishing novel therapeutic techniques to enhance global health outcomes.

Saifullah
Department of Botany, University of Agriculture, Faisalabad, Pakistan

I. U. Haq · N. A. Khan · A. Nisar · Abdullah
Department of Plant Pathology, University of Agriculture, Faisalabad, Pakistan

S. Ijaz (✉)
Centre of Agricultural Biochemistry and Biotechnology (CABB), University of Agriculture, Faisalabad, Pakistan

M. Aslam
Department of Plant Breeding and Genetics, University of Agriculture, Faisalabad, Pakistan

J. Fatima
Department of Environmental Sciences and Safety Engineering, Jiangsu University, Zhenjiang, China

I. U. Haq et al. (eds.), *Mushrooms*, https://doi.org/10.1007/978-981-95-2539-3_4

Keywords Threatened · Contamination · Consequences · Immunomodulatory · Metabolites · Therapeutic techniques

4.1 Introduction

Recent studies indicate that fungi encompass 2.2–3.8 million species globally, including all types of fungi, such as mushrooms (Hawksworth and Lücking 2017). Mushrooms, the reproductive structures of specific fungi, contribute significantly to fungal diversity. It is estimated that there are approximately 140,000–160,000 species of mushrooms worldwide, but detailed studies have been conducted on only about 10% of them (Chang and Wasser 2017). Mushrooms are terrestrial fungi that produce fruiting bodies above the earth. Because they lack photosynthetic pigments, they have a distinct lifecycle compared with other nonmotile organisms, such as plants. Many mushroom species are classified as Basidiomycota, with Ascomycota accounting for a smaller fraction. Mushrooms can grow below ground (hypogeous) or above ground (epigeous), and they are often large enough to be seen with the human eye, enabling hand harvesting (Chang and Wasser 2017). Mushrooms are widely consumed as food and medicine because of their good nutritional content and medicinal activities. They contain physiologically active chemicals and nutrients useful in industries like food, pharmacology, cosmetics, and agriculture (Venturella et al. 2021; El-Ramady et al. 2022). Mushrooms get nutrients via symbiotic partnerships (as mycorrhiza), saprophytes, or parasitism, and they have two life stages: vegetative mycelial and reproductive fruiting phase. Over 2000 mushroom species have been identified as edible worldwide, representing approximately 31 genera (Li et al. 2021). Such edible mushrooms, as well as medicinal species, thrive in rich, moist wood, organic soils, animal manure, and other substrates, especially after heavy rains or sudden temperature changes, whereas their fruiting bodies frequently vanish, leaving only the mycelium remains (El-Ramady et al. 2022).

4.2 Medicinal Importance

The estimated number of fungus species worldwide is between 2.2 and 3.8 million, with only approximately 150,000 completely documented. Around 15,000 different forms of macrofungi have been named, with approximately 2000 species designated as edible mushrooms. Notably, over 200 kinds of wild mushrooms have been identified for their medicinal characteristics, offering various benefits to health and medicinal aspects (Hyde 2022). These medicinal mushrooms and other fungi have around 130 reported therapeutic benefits, which include anticancer, immunomodulatory, antioxidant, cardiovascular, antiviral, antibacterial, antiparasitic, and antifungal characteristics (Chugh et al. 2022). Medicinal mushrooms are edible or inedible macrofungi that have health or medicinal properties. Because of their evolutionary partition from plants and animals, fungi possess distinct chemical compounds that other kingdoms may not, such as penicillin generated from Penicillium species. These mushrooms' unique bioactive compounds, including β-glucans and triterpenoids, have sparked global scientific attention, especially in East Asian countries, as well as in China.

Ganoderma and other mushrooms have traditionally been used for over 2000 years in Chinese medicine to treat different health problems (Venturella et al. 2021; Łysakowska et al. 2023). The Basidiomycetes class of medicinal mushrooms is distinguished by the presence of metabolically active compounds in the fruiting bodies, broth, and mycelium culture (Sandargo et al. 2019). Mushroom secondary metabolites and polysaccharides are particularly interesting since they have anticancer and immune-boosting characteristics. The data on the biological activity of these compounds are documented for around 700 species belonging to both hetero- and homobasidiomycetes. Polysaccharide-protein complexes and simple Polysaccharides produced from these fungi have been documented in preclinical and clinical trials to induce innate and cell-mediated immune responses, with potential anticancer characteristics in both animals and humans. While the particular mechanisms of their anticancer actions have yet to be fully understood, immune regulation appears to be a substantial contributor. Furthermore, these fungi create low-molecular-weight secondary metabolites, which increase their therapeutic potential. Several drugs are now undergoing clinical studies (Phases I–III) and are widely utilized in Asia for cancer therapy and disease management (Wasser 2014).

4.3 Morphology of Medicinal Mushrooms

Mushrooms are frequently recognized based on their visual and chemical features. These include form, size, colors, stalk and cap texture, spore coloration, and gill attachment pattern to the stalk (Bhambri et al. 2022). Molecular approaches, notably the utilization of internal transcribed spacer (ITS) sections in the sequencing of genes, are also commonly used for mushroom identification. Because of its conserved sequence and diversity, the ITS region is a very reliable molecular marker for differentiating fungal species, allowing discrimination at both the genus and species level (Kauserud 2023). Other modern approaches supplement conventional and molecular techniques of mushroom identification. These consist of microscopy, chemical analysis (including thin-layer chromatography, nuclear magnetic resonance spectroscopy, and chemical fingerprinting), and chromatography techniques, which include high-performance liquid chromatography, gas chromatography-mass spectrometry, and ultra-performance liquid chromatography (Toh Choon et al. 2012; Hennicke et al. 2016; Awadasseid et al. 2017). More advanced techniques, such as liquid chromatography, quadrupole time-of-flight mass spectrometry, and Fourier transform infrared spectroscopy also used. One of the most notable recent developments is matrix-assisted laser desorption ionization-time of flight mass spectrometry. Because of its quick processing time, cost-effectiveness, and high-throughput capabilities, this method is currently often employed in clinical laboratories for fungal identification, making it a promising tool for fungal diagnostics (Chen et al. 2021a, b).

4.4 Historical Overview of Medicinal Mushrooms

Mushrooms have been commonly used in traditional medicine throughout cultures for many years due to their reported medicinal characteristics (Stamets and Zwickey 2014). One of the earliest documented applications of therapeutic mushrooms goes back to around 3300 BCE. Otzi the Iceman, a well-preserved natural mummy found frozen in ice, was carrying *Fomitopsis betulina* (birch polypore), a mushroom known for its antiparasitic and antibacterial qualities. Similarly, evidence of mushroom usage has been found in Egyptian hieroglyphics, indicating its cultural and therapeutic value in ancient communities (Stamets and Zwickey 2014). The Reishi mushroom (*Ganoderma lucidum*) is well known in traditional Chinese medicine, with documented therapeutic usage dating back over 2000 years. Reishi, known as the "mushroom of immortality," was highly regarded for its ability to improve lifespan and general health. Because of its rarity and prestige, it was traditionally reserved for emperors and nobles (El Sheikha 2022). Medicinal mushrooms are also mentioned in the works of ancient Greek and Roman intellectuals like Seneca, Pliny the Elder, and Dioscorides, who disputed their possible health advantages. By 100 BCE, Chinese medical books described fungi's medicinal benefits, including their use to cure respiratory diseases and perhaps cancer. Mushrooms such as chaga (*Inonotus obliquus*) were widely employed by indigenous cultures throughout North America and Siberia for their alleged medicinal advantages, with records dating back to at least the sixteenth century. In Mesoamerican societies, fungus served a dual purpose, both medical and spiritual (Shikov et al. 2014).

4.5 Health Benefits of Mushroom Medicines

4.5.1 Cognitive Enhancement

Among the various medicinal mushrooms, *Hericium erinaceus* has received substantial attention for its possible advantages to cognitive and neurological health. Several preclinical researches has have investigated its effects, resulting in extensive information on its function in animal models and cell lines (Sabaratnam et al. 2013). Clinical studies have also been done, including a modest double-blind, placebo-controlled experiment on patients aged 50–80 with moderate cognitive impairment. In this trial, 30 people were randomly assigned to treat either *H. erinaceus* medicine (three times daily–250 mg) or a placebo for 4 months, followed by a 4-week observation phase (Mori et al. 2009). Participants in the *H. erinaceus* group exhibited significantly better cognitive function, as observed by the Rrevised Ddementia Sscale, at 2, 3, and 4 months compared to the placebo group. However, cognitive enhancements faded the hepatoprotective activity of the mushroom.

The liver contributes significantly to the metabolism and detoxification of harmful chemicals by minimizing exposure to contaminants of chemicals, such as medicine, during diseases (Messelmani et al. 2022; Barouki et al. 2023). It is thus vital to find new medicines for the management of liver illnesses. The compounds of medicinal mushrooms, such as terpenes, steroids, and phenolics are protective functions against liver disease. When tested against exogenous compounds, mycelial extracts from medicinal mushrooms caused liver injury (Soares et al. 2013; Sun et al. 2024). *G. lucidum* prevents Bacillus Calmette-Guerin vaccine-induced liver damage in rats (Chaturvedi et al. 2018). In other trials, four individuals with liver damage and hepatitis B kept going for more than 3 years were clinically evaluated for 12 months. Throughout this period, they received no drugs other than a daily dosage of 1500 mg of *A. blazei* Murill (ABM) extract. The results showed that ABM extract was effective in increasing liver function. Additionally, the levels of aspartate aminotransferase and alanine aminotransferase were reduced when compared to other hepatitis B patients, confirming the potential medicinal effects of ABM extract in liver management (Hsu et al. 2008). Radiation exposure causes the liver to produce reactive oxygen species, which include the peroxyl radical and hydroxyl radical, both of which can cause the peroxidation of lipids. The extracts from edible and medicinal mushrooms *Ganoderma lucidum, Pleurotus florida, Phellinus rimosus*, and *Pleurotus sajor-caju*, showed considerable activity as antioxidants in the liver, particularly evaluated by radical scavenging (Chaturvedi et al. 2018; Abo Nahas et al. 2023).

4.5.2 Antimicrobial Properties

The extracts from medicinal mushroom *Osmoporus odoratus* contain different compounds like petroleum ether, chloroform, acetone, and water that have antibacterial properties toward *Escherichia coli, Staphylococcus aureus, Pseudomonas aeruginosa, Streptococcus pyogenes*, and *Bacillus subtilis* (Sai et al. 2023). The *Osmoporus odoratus* aqueous extract exhibits antibacterial action toward organisms, and it is compared to ampicillin instead of chloramphenicol (Temesgen 2018). Many solvent extracts of *G. lucidum* mushroom, including ethanol, acetone, methanol, and an aqueous extract, have antibacterial action against the six species of bacteria: *P. aeruginosa, S. aureus, B. subtilis, E. coli, Klebsiella pneumonia, and Salmonella typhi* (Agarwal et al. 2023). The methanol extracts of six species of wild medicinal mushrooms (*Cantharellus* cibarius, *Lycoperdon perlatum, Ramaria formosa, Cantharellus vermiculris, Pleurotus pulmonarius,* and *Marasmius oreades*) reported in the west of India showed considerable antibacterial effects against *B. subtilis, P. aeruginosa, E. coli, Candida albicans, and S. aureus* (Al-Faqeeh et al. 2022). The water extract and methanol from the fruiting body of Phellinus are utilized against the five pathogenic bacteria, which include *S. aureus, S. typhi, P. aeruginosa, Streptococcus mutans, and E. coli*, as well as five kinds of fungi, including *Aspergillus flavus*, Penicillium spp., *Aspergillus fumigatous, Mucor indicus*, and *Aspergillus niger* (Vazquez-Armenta et al. 2022; Sułkowska-Ziaja et al. 2023).

4.5.3 Antiasthmatic Properties of Mushrooms

Asthma is a chronic respiratory disorder that affects both adults and children. It is distinguished by airway inflammation and hyperresponsiveness. Allergens are a primary cause of asthma symptoms (Wypych-Ślusarska et al. 2022). Medicinal mushrooms, such as *Flammulina velutipes*, have received attention for their possible medicinal benefits. Extracts of *F. velutipes* include a fungal immunomodulatory protein (FIP-fve), which has shown promise in modifying immunological responses. Using a mouse model of allergen-induced asthma, researchers discovered that oral treatment of isolated FIP-fve significantly prevented chronic airway inflammation produced by ovalbumin. This impact is achieved by a shift in the proportion of type 1 T helper cells (Th1) and type 2 T helper cells (Th2), which favor a Th1-biased immune response (Wu et al. 2020; Hu et al. 2022; Song et al. 2024). Intranasal OVA administration in mice resulted in increased airway inflammation and hyperresponsiveness, as seen by BALF analysis and ELISA tests. Oral FIP-fve demonstrated anti-inflammatory characteristics by reducing airway inflammation, indicating that it might be used to treat allergic asthma (Tsai et al. 2023). Another research effort looked at the antiasthmatic characteristics of a Cordyceps sphecocephala mycelial extract. The extract was given to OVA-induced asthmatic mice to see how it affected lung immune responses (Tsai et al. 2023). The results revealed that pro-inflammatory cytokines such as IL-25, IL-13, and IL-4 were dramatically decreased. This reduction in cytokine expression helped to regulate overactive immune responses and alleviate asthma symptoms (Lee 2010; Wu et al. 2020; Hu et al. 2022; Song et al. 2024).

4.5.4 Mushroom Applications for Bone Health

Medicinal mushrooms consist of different metabolites that can alter the metabolism of bones in humans, potentially lowering the incidence of osteoporosis (Lindequist and Haertel 2021; Riaz et al. 2022). The water extract of the medicinal mushroom *Grifola frondosa* increases ALP and mineralization activity in human osteoblastic cell cultures (SaOS-2 and HOS58). The entire extract of *G. frondosa* promotes the growth of bones and mineralization while also increasing the osteogenicity of growing bone cells (Lindequist and Haertel 2021; Lindequist 2024). The process of mineralization and alkaline phosphate activity is considered a sign of bone maturation and is dependent on the maturation of bone cells and vitality. Mushrooms play an important function in osteoclastogenesis, or delayed bone cell breakdown. *Ganoderma lucidum* ethanol extracts are protective against ovariectomized rats. The tumor necrosis factor (TNF) superfamily includes NF-κB ligand (RANKL), also referred to as type II membrane protein (Erjavec et al. 2016). The receptor activator of nuclear factor kappa-B ligand (RANKL) can enhance bone growth as well as density while also participating in the production of osteoclasts. *Ganoderma lucidum* ethanol extracts activate NF-κB ligand (RANKL) and block TNF-α signaling (Rashed et al. 2021; Boyce et al. 2023).

4.5.5 Anti-Obesity Effects

Obesity is a common worldwide health issue that is connected to the development of a variety of chronic illnesses, including diabetes, hyperlipidemia, and atherosclerosis (Henning 2021). According to research, *Pleurotus florida* extracts may have anti-obesity qualities. In animal trials, feeding these extracts to rats on a high-cholesterol diet has been found to limit weight growth, lower low-density lipoprotein, lower total cholesterol, and triglyceride levels, and reduce fat deposition in high-fat diet-induced obese mice (Chaturvedi et al. 2018). *Pleurotus sajor-caju* contains β-glucan, a bioactive molecule that has been shown to effectively reduce obesity in mice fed a high-fat diet (Ahmad Tarmizi et al. 2022). A comparative analysis of eight medicinal mushroom species indicated hypolipidemic and anti-obesity benefits in mice with fat-induced metabolic abnormalities (Chaturvedi et al. 2018). In vitro tests provide additional support for these findings. *Pleurotus eryngii* water extract has been demonstrated to lower pancreatic lipid levels, but *Grifola frondosa* inhibits pancreatic lipase activity, preventing the hydrolysis of substrates such as 4-methylumbelliferyl oleate and trioleoylglycerol emulsified with lecithin (Palanisamy et al. 2012; Chaturvedi et al. 2018).

4.5.6 Antiaging Effects

Advanced research has shown that the fungus *Tricholoma lobayense* generates a variety of bioactive metabolites with possible antiaging properties. In tests with a D-galactose-induced aging mice model, the polysaccharide TLH-3, derived from the young fruiting body of *Tricholoma lobayense*, showed considerable antiaging efficacy (Ding et al. 2016). The *G. lucidum*-derived methanol extract contains two unique antiaging ergosterols, Ganoderma sidase (A & B); these compounds are believed to increase the replicative cycle of yeast via modulating the expression of oxidative stress-responsive genes, such as UTH1 (Weng et al. 2010; Lian et al. 2024). Table 4.1 shows some patents that have confirmed antiaging/menopause along with additional treatment using *Ganoderma lucidum* pores, among other medicinal applications (Oke et al. 2022), whereas Table 4.2 shows metabolites of

Table 4.1 Reported patents of different medicinal mushrooms with their biological functions

Serial No.	Name of extract/product	Source of product/ mushroom name	Patent ids	Biological Functions	References
1	Spores	*Ganoderma lucidum*	HK1056689A1	Delay aging/ antiaging properties	Chung and Tong (2004)
2	Spores	*G. lucidum*	US-(7087233)	Antimutagenic activities	Chung and Tong (2006)
3	Glycoproteins	*G. frondosa*	US-(7214778)	Antidiabetic activities	Zhuang et al. (2007)
4	Myco-nutrient Compositions	*Grifola frondosa, Pleurotus ostreatus, Lentinula edodes*, and *Agaricus bisporus,*	WO-2007-US-63984 20070314	Prevention of radiation damage and neuro-degenerative treatment	Beelman et al. (2007)
5	Prepared food supplement	*Hericium erinaceus, Grifola frondosa, Pleurotus eryngii,*	CN-101292726-A-20081029	Regulation of lipid and sugar level in blood	Liu et al. (2008)
6	Mushroom extract	*Hericium erinaceus*	US-2009274720-(A1)	Promote the synthesis of nerve growth factors and reduce neuronal toxicity	Zhuang et al. (2009)
7	Polysaccharides	*Ganoderma lucidum*	US-9758595	Function of antiobesity	Ko et al. (2017)
8	Terpenoids	Agaricus spp and *Agaricus subrufescens*	EP-2468 253-(A1)	Antibacterial, antiviral, and agonists function for liver –x-receptor	Grothe et al. (2012)
9	Mushroom extracts, stimulators (insulin), and food products	Agaricus spp, *Hericium erinaceum*, and *Hypsizigus marmoreus*	JP-2012077004-A	Antidiabetic functions	Takeshi et al. (2012)
10	Nutriceuticals agent/extracts	higher fungi-mushroom (basidio-mycetes)	US-6383799 US-7517682	Prevention of hyper-glycaemia Immune regulations	Wasser and Reshetnikov (2002), Watanabe (2009)

Table 4.2 Reported Metabolites of different medicinal mushrooms with their biological applications

Serial No.	Name of mushroom	Different metabolites	Biological mechanism/ pathways	Biological application/functions	References
1	*Lentinus Edodes*	Lipids, polysaccharides, terpenoids, sterols, and phenolic compounds	Inhibition of TNF-α and IL-1β	Diabetes, fungal infection, hypertension, bronchial inflammation, flu and cold, hyperlipidemia, heart disease, hepatitis, cancer, depressed immune function, infectious disease, urinary inconsistencies, environmental allergies,	Xu et al. (2024)
2	*Grifola frondosa*	Sesquiterpene, polysaccharide, glycoproteins, glucans, etc.	Immunity stimulator	Antiviral, antioxidation, hyperglycemia, immunomodulation, antitumor, liver diseases, hypertension, antidiabetic	Su et al. (2020)
3	Ganoderma lucidum	Polysaccharides, triterpenes, terpenoids, proteins, phenols, glycoproteins, amino acids, ganodermic acids, steroids, peptidoglycans	Immune regulator, activation of nitric oxide synthase	Antioxidant, antibacterial, insecticidal, anticancer, antiproliferative, antifungal, blood regulation, asthma, antiaging, immune regulation, diabetic, blood pressure, etc.	Sanodiya et al. (2009)
4	*Trametes versicolor*	Lipids, carbohydrates, proteins, bioflavonoids, flavanols, flavonoids, flavones, flavanone, coriolin, ergosterol, stigmast, methylquinoline, β-sitosterol, oxycoriolic acid and sesquiterpene	Immunity stimulant, natural killer cell toxicity, and activation of macrophage	Antioxidant, anticoagulant, antibacterial, insecticidal, antiproliferative, anticancer, antifungal, antiparasitic, antidiabetic, hepatoprotective, antiviral, respiratory, urinary tract infections, anti-inflammatory, chronic hepatitis	Habtemariam (2020)
5	Cordyceps Mushrooms	Carbohydrates, proteins, alkaloids, flavonoids, gums, phenols, saponins, mucilage, cordycepic acid, amino acid, cordycepin compounds	Immunity stimulant, regulation of neuromuscular functions, enhancing endurance activity	Heart issues, liver diseases, protects kidneys, oxygen uses, generation of ATP, regulation of sugar level, improve reproductive cycle	Jędrejko et al. (2021)

(continued)

Table 4.2 (continued)

Serial No.	Name of mushroom	Different metabolites	Biological mechanism/ pathways	Biological application/functions	References
6	Tremella fuciformis	Proteins, polysaccharides, flavonoids, phenolics, trace elements, fatty acids, dietary fibers	Inhibition of enzymes	Cytotoxic, anti-inflammatory, antimicrobial, wound healing, antioxidant, Fe chelation, anti-hypoxic	Mingyi et al. (2019)
7	Cantharellus cibarius	Tannins, phenolic compounds, flavonoids, steroids, terpenes, lectins, proteins, polysaccharides	Cytotoxicity for regulation angiotensin responsive enzyme	Antioxidant, wound healing, Fe chelation, antimicrobial, anti-inflammatory effects, cytotoxicity, antihyperglycemic	Kozarski et al. (2015a, b)
8	Coprinus comatus	Phenolic compounds, carbohydrates, proteins, dietary fibers	Regulation of homeostasis by antioxidative mechanism	Antibacterial, immunity stimulant, antioxidant, antitumor, hypolipidemic, hypoglycemic, regulation of glucose level in blood	Stilinović et al. (2020)
9	Pleurotus ostreatus	Tannin, terpenoids, glycosidase, steroidal, polysaccharides	Inhibition of antherogenesis and hypocholesterolemic	Antibacterial, anticancer, anti-cholestrolic, antiviral, antioxidant, antidiabetic, antiarthritic and eye health	Deepalakshmi and Mirunalini (2014)
10	Agaricus	Sterols, unsaturated fatty acids, phenolics, sugar, organic acids, nutraceuticals, amino acids, lectins, indole compounds, polyols	Antioxidant and anti-inflammatory activities inhibit growth of cancer cell line	Stomach ulcer, heart disease, anticancerous, digestive issues, cholesterol level, diabetic problems, arteriosclerosis, antiviral, bloodstream	Usman et al. (2021)

different medicinal mushrooms with their biological applications (Xu et al. 2024; Su et al. 2020; Sanodiya et al. 2009; Habtemariam 2020; Jędrejko et al. 2021; Mingyi et al. 2019; Kozarski et al. 2015a, b; Stilinović et al. 2020; Deepalakshmi and Mirunalini 2014; Usman et al. 2021). Several derived compounds from the mycelium and fruiting body of different medicinal mushrooms, including terpenoids, polysaccharides, polyphenolics, vitamins, phenolics, volatile organic compounds, and selenium have powerful, antiaging, antioxidant, skin whitening, antiwrinkle, and moisturizing properties (Wu et al. 2016; Dhanapal and Rajoo 2023). Auricularia auricular-derived hydrophilic polysaccharide (AAP I-a) protects aged mice from oxidative damage caused by d-galactose. In the industry, the paste polysaccharide is utilized to form antiaging lotions or creams, as well as in many skin-care products (Xiang et al. 2021). These polysaccharides (AAP I-a) can slow aging by inhibiting the activity of functional enzymes that cause skin aging, including hyaluronidase, tyrosinase, elastase, and MMP-1 enzymes. These polysaccharides additionally contribute to recovering the elasticity of the skin by boosting collagen formation (Chaturvedi et al. 2018)

4.6 Mushrooms in Modern and Traditional Medicine Use

4.6.1 Traditional Use

Of the total 14,000–22,000 estimated mushroom species, only 35–45 are deemed safe for human consumption. Among these, 20–30 species are being grown, with around 15 species collected from the wild. Mushrooms have a wide range of uses, including culinary use, tea brewing, and the creation of powdered extracts or tinctures. Culinary mushrooms, such as shiitake, are popular for their flavor and are widely available in grocery shops. These mushrooms are versatile and frequently used in soups, stir-fries, pasta dishes, and other recipes. Shelf mushrooms, a form of therapeutic fungus, grow on the sides of trees and aid in the decomposition of their hosts. While these mushrooms are too woody to consume directly, they are commonly used to make mushroom tea or processed into powder for medicinal extracts and tinctures. To make mushroom tea, soak 3–10 g of dried mushroom pieces in boiling water for 5–10 min. However, woody shelf mushrooms usually require longer steeping durations or boiling to extract their therapeutic components. Sweeteners such as honey might be added after the dish has been prepared to improve the flavor (Milenge Kamalebo et al. 2018; Hussain et al. 2023).

4.6.2 Mushroom-Derived Pharmaceuticals/Modern Use

Mushrooms have therapeutic characteristics that are still utilized today. Alexander Fleming developed penicillin, a mold-derived antibiotic, in 1928. This resulted in the identification of additional antiviral and antibacterial properties in a diversity of fungi. Even now, researchers are continuously discovering novel medicinal properties in fungi and mushrooms. This consists of antiparasitic, antiviral, antifungal, and antiprotozoan properties. Medicinal Mushrooms are frequently used in different cancer treatments. Lentinan, derived from shiitake mushrooms, is one such example. It is the third most commonly given supplementary cancer treatment globally (Ayoub and Mehta 2024).

4.6.3 Cultivation and Sustainability

Many therapeutic mushrooms are rare in natural environments, therefore, production of fungal fruiting bodies frequently falls short of satisfying the expanding demands of the food and pharmaceutical sectors. Over the previous four decades, there has been a significant increase in the cultivation and production of mushroom fruiting bodies using diverse farming techniques. These methods include conventional growing on wood logs and beds and current substrate-based systems that use bags, bottles, trays, and other media. While farming is inexpensive, it may be time consuming when scaled up to large-scale production (Khan et al. 2024). Common substrates include agricultural, wood, and food industry by-products that have been processed with delignification and enrichment to increase protein content and output of bioactive pharmaceutical chemicals. Mushroom farming is expected to increase in response to increased worldwide agricultural waste output and market demand for both medicinal and gourmet mushrooms. This increase is likely to be driven by innovations in substrate selection, particularly the utilization of waste materials, as well as advances in cultivating techniques, allowing for more efficient and sustainable production (Nikšić et al. 2022).

4.7 Side Effects/Safety and Toxicity of Mushroom Medicines

4.7.1 Positive Side Effects of Mushroom Medicine

According to several research findings, mushroom medicine has significant health benefits and a wide range of pharmacological uses, including antiallergic, antifungal, antibacterial, anti-inflammatory, antiviral, antioxidative, antidiabetic, immunomodulating, antidepressive, antihyperlipidemic, hepatoprotective, neuroprotective, digestive, nephroprotective, cytotoxic, hypotensive, and osteoprotective properties (Venturella et al. 2021). Figure 4.1 shows some important bioactive compounds derived from different medicinal mushrooms, whereas Fig. 4.2 shows different

Fig. 4.1 Some important bioactive compounds are derived from different medicinal mushrooms

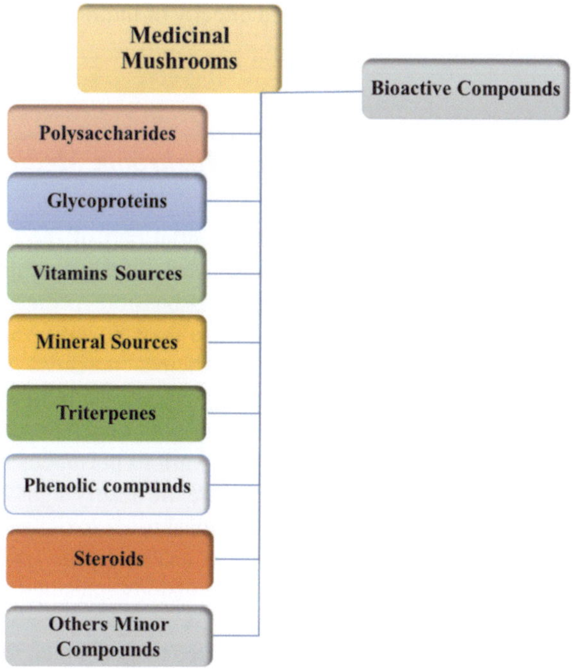

Fig. 4.2 Different biological activities are performed directly or indirectly by the metabolites from different medicinal mushrooms

biological activities performed directly or indirectly by the metabolites from different medicinal mushrooms.

4.7.2 *Negative Effects of Mushroom Medicine*

Mushrooms are considered an essential part of the human diet, known for their medicinal and nutritional benefits. Many mushroom species are known worldwide for their health advantages, including possible medical uses. However, certain species pose significant concerns due to their toxicity. Various mushrooms have been intensively studied for their medicinal potential and harmful consequences (Nieminen and Mustonen 2020). Every year, multiple incidents of serious human poisoning are documented, the majority of which are caused by misidentifying poisonous mushrooms as edible types. Various poisonous mushrooms have been found to have different chemicals that cause the deadly signs and symptoms of mushroom intoxication. Research studies on mushroom toxicity have found that exposure to some poisonous species can cause serious health problems such as bradycardia, liver damage, chest discomfort, convulsions, intestinal fibrosis, gastroenteritis, renal failure, rhabdomyolysis, and erythromelalgia. Accurate identification and knowledge of mushroom species are required to assure their safety as functional foods and their efficacy in medicinal applications (Jo et al. 2014; Nieminen and Mustonen 2020; Yao et al. 2023).

4.8 Future Research Directions

The therapeutic potential of mushroom metabolites is largely unknown, specifically in terms of polysaccharide–protein complexes and their involvement in treating viral and bacterial infections, metabolic disorders, cancer, and cholesterol. Advanced methodologies and cosmeceutical uses require more research, as do efforts to uncover new species with medicinal characteristics. To establish mushroom metabolites' effectiveness and safety, clinical research should focus on long-term, placebo-controlled trials. Untapped prospects include agricultural applications such as pesticides and antiviral medicines, as well as farm animal studies. Intellectual property management of mushroom genetic resources is critical for driving innovation and increasing their commercial worth. Public education on the health advantages of mushroom metabolites, as well as access to research data, are critical for increasing society's understanding and adoption of these potential bioresources. Furthermore, advances in biotechnology for medicinal mushrooms, such as the creation of effective CRISPR-Cas9 gene editing techniques, are critical for increasing bioactive chemical synthesis. CRISPR technology can enhance the targeted editing of mushroom genomes, accelerating the development of strains with improved medicinal qualities. This is especially important for the industrial-scale production and optimization of bioactive metabolites like polysaccharides and alkaloids, which are essential in therapeutic applications. Furthermore, the sustainable production of medicinal mushrooms to meet the demand for mushroom medicine, as well as the utilization of mushroom medicine in modern healthcare, are major issues at present.

4.9 Conclusion

Mushrooms possess a wide range of qualities that have high medical significance across the world. More than 85% of therapeutic mushroom products come from their fruiting bodies. Mushrooms are utilized for both food and medicine due to their properties. They are used to treat many infections and disorders. Even still, many therapeutic mushroom qualities remain unknown. More study is needed on therapeutic mushrooms and their qualities.

References

Abo Nahas HH, Abdel-Rahman MA, Gupta VK, Abdel-Azeem AM. Myco-antioxidants: insights into the natural metabolic treasure and their biological effects. Sydowia. 2023;75:151–79. https://doi.org/10.12905/0380.sydowia75-2023-0151.

Agarwal A, Gupta V, Yadav AN, Sain D, Rahi RK, Bera SP, Neelam D. Aspects of mushrooms and their extracts as natural antimicrobial agents: microbiology. J Microbiol Biotechnol Food Sci. 2023;e9191 https://doi.org/10.55251/jmbfs.9191.

Ahmad Tarmizi NAK, Kushairi N, Phan CW, Sabaratnam V, Naidu M, David P. β -Glucan-Rich extract of gray oyster mushroom, *Pleurotus pulmonarius*, improves object recognition memory and hippocampus morphology in mice fed a high-fat diet. J Med Food. 2022;25(3):230–8. https://doi.org/10.1089/jmf.2021.K.0121.

Al-Faqeeh LAS, Naser R, Kagne SR, Khan SW. Review on anticancer and antimicrobial activities of mushrooms. World J Pharm Pharmaceut Sci. 2022; https://doi.org/10.17605/OSF.IO/67CNG.

Awadasseid A, Hou J, Gamallat Y, Xueqi S, Eugene KD, Musa Hago A, Bamba D, Meyiah A, Gift C, Xin Y. Purification, characterization, and antitumor activity of a novel glucan from the fruiting bodies of Coriolus versicolor. PLoS One. 2017;12(2):e0171270. https://doi.org/10.1371/journal.pone.0171270.

Ayoub Z, Mehta A. Use of fungi in pharmaceuticals and production of antibiotics. In: Entrepreneurship with microorganisms. Elsevier; 2024. p. 69–87. https://doi.org/10.1016/B978-0-443-19049-0.00017-7.

Barouki R, Samson M, Blanc EB, Colombo M, Zucman-Rossi J, Lazaridis KN, Miller GW, Coumoul X. The exposome and liver disease—how environmental factors affect liver health. J Hepatol. 2023;79(2):492–505. https://doi.org/10.1016/j.jhep.2023.02.034.

Beelman RB, Dubost NJ, Peterson DG, Hausman M. Phytonutrient compositions from mushrooms or filamentous fungi comprising Ergothioneine for treatment of neurodegenerative diseases and radiation damage database: CAPLUS. WO 2007106859 A2 20070920. Cham: Springer International Publishing AG; 2007.

Bhambri A, Srivastava M, Mahale VG, Mahale S, Karn SK. Mushrooms as potential sources of active metabolites and medicines. Front Microbiol. 2022;13:837266. https://doi.org/10.3389/fmicb.2022.837266.

Boyce BF, Li J, Yao Z, Xing L. Nuclear factor-Kappa B regulation of osteoclastogenesis and osteoblastogenesis. Endocrinol Metab. 2023;38(5):504–21. https://doi.org/10.3803/EnM.2023.501.

Chang ST, Wasser SP. The cultivation and environmental impact of mushrooms. In: Chang ST, Wasser SP, editors. Oxford research encyclopedia of environmental science. Oxford University Press; 2017. https://doi.org/10.1093/acrefore/9780199389414.013.231.

Chaturvedi VK, Agarwal S, Gupta KK, Ramteke PW, Singh MP. Medicinal mushroom: boon for therapeutic applications. 3 Biotech. 2018;8(8):334. https://doi.org/10.1007/s13205-018-1358-0.

Chen X, Wang Y, Shen M, Yu Q, Chen Y, Huang L, Xie J. The water-soluble non-starch polysaccharides from natural resources against excessive oxidative stress: a potential health-promoting effect and its mechanisms. Int J Biol Macromol. 2021a;171:320–30. https://doi.org/10.1016/j.ijbiomac.2021.01.022.

Chen X-F, Hou X, Xiao M, Zhang L, Cheng J-W, Zhou M-L, Huang J-J, Zhang J-J, Xu Y-C, Hsueh P-R. Matrix-Assisted Laser Desorption/Ionization Time of Flight Mass Spectrometry (MALDI-TOF MS) analysis for the identification of pathogenic microorganisms: a review. Microorganisms. 2021b;9(7):1536. https://doi.org/10.3390/microorganisms9071536.

Chugh RM, Mittal P, Mp N, Arora T, Bhattacharya T, Chopra H, Cavalu S, Gautam RK. Fungal mushrooms: a natural compound with therapeutic applications. Front Pharmacol. 2022;13:925387. https://doi.org/10.3389/fphar.2022.925387.

Chung C-L, Tong SK. Anti-aging/menopause symptoms relief using ganoderma lucidum spores (Patent HK1056689A1). 2004. https://patents.google.com/patent/HK1056689A1/en.

Chung C, Tong SK. Antimutagenic effects of Ganoderma lucidum spores (Patent 7087233). 2006. https://www.freepatentsonline.com/7087233.html.

Deepalakshmi K, Mirunalini S. Pleurotus ostreatus: an oyster mushroom with nutritional and medicinal properties. J Biochem Technol. 2014;5(2):718–26.

Dhanapal D, Rajoo B. Value addition of mushrooms by incorporation in the food products: an overview. Int J Food Eng. 2023;19(12):573–91. https://doi.org/10.1515/ijfe-2023-0073.

Ding Q, Yang D, Zhang W, Lu Y, Zhang M, Wang L, Li X, Zhou L, Wu Q, Pan W, Chen Y. Antioxidant and anti-aging activities of the polysaccharide TLH-3 from Tricholoma lobayense. Int J Biol Macromol. 2016;85:133–40. https://doi.org/10.1016/j.ijbiomac.2015.12.058.

El Sheikha AFE. Nutritional profile and health benefits of Ganoderma lucidum "Lingzhi, Reishi, or Mannentake" as functional foods: current scenario and future perspectives. Foods. 2022;11(7):1030. https://doi.org/10.3390/foods11071030.

El-Ramady H, Abdalla N, Badgar K, Llanaj X, Törős G, Hajdú P, Eid Y, Prokisch J. Edible mushrooms for sustainable and healthy human food: nutritional and medicinal attributes. Sustainability. 2022;14(9):4941. https://doi.org/10.3390/su14094941.

Erjavec I, Brkljacic J, Vukicevic S, Jakopovic B, Jakopovich I. Mushroom extracts decrease bone resorption and improve bone formation. Int J Med Mushrooms. 2016;18(7):559–69. https://doi.org/10.1615/IntJMedMushrooms.v18.i7.10.

Grothe T, Stadler M, Köpcke B, Roemer E, Bitzer J, Wabnitz P. Terpenoid spiro ketal compounds with LXR agonists activity, their use and formulations with them. EP2468253 A1. Dortmund: World International Property Organization (WIPO); 2012.

Habtemariam S. Trametes versicolor (Synn. Coriolus versicolor) polysaccharides in cancer therapy: targets and efficacy. Biomedicine. 2020;8(5):135. https://doi.org/10.3390/biomedicines8050135.

Hawksworth DL, Lücking R. Fungal diversity revisited: 2.2 to 3.8 million species. In: Heitman J, Howlett BJ, Crous PW, Stukenbrock EH, James TY, Gow NAR, editors. The fungal kingdom. ASM Press; 2017. p. 79–95. https://doi.org/10.1128/9781555819583.ch4.

Hennicke F, Cheikh-Ali Z, Liebisch T, Maciá-Vicente JG, Bode HB, Piepenbring M. Distinguishing commercially grown Ganoderma lucidum from Ganoderma lingzhi from Europe and East Asia on the basis of morphology, molecular phylogeny, and triterpenic acid profiles. Phytochemistry. 2016;127:29–37. https://doi.org/10.1016/j.phytochem.2016.03.012.

Henning RJ. Obesity and obesity-induced inflammatory disease contribute to atherosclerosis: a review of the pathophysiology and treatment of obesity. Am J Cardiovasc Dis. 2021;11(4):504–29.

Hsu C-H, Hwang K-C, Chiang Y-H, Chou P. The mushroom *Agaricus blazei* Murill extract normalizes liver function in patients with chronic hepatitis B. J Altern Complement Med. 2008;14(3):299–301. https://doi.org/10.1089/acm.2006.6344.

Hu R-H, Wu C-T, Wu T-S, Yu F-Y, Ko J-L, Lue K-H, Liu Y-F. Systematic characterization of the group 2 house dust mite allergen in Dermatophagoides microceras. Front Cell Infect Microbiol. 2022;11:793559. https://doi.org/10.3389/fcimb.2021.793559.

Hussain S, Sher H, Ullah Z, Elshikh MS, Al Farraj DA, Ali A, Abbasi AM. Traditional uses of wild edible mushrooms among the local communities of Swat, Pakistan. Foods. 2023;12(8):1705. https://doi.org/10.3390/foods12081705.

Hyde KD. The numbers of fungi. Fungal Divers. 2022;114(1):1–1. https://doi.org/10.1007/s13225-022-00507-y.

Jędrejko KJ, Lazur J, Muszyńska B. Cordyceps militaris: an overview of its chemical constituents in relation to biological activity. Foods. 2021;10(11):2634. https://doi.org/10.3390/foods10112634.

Jo W-S, Hossain MA, Park S-C. Toxicological profiles of poisonous, edible, and medicinal mushrooms. Mycobiology. 2014;42(3):215–20. https://doi.org/10.5941/MYCO.2014.42.3.215.

Kauserud H. ITS alchemy: on the use of ITS as a DNA marker in fungal ecology. Fungal Ecol. 2023;65:101274. https://doi.org/10.1016/j.funeco.2023.101274.

Khan A, Murad W, Salahuddin A, Shah SS, Halim SA, Khalid A, Kashtoh H, Khan A, Al-Harrasi A. Contribution of mushroom farming to mitigating food scarcity: current status, challenges and potential future prospects in Pakistan. Heliyon. 2024;10(23):e40362. https://doi.org/10.1016/j.heliyon.2024.e40362.

Ko Y, Martel J, Liau J, Chang I, Jian WT, Lin M, Yang C, Chiu C, Chang C, Lin C, Wu T, Lu C, Ojcius DM, Lai H, Young JD. Method to prepare Ganoderma lucidum polysaccharides possessing anti-obesity properties and uses thereof (Patent 9758595). 2017. https://www.freepatentsonline.com/9758595.html.

Kozarski M, Klaus A, Jakovljevic D, Todorovic N, Vunduk J, Petrović P, Niksic M, Vrvic M, Van Griensven L. Antioxidants of edible mushrooms. Molecules. 2015a;20(10):19489–525. https://doi.org/10.3390/molecules201019489.

Kozarski M, Klaus A, Vunduk J, Zizak Z, Niksic M, Jakovljevic D, Vrvic MM, Van Griensven LJLD. Nutraceutical properties of the methanolic extract of edible mushroom Cantharellus cibarius (fries): primary mechanisms. Food Funct. 2015b;6(6):1875–86. https://doi.org/10.1039/C5FO00312A.

Lee. Anti-asthmatic activities in mycelial extract and culture filtrate of Cordyceps sphecocephala J201. Int J Mol Med. 2010; https://doi.org/10.3892/ijmm_00000472.

Li H, Tian Y, Menolli N, Ye L, Karunarathna SC, Perez-Moreno J, Rahman MM, Rashid MH, Phengsintham P, Rizal L, Kasuya T, Lim YW, Dutta AK, Khalid AN, Huyen LT, Balolong MP, Baruah G, Madawala S, Thongklang N, et al. Reviewing the world's edible mushroom species: a new evidence-based classification system. Compr Rev Food Sci Food Saf. 2021;20(2):1982–2014. https://doi.org/10.1111/1541-4337.12708.

Lian W, Yang X, Duan Q, Li J, Zhao Y, Yu C, He T, Sun T, Zhao Y, Wang W. The biological activity of Ganoderma lucidum on neurodegenerative diseases: the interplay between different active compounds and the pathological hallmarks. Molecules. 2024;29(11):2516. https://doi.org/10.3390/molecules29112516.

Lindequist U. Medicinal mushrooms as multicomponent mixtures—demonstrated with the example of Lentinula edodes. J Fungi. 2024;10(2):153. https://doi.org/10.3390/jof10020153.

Lindequist U, Haertel B. Medicinal mushrooms for prevention and therapy of osteoporosis (review). Int J Med Mushrooms. 2021;23(4):13–22. https://doi.org/10.1615/IntJMedMushrooms.2021038084.

Liu Z, Cang L, Chen C, Zhou G. Method for preparing gold needle mushroom canned food for decreasing blood sugar and adjusting blood fat database: CAPLUS. CN 101292726 A 20081029. Cham: SpringerNature; 2008.

Łysakowska P, Sobota A, Wirkijowska A. Medicinal mushrooms: their bioactive components, nutritional value and application in functional food production—a review. Molecules. 2023;28(14):5393. https://doi.org/10.3390/molecules28145393.

Messelmani T, Morisseau L, Sakai Y, Legallais C, Le Goff A, Leclerc E, Jellali R. Liver organ-on-chip models for toxicity studies and risk assessment. Lab Chip. 2022;22(13):2423–50. https://doi.org/10.1039/D2LC00307D.

Milenge Kamalebo H, Nshimba Seya Wa Malale H, Masumbuko Ndabaga C, Degreef J, De Kesel A. Uses and importance of wild fungi: traditional knowledge from the Tshopo province in The Democratic Republic of the Congo. J Ethnobiol Ethnomed. 2018;14(1):13. https://doi.org/10.1186/s13002-017-0203-6.

Mingyi Y, Belwal T, Devkota HP, Li L, Luo Z. Trends of utilizing mushroom polysaccharides (MPs) as potent nutraceutical components in food and medicine: a comprehensive review. Trends Food Sci Technol. 2019;92:94–110. https://doi.org/10.1016/j.tifs.2019.08.009.

Mori K, Inatomi S, Ouchi K, Azumi Y, Tuchida T. Improving effects of the mushroom Yamabushitake (*Hericium erinaceus*) on mild cognitive impairment: a double-blind placebo-controlled clinical trial. Phytother Res. 2009;23(3):367–72. https://doi.org/10.1002/ptr.2634.

Nieminen P, Mustonen A-M. Toxic potential of traditionally consumed mushroom species—a controversial continuum with many unanswered questions. Toxins. 2020;12(10):639. https://doi.org/10.3390/toxins12100639.

Nikšić M, Podgornik BB, Berovic M. Farming of medicinal mushrooms. In: Berovic M, Zhong J-J, editors. Biochemical engineering and biotechnology of medicinal mushrooms, vol. 184. Springer International Publishing; 2022. p. 29–76. https://doi.org/10.1007/10_2021_201.

Oke MA, Afolabi FJ, Oyeleke OO, Kilani TA, Adeosun AR, Olanbiwoninu AA, Adebayo EA. Ganoderma lucidum: unutilized natural medicine and promising future solution to emerging diseases in Africa. Front Pharmacol. 2022;13:952027. https://doi.org/10.3389/fphar.2022.952027.

Palanisamy M, Gil-Ramírez A, Ruiz-Rodríguez A, Marín FR, Reglero G, Soler-Rivas C. Testing edible mushrooms to inhibit the pancreatic lipase activity by an *in vitro* digestion model. Int J Food Sci Technol. 2012;47(5):1004–10. https://doi.org/10.1111/j.1365-2621.2011.02934.x.

Rashed F, Kamijyo S, Shimizu Y, Hirohashi Y, Khan M, Sugamori Y, Murali R, Aoki K. The effects of receptor activator of NF-κB ligand-binding peptides on bone resorption and bone formation. Front Cell Develop Biol. 2021;9:648084. https://doi.org/10.3389/fcell.2021.648084.

Riaz S, Ahmad A, Farooq R, Ahmed M, Shaheryar M, Hussain M. Edible mushrooms, a sustainable source of nutrition, biochemically active compounds and its effect on human health. In: Shiomi N, Savitskaya A, editors. Current topics in functional food. IntechOpen; 2022. https://doi.org/10.5772/intechopen.102694.

Sabaratnam V, Kah-Hui W, Naidu M, David PR. Neuronal health—can culinary and medicinal mushrooms help? J Tradit Complement Med. 2013;3(1):62–8. https://doi.org/10.4103/2225-4110.106549.

Sai I, Basavaraju R, Thakur M, Lakhanpal TN. Health benefits of mushrooms: an appraisal. In: Singh I, Rajpal VR, Navi SS, editors. Fungal resources for sustainable economy: current status and future perspectives. Springer Nature; 2023. p. 179–226. https://doi.org/10.1007/978-981-19-9103-5_7.

Sandargo B, Chepkirui C, Cheng T, Chaverra-Muñoz L, Thongbai B, Stadler M, Hüttel S. Biological and chemical diversity go hand in hand: Basidiomycota as source of new pharmaceuticals and agrochemicals. Biotechnol Adv. 2019;37(6):107344. https://doi.org/10.1016/j.biotechadv.2019.01.011.

Sanodiya B, Thakur G, Baghel R, Prasad G, Bisen P. Ganoderma lucidum: a potent pharmacological macrofungus. Curr Pharm Biotechnol. 2009;10(8):717–42. https://doi.org/10.2174/138920109789978757.

Shikov AN, Pozharitskaya ON, Makarov VG, Wagner H, Verpoorte R, Heinrich M. Medicinal plants of the Russian pharmacopoeia; their history and applications. J Ethnopharmacol. 2014;154(3):481–536. https://doi.org/10.1016/j.jep.2014.04.007.

Soares A, De Sá-Nakanishi A, Bracht A, Da Costa S, Koehnlein E, De Souza C, Peralta R. Hepatoprotective effects of mushrooms. Molecules. 2013;18(7):7609–30. https://doi.org/10.3390/molecules18077609.

Song W, Li M, Yue X, Zheng Y. Effect of extraction and purification on the structure and activity of *Flammulina velutipes* polysaccharides: a review. Int J Food Sci Technol. 2024;59(12):9571–91. https://doi.org/10.1111/ijfs.17393.

Stamets P, Zwickey H. Medicinal mushrooms: ancient remedies meet modern science. Integrat Med (Encinitas, Calif). 2014;13(1):46–7.

Stilinović N, Čapo I, Vukmirović S, Rašković A, Tomas A, Popović M, Sabo A. Chemical composition, nutritional profile and *in vivo* antioxidant properties of the cultivated mushroom *Coprinus comatus*. R Soc Open Sci. 2020;7(9):200900. https://doi.org/10.1098/rsos.200900.

Su C-H, Lu M-K, Lu T-J, Lai M-N, Ng L-T. A (1→6)-Branched (1→4)-β- d -Glucan from *Grifola frondosa* inhibits lipopolysaccharide-induced cytokine production in RAW264.7 macrophages by binding to TLR2 rather than Dectin-1 or CR3 receptors. J Nat Prod. 2020;83(2):231–42. https://doi.org/10.1021/acs.jnatprod.9b00584.

Sułkowska-Ziaja K, Trepa M, Olechowska-Jarząb A, Nowak P, Ziaja M, Kała K, Muszyńska B. Natural compounds of fungal origin with antimicrobial activity—potential cosmetics applications. Pharmaceuticals. 2023;16(9):1200. https://doi.org/10.3390/ph16091200.

Sun X, Shi Y, Shi D, Tu Y, Liu L. Biological activities of secondary metabolites from the edible-medicinal macrofungi. J Fungi. 2024;10(2):144. https://doi.org/10.3390/jof10020144.

Takeshi I, Hiroshi H, Satoshi I, Aya K. Mushroom extracts from Agaricus, Hericiumerinaceum, and Hypsizigusmarmoreus as insulin secretion stimulators and health foods for prevention and therapy of diabetes mellitus. JP 2012077004A. Cham: SpringerNature; 2012.

Tang C, Hoo PC-X, Tan LT-H, Pusparajah P, Khan TM, Lee L-H, Goh B-H, Chan K-G. Golden needle mushroom: a culinary medicine with evidenced-based biological activities and health promoting properties. Front Pharmacol. 2016;7 https://doi.org/10.3389/fphar.2016.00474.

Temesgen T. Application of mushroom as food and medicine. Adv Biotechnol Microbiol. 2018;11(4) https://doi.org/10.19080/AIBM.2018.11.555817.

Toh Choon RL, Sariah M, Siti Mariam MN. Ergosterol from the soilborne fungus *Ganoderma boninense*. J Basic Microbiol. 2012;52(5):608–12. https://doi.org/10.1002/jobm.201100308.

Tsai J-J, Yen C-Y, Hsu C-H, Yu S-J, Chen C-H, Liao E-C. Immunomodulatory effects of modified Liu-Wei-Di-Huang-Wan traditional Chinese medicine on allergic asthmatic mice. Allergy, Asthma Clin Immunol. 2023;19(1):35. https://doi.org/10.1186/s13223-023-00792-5.

Usman M, Murtaza G, Ditta A. Nutritional, medicinal, and cosmetic value of bioactive compounds in button mushroom (Agaricus bisporus): a review. Appl Sci. 2021;11(13):5943. https://doi.org/10.3390/app11135943.

Vazquez-Armenta FJ, Leyva JM, Mata-Haro V, Gonzalez-Aguilar GA, Cruz-Valenzuela MR, Esqueda M, Gutierrez A, Nazzaro F, Fratianni F, Gaitán-Hernández R, Ayala-Zavala JF. Phenolic compounds of Phellinus spp. with antibacterial and antiviral activities. Braz J Microbiol. 2022;53(3):1187–97. https://doi.org/10.1007/s42770-022-00745-x.

Venturella G, Ferraro V, Cirlincione F, Gargano ML. Medicinal mushrooms: bioactive compounds, use, and clinical trials. Int J Mol Sci. 2021;22(2):634. https://doi.org/10.3390/ijms22020634.

Wasser S. Medicinal mushroom science: current perspectives, advances, evidences, and challenges. Biom J. 2014;37(6):345. https://doi.org/10.4103/2319-4170.138318.

Wasser SP, Reshetnikov SV. Process for producing, methods and compositions of glucuronoxylomannan as nutriceutical agent from higher basidiomycetes mushroom (Patent 6383799). 2002. https://www.freepatentsonline.com/6383799.html.

Watanabe T. Basidiomycetes, basidiomycetes extract composition, health foods, and immunopotentiators (Patent 7517682). 2009. https://www.freepatentsonline.com/7517682.html.
Weng Y, Xiang L, Matsuura A, Zhang Y, Huang Q, Qi J. Ganodermasides A and B, two novel anti-aging ergosterols from spores of a medicinal mushroom Ganoderma lucidum on yeast via UTH1 gene. Bioorg Med Chem. 2010;18(3):999–1002. https://doi.org/10.1016/j.bmc.2009.12.070.
Wu Y, Choi M-H, Li J, Yang H, Shin H-J. Mushroom cosmetics: the present and future. Cosmetics. 2016;3(3):22. https://doi.org/10.3390/cosmetics3030022.
Wu C-T, Lee Y-T, Ku M-S, Lue K-H. Role of biomarkers and effect of FIP-fve in acute and chronic animal asthma models. J Microbiol Immunol Infect. 2020;53(6):996–1007. https://doi.org/10.1016/j.jmii.2020.07.006.
Wypych-Ślusarska A, Grot M, Kujawińska M, Nigowski M, Krupa-Kotara K, Oleksiuk K, Głogowska-Ligus J, Grajek M. Respiratory symptoms, allergies, and environmental exposures in children with and without asthma. Int J Environ Res Public Health. 2022;19(18):11180. https://doi.org/10.3390/ijerph191811180.
Xiang H, Sun-Waterhouse D, Cui C. Hypoglycemic polysaccharides from Auricularia auricula and Auricularia polytricha inhibit oxidative stress, NF-κB signaling and proinflammatory cytokine production in streptozotocin-induced diabetic mice. Food Sci Human Wellness. 2021;10(1):87–93. https://doi.org/10.1016/j.fshw.2020.06.001.
Xu X, Yu C, Liu Z, Cui X, Guo X, Wang H. Chemical composition, antioxidant and anti-inflammatory activity of shiitake mushrooms (Lentinus edodes). J Fungi. 2024;10(8):552. https://doi.org/10.3390/jof10080552.
Yao Q, Wu Z, Zhong J, Yu C, Li H, Hu Q, He J, Du J, Sun C. A network system for the prevention and treatment of mushroom poisoning in Chuxiong autonomous prefecture, Yunnan Province, China: implementation and assessment. BMC Public Health. 2023;23(1):1979. https://doi.org/10.1186/s12889-023-16042-7.
Zhuang C, Kawagishi H, Preuss HG. Glycoprotein with antidiabetic, antihypertensive, antiobesity and antihyperlipidemic effects from Grifola frondosa, and a method for preparing same (Patent 7214778). 2007. https://www.freepatentsonline.com/7214778.html.
Zhuang C, Kawagishi H, Zhang L, Anzai H. Anti-dementia substance from hericium erinaceum and method of extraction (Patent 20090274720). 2009. https://www.freepatentsonline.com/y2009/0274720.html.

Chapter 5
Mushrooms to Avoid

Asifa Kausar, Khadija Javed, Muhammad Zaid Salar, Naila Ghafoor, Quratulann Sattar, Amara Tahir, Ayesha Ghafoor, and Arooba Farooq

Abstract Humans have used mushrooms for centuries, serving as a vital food source and in traditional medicine, culinary practices, and cultural rituals. At the same time, their modern applications extend to pharmaceuticals, biotechnology, and sustainable industries. However, not all mushrooms are edible; some inedible mushrooms are also present and can be poisonous. The consumption of poisonous mushrooms has led to many health problems and even deaths throughout the world. Therefore, this chapter discusses mushrooms that should be avoided. In our study, we reviewed the comparison between inedible and edible mushrooms, as well as the taxonomy of the latter. Moreover, we have also discussed the toxic compounds in inedible mushrooms and their effects on living organisms. In addition to this, the medicinal and nutritional importance of inedible mushrooms is also provided. Finally, protective measures against mushroom poisoning are presented as precautions against mushroom toxicity.

Keywords Mushrooms · Inedible mushrooms · Toxicity · Mortality · Medicine

A. Kausar (✉)
National Institute of Food Science and Technology (NIFSAT), University of Agriculture, Faisalabad, Pakistan

K. Javed
School of Natural Sciences, University of Chester, Chester, UK

M. Z. Salar · N. Ghafoor · Q. Sattar · A. Tahir
Department of Zoology, Wildlife and Fisheries, University of Agriculture, Faisalabad, Pakistan

A. Ghafoor
Department of Zoology, Government College University Faisalabad, Faisalabad, Pakistan

A. Farooq
Department of Chemistry, University of Agriculture, Faisalabad, Pakistan

I. U. Haq et al. (eds.), *Mushrooms*, https://doi.org/10.1007/978-981-95-2539-3_5

5.1 Introduction

The term mushroom has been used across various eras and nations. Mushroom often denotes all larger fungi characterized by stalks and caps. The limited application of mushrooms refers specifically to larger edible fungi, which may include medical properties. The term mushroom is often used specifically to refer solely to the edible species of Agaricus (Hawksworth 2001). Mushrooms are characterized as macrofungi by a prominent fruiting body, located either above or below the earth. They attain sufficient size for easy recognition by the naked eye and can be manually harvested (Chang and Miles 1992). This implies that mushrooms are not necessarily Basidiomycetes, airborne, or edible. They may belong to the Ascomycetes, develop subterraneously, possess a nonfleshy texture, and are not always edible (Ukwuru et al. 2018). This concept is broadly relevant in the present context of mushroom cultivation, where various species of mushrooms are cultivated globally with the emergence of new technology in mushroom production.

Mushrooms have been incorporated into the food for humans for millennia (Zhang et al. 2016). In certain nations, mushrooms are referred to as "*meat for the poor*." Mushrooms are also extensively utilized for medicinal uses (Filippova 2018; VanderMolen et al. 2017). The therapeutic applications of mushrooms date back over 5000 years (Halpern 2010). Taofiq et al. (2016) also elucidated the application of mushroom extracts and chemicals in cosmetics. Approximately 14,000 species of mushrooms have been identified; nevertheless, scientists concur that numerous other species remain undiscovered (Zahan et al. 2021).

Mushrooms are a nutritious food source; however, not all mushrooms are edible, as some are inedible and potentially toxic. Of the identified mushroom species, 25% are edible, 50% are inedible, 20% are toxic, 4% may be palatable, and the remaining 1% are lethal (Miles and Chang 1997). Certain edible mushrooms may lack palatability, yet several justifications exist for their consumption. Edible mushrooms have notable attributes, including enhancing the immune system, combating infections, and eradicating cancer cells. Inedible mushrooms are too tough to contemplate chewing, akin to certain woody rack mushrooms (Zahan et al. 2021). It is also indigestible and may evacuate the human gastrointestinal tract. Moreover, ignorance regarding toxic mushrooms results in several fatalities from their consumption. Mushrooms exhibiting intact cells, vibrant colors, and a lack of avian and insect interaction are likely toxic. The National Poison Data System (NPDS) reported 133,700 instances of mushroom exposure from 1999 to 2016, averaging 7248 cases per year (Brandenburg and Ward 2018). Between 2004 and 2014 in China, there were 576 recorded incidents involving 3701 patients, of which 786 resulted in fatalities (Zhou et al. 2016). In 2021, there were 327 documented occurrences of mushroom poisoning across 25 provincial-level administrative units, affecting 923 individuals and resulting in 20 fatalities. The mortality rate was 2.17% (Li et al. 2022).

The applications of mushrooms in people's daily lives are expanding at an exceptional rate. Therefore, there is a dire need to educate people about inedible mushrooms, their applications, pros and cons, and the methods to distinguish between edible and inedible mushrooms. This chapter discussed the mushrooms we should avoid and the methods to distinguish between edible and inedible mushrooms. Moreover, it also discussed the potential benefits of edible mushrooms.

5.2 Historical Perspective

The history of mushroom consumption likely parallels the history of food collecting, a fundamental aspect of early human existence (Boyden 1973). Numerous experts presently advocate for the significance of fungi in developing early religious activities (Guzmán 2009). Numerous instances indicate the historical utilization of fungal fruiting structures carrying psychoactive substances to attain mystical experiences. Robert Gordon Wasson, a pioneer in ethnomycological study, believed that fly agaric [*Amanita muscaria* (L.) Lam.] was the inaugural fungal species employed as an entheogen. It is hypothesized that it was utilized by inhabitants of the northern Eurasian woodlands around 9000 BCE (Wasson and Wasson 1957; Hajicek-Dobberstein 1995). In addition to Siberia, this fungal species was utilized for religious purposes throughout the Palaeolithic age in the Saharan Region and Northern Europe, where macrofungi are presently regarded as a rarity (Samorini 2001; Wexler 2014; Niksic et al. 2015).

The mushroom cult existed in Mayan civilization, which referred to psychoactive fungus (mostly species from the Psilocybe genus) as Teonanácatl, signifying the body and sustenance of deities (Wexler 2014; Niksic et al. 2015). The fly agaric is also considered an element in the Vedic ritual drink Soma, intended to produce hallucinations and enable individuals to perform heroic feats (Hajicek-Dobberstein 1995). In numerous cultures, the fly agaric was referred to as a single eye, allegedly due to its likeness to a human eye in the early stages of development (Wasson 1971). Consequently, it was linked, among others, to the one-eyed Odin, the supreme god of the Norse pantheon (Hajicek-Dobberstein 1995; Motyka and Marcinkowski 2014). In Grímnismál, a poem from the earliest extant manuscript of Icelandic literature, the Poetic Edda, Odin is referred to as Sithhott or Sidhhottr, which translates as *long hood or broad hat* (Hajicek-Dobberstein 1995). This undeniably elicits further mycological associations.

In Ancient Egypt, mushrooms were considered plants of immortality, bestowed upon individuals by the deity Osiris. Mushrooms were designated as a delicacy exclusively for Egyptian nobility due to their distinctive flavor. Common individuals were prohibited from consuming them and prevented from touching them (Niksic et al. 2015). The Romans likewise held mushrooms in high regard. The ancient Romans referred to them as the sustenance of the Gods (Niksic et al. 2015). Among the extant records of mushroom utilization in ancient Roman times, the initial references to awareness of the toxic effects of mushrooms are also included. The rulers' fondness for mushroom ingestion was used by poisoners since mushrooms were an ideal means to usurp emperors. The demise of Claudius, a Roman emperor, attributed to mushroom poisoning, is widely recognized. Numerous historians have documented the demise of Claudius (Cary 1925; Mottershead 1969). Agrippina, the spouse of Claudius, shared a dish of mushrooms, one of which was poisoned, with her husband (Feldman 1965). The toxic mushroom resulted in the emperor's demise. It is reported that the toxic mushroom was a death cap (*Amanita phalloides*), which resulted in Claudius's demise following a night of intense suffering (Eden 1990). The aforementioned historical context of mushrooms offers insights into their usage, applications, and understanding by ancient humans.

5.3 Comparison of Edible and Inedible Mushrooms

Numerous studies have been undertaken to compare edible and inedible mushrooms. Due to the absence of a definitive method for promptly determining the edibility of a mushroom, it should not be consumed unless it has been correctly identified and its species' edibility is established (Ukwuru et al. 2018). Nonetheless, certain characteristics can still be utilized to evaluate the edibility of a mushroom based on its form and nutritional or poisonous properties. Edible mushrooms are fungi that produce fruiting structures visible to the naked eye. They can be epigeous or hypogeous and are harvestable by hand. The edibility of mushrooms is highlighted by the lack of toxic effects on humans and their appealing taste and aroma (Arora 1996). The nutritional makeup of edible mushrooms is substantial, which accounts for their consumption due to their nutritional and purported medical benefits. Practitioners of folk medicine utilize medicinal mushrooms for their nutritional properties. Psilocybin mushrooms, a type of hallucinogenic fungus, are ingested for recreational or spiritual reasons. However, they can induce significant disorientation and nausea, so they are not usually called edible mushrooms (Boa 2004). Inedible or poisonous mushrooms account for less than 1% of the world's known species, representing harmful and potentially lethal varieties. A qualified mycological expert must perform mushroom identification before eating (Ukwuru et al. 2018). Consequently, it is imperative to ascertain whether a particular mushroom is edible before eating. The poisons in different species exhibit significant variation in chemical makeup, resulting in markedly diverse poisoning effects depending on the species in question (Shepherd and Totterdell 1988).

Traditionally, ethnomycological information about edible mushrooms is confined to their observable fruiting bodies. Identification based on morphology relies on characteristics such as shape, cap color and form, gill size, stipe color, and fruiting bodies, which can significantly fluctuate within a species due to environmental conditions, often resulting in species misidentification. A study indicates that the Bado community in Kokrajhar district, Bodoland Territorial Region (BTR) Assam, India, has considerable mycological knowledge for distinguishing the edibility of wild macrofungi (Basumatary and Gogoi 2016). The indigenous understanding of edible mushrooms and their use by local populations is a significant aspect of ethnomycology (Das et al. 2014). Palapala et al. (2002) documented distinctions between edible and inedible mushrooms. He asserted that inedible and toxic mushrooms exhibit a green or purple hue upon being cut and cause a burning or stinging sensation on the tongue.

Furthermore, they possess a pungent odor and an unpleasant flavor, exhibit scales on the cap, and are devoid of worms. Conversely, edible mushrooms possess entirely contrasting qualities. Table 5.1 presents the catalog of edible and inedible/poisonous mushrooms documented by the Missouri Department of Conservation (Briggler 2018).

Mushroom-Type Description Size Growth Examples
Amanitas numerous species of this genus possess amanitin, the most deadly toxins on earth. A single cap of the destroying angel can be lethal to a human.

Amanita begins as an ovate "button" that may resemble a little puffball-shaped mushroom. The button ruptures as the mushroom expands. Mature amanitas are fungi with gills characterized by parasol-shaped caps that can be yellow, brown, white, or red. Their sizes vary significantly, contingent upon species and growth conditions. Amanitas are typically located on the ground in open areas and woodlands during the fall and summer. Destroying angel (*Amanita bisporigera*), Fly agaric (*A. muscaria*), death cap (*A. phalloides*), and fool's mushroom (*Amanita verna*).

False morels resemble true morels, although notable distinctions exist among them.

Their cap surface has flaps, folds, wrinkles, or lobes, but it does not contain ridges and pits like true morels. They also lack the hollow center, a characteristic of a true morel. False morels vary in size from 2 to 8 in. These mushrooms are found in woodlands during fall, spring, and summer. An example of a false morel is *Gyromitra caroliniana.*

Little brown mushrooms (LBMs) encompass all small- to medium-sized, difficult-to-distinguish brownish mushrooms with spores of various colors.

Numerous LBMs are nontoxic; however, a few are slightly to severely toxic. The seemingly benign small mushrooms of the *Galerina* genus are likely the most perilous of the LBMs. They are found in all seasons and all habitats. Galerina mushrooms possess a brownish spore print, and they grow on wood in the form of clusters. Deadly galerina (*Galerina autumnalis*).

Jack-o'-Lantern: A vibrant orange to yellowish-orange mushroom, aptly recognized because of its pumpkin color, seasonal look during fall, and occasional emission of a greenish luminescence seen at night or in dimly lit environments.

The gills are sharply defined and extend along the stalk, while the cap is initially convex, transitioning to a flat or funnel shape with a central depression. They are normally 2–8 in tall, and their caps are 2–8 in wide. These mushrooms appear in fall and summer (July–October) on stumps, buried wood at the base of trees. An example of Jack-o'-Lantern is *Omphalotus olearius*.

Table 5.1 Insights into different types of edible and inedible mushrooms as reported by the Missouri Department of Conservation

Mushroom type	Description	Size	Growth	Examples
Edible mushrooms				
The Puffballs	They are white, rounded and may have or have not spiny warts As they mature, the color becomes yellowish, to pinkish and to brownish	Puffballs are usually 1–2.5 in expended and 1–3 in long. Some of them can reach a size of 20 in wide and tall	They grow in end of summer or fall (July–October) and grow in barren areas, lawns pastures, decaying wood, soil, and open woods	Gem-studded puffball (*Lycoperdon perlatum*)
Shaggy Mane	It is mushroom with white color and cylindrical, shaggy cap that looks inky and black at different growing stages The cap become bell shaped and white scales turn brownish, as it matures	Shaggy mane is 4–6 in tall	They are mostly found in fall (September–October) in the lawns and along the roadsides	Lawyer's wig or Shaggy ink cap (*Coprinus comatus*)
Coral Fungi/ Club Fungi	They possess clumps of branching stems that point upward, forming a structure similar to corals Most are yellowish, tan, or whitish, while some are purple or pink	Coral fungi can grow more than 8 in long	In spring and summer (June–September) coral mushrooms are found and grow in woodedland on the decaying logs and on ground	Crown-tipped coral (*Artomyces pyxidatus*)
Morels	They have honeycomb like cap containing ridges that are brownish black in color and yellowish colored pits Spores are present inside the pits and are white to cream-colored	Their size ranges from 2–12 in tall	Morels typically grows in March to early May time period. They are particularly found in moist woodlands and in river bottoms	Common morel (*Morchella americana*) and Black/smoky morel (*Morchella angusticeps*)

(continued)

Table 5.1 (continued)

Mushroom type	Description	Size	Growth	Examples
Chanterelles	They possess wavy cap edges and have shape like small funnels. Most of chanterelles are bright orange or yellow. Some fresh chanterelles possess pleasant, fruity fragrance Their spore print is pinkish-yellow to cream	Chanterelles can vary in size from 0.5–6 in tall and 0.5–6 in wide	They are found during the summer to fall season (May–September) Chanterelles are found growing in the leaf litter in hardwood forests and ground in grass	Common chanterelle (*Cantharellus cibarius*)
Oyster Mushrooms	They have grayish, tan cap along with an off center, stubby stalk. The cap is elongated to semicircular (oyster shaped), shell shaped, and having smooth, occasionally wavy margins Their gills extends downward along a short, robust stalk and are narrow. They have grayish lilac to white spores	The cap of oyster mushrooms is 1 to 7 in expended with a stalk of 1.25 in long and 0.75 in expanded	Oyster mushrooms found throughout the year and exclusively on wood. They occasionally seem to emerge from the earth, however they are connected to tree roots below the soil surface	Oyster mushroom (*Pleurotus ostreatus*) and Italian Oyster (*P. pulmonarius)*

(continued)

Table 5.1 (continued)

Mushroom type	Description	Size	Growth	Examples
Inedible/poisonous mushrooms				
Amanitas	Numerous species of this genus possess amanitin, the most deadly toxins on earth. A single cap of the destroying angel can be lethal to a human Amanita begins as an ovate "button" that may resemble a little puffball-shaped mushroom. The button ruptures as the mushroom expands. Mature amanitas are fungi with gills characterized by parasol shaped caps that can be yellow, brown, white, or red	Their sizes vary significantly, contingent upon species and growth conditions	Amanitas are typically located on the ground in open areas and woodlands during the fall and summer	Destroying angel (*Amanita bisporigera*), Fly agaric (*A. muscaria*), death cap (*A. phalloides*), and Fool's mushroom (*Amanita verna*)
False morels	They resemble true morels, although there are notable distinctions among them Their cap surface has flaps, folds, wrinkles or lobes, but does not contain ridges and pits like true morels. They also lack the hollow center, a characteristic of a true morel	False, morels vary in size between 2 and 8 in	These mushrooms are found in woodlands during fall, spring, and summer seasons	False morel (*Gyromitra caroliniana*)

(continued)

Table 5.1 (continued)

Mushroom type	Description	Size	Growth	Examples
Little brown mushrooms (LBMs)	It encompasses all small to medium sized, difficult-to-distinguish brownish mushrooms with spores of various colors Numerous LBMs are nontoxic, however a few are slightly to severely toxic. The seemingly benign small mushrooms of the *Galerina* genus are likely the most perilous of the LBMs	–	They are found in all seasons and in all habitats. Galerina mushrooms possess a brownish spore print and they grow on wood in the form of clusters	Deadly galerina (*Galerina autumnalis*)
Jack-o'-Lantern	The jack-o'lantern is a vibrant orange to yellowish-orange mushroom, aptly recognized because of its pumpkin color, seasonal look during fall, and occasional emission of a greenish luminescence seen at night or in dimly lit environments The gills are sharply defined and extend along the stalk, while the cap is initially convex, transitioning to a flat or funnel shape with a central depression	They are normally 2–8 in tall, and their caps are 2–8 in wide	These mushrooms are appear in fall and summer (July–October) on stumps, buried wood at the base of trees	Jack-o'-Lantern (*Omphalotus olearius*)

5.4 Distribution and Dispersal of Inedible Mushrooms

In the last decade, China has documented over 10,000 cases of mushroom poisoning, resulting in at least 788 fatalities. Furthermore, during 2010 and 2020, the incidence of documented mushroom poisoning escalated markedly, rising from 37 cases in 2010 to 2705 cases in 2020. China has 480 recognized toxic mushroom taxa

(Yu et al. 2019; Nieminen and Mustonen 2020). Among these numerous macro-fungi, over 1000 species are believed to be poisonous mushrooms (Mao 2006). China has recorded up to 480 types of toxic mushrooms. The State of the World's Fungi 2018 reports that over 350 edible mushroom species have been discovered globally (Wu et al. 2019).

Poisonous mushroom species mostly belong to the orders Russulales (Basidiomycota), Boletales, Agaricales, and Pezizales (Ascomycota), since the majority of species within these orders produce large fruiting bodies (Hall et al. 2003; Wu et al. 2019). The deadly species mostly belong to Agaricales, such as amanitin-producing mushrooms: *Amanita bisporigera* G.F. Atk. *A. phalloides* (Vaill. ex Fr.) Link, *Lepiota brunneoincarnata* Chodat and C. Martín, and *Galerina marginata* (Batsch) Kühner. Species of Amanita, regarded as the most perilous mushrooms, account for 90% of fatal mushroom poisoning incidents (Bresinsky and Besl 1990; Chen et al. 2014a). The death cap, *A. phalloides*, is one of the most notorious species, having caused the deaths of thousands over history, hence earning a position on the list of the world's ten most feared fungi (Hyde et al. 2018). In addition to amanitin-producing mushrooms, deadly species include genera such as Trogia, Paxillus, Gyromitra, and Russula.

The ingestion of deadly mushrooms, erroneously gathered in place of edible varieties, results in poisoning. The prevalence of mushroom poisoning significantly differs across many regions globally, influenced by local customs, lifestyle, dietary variables, climate, and the availability of wild mushrooms. Table 5.2 presents data on the prevalence of deadly mushrooms throughout several nations, regions, and globally. Table 5.2 illustrates a significant disparity among various studies on this subject.

Mushroom poisoning is a significant issue in the following countries: Bulgaria (Marinov et al. 2018); Krenova et al. (2007); Czechia (Kieslichova et al. (2018); China (Zhou et al. 2012; Sun et al. 2018); Iran (Dadpour et al. 2017); Mexico (Ruiz-Gonzalez et al. 2017); Japan (Matsuura et al. 2009); Italy (Cervellin et al. 2018); S. Korea (Kim et al. 2017; Sohn 2015); Nepal (Chaudhary et al. 2013); Poland (Krakowiak et al. 2017); Romania (Precup et al. 2012); Russia (Shen et al. 2013); Thailand (Trakulsrichai et al. 2017); Turkey (Erenler et al. 2016); and Ukraine (Kurdil et al. 2016). Isolated instances and case series have been documented from Argentina (Romano et al. 2013); Australia (Nicholson and Korman 1997); Brazil (Vargas et al. 2011); and Canada (Berch et al. 2017); Colombia (Meijer et al. 2007); Guatemala (Logemann et al. 1987); Hungary (Vetter and Vetter 2014); France (Varvenne et al. 2015); India (Latha et al. 2018); Ireland (Cassidy et al. 2011); Taiwan (Yang et al. 2006); the United States (Vo et al. 2017); Zambia (Tagwireyi et al. 2016); and Zimbabwe (Tagwireyi et al. 2016). Numerous findings demonstrated mycotoxicity in animals. The poisonous impacts of mushrooms in mammals are thought to be similar to those in humans; however, other animal taxa exhibit reduced sensitivity (Vetter and Vetter 2014).

Table 5.2 World and country reports of hazardous mushroom species as reported by Govorushko et al. (2019a, b)

Region	Toxic species	Potentially fatal species	References
China	435		Chen et al. (2014a, b), Sun et al. (2017)
Norway	60–100		Holsen and Aarebrot (1997)
European region	32	11	Schmutz et al. (2018)
European region	>100	–	Karaseva and Butvilovsky (2014)
Russia	>150		Vishnevsky (2017)
Russia	40		Suliga and Yarovova (2017)
Japan	30	–	Gonmori and Yoshioka (2005)
Türkiye	<30		Sönmez et al. (2015)
North America	<100	–	Broussard et al. (2001)
Iran	50		Soltaninejad (2018)
Worldwide	50–100	–	Diaz (2016a, b), Kintziger et al. (2011) and Cervellin et al. (2018)
Worldwide	170	–	Boa (2004)
Worldwide	150	< 10	Smith and Davis (2016)
Worldwide	150	–	Erenler et al. (2016)
Worldwide	<100	–	Eren et al. (2010a, b)
Worldwide	400	–	Horowitz (2015)
Worldwide	100	15–20	Rajarathnam and Shashirekha (2003), Gawlikowski et al. (2015)
Worldwide	25–55	10	Pajoumand et al. (2005)
Worldwide	70–100	45	Sönmez et al. (2015)
Worldwide	50–100	10	Barman et al. (2018), Graeme (2014)

5.5 Taxonomy of Inedible Mushrooms

Wild mushrooms are cultivated and consumed due to the knowledge amassed in our cultural memory, rooted in our worldview. The accumulation of cultural legacy is facilitated by identifying unique traits that enable individuals to distinguish between nonedible and edible mushrooms. Conceptual representations are constructed from group and individual experiences, which evolve into logical, collective understandings that reside in the collective unconscious (Ford 2011). Prior studies indicated that nonedible fungal species are essential components in the cosmos, corpus, and praxis of traditional knowledge on wild mushrooms. Literature has found over 100 nonedible taxa; however, other ethnomycological studies often cite a maximum of 17 species acknowledged as toxic or poisonous (Agnihotri and Si 2012). Furthermore, the definition of the local understanding of nonedible mushrooms has been cursory and typically stems from assumptions concerning edible mushrooms (Zamudio and Hilgert 2015).

Mushrooms are recognized and categorized by several approaches. The Mushroom Diagnosis Assistance System (MDAS) is primarily a mobile device that incorporates three components: a web application (server), a unified database, and a mobile application. This study mainly focused on the classification rates of the Naïve Bayes and decision tree classifiers. The proposed approach first identified the characteristics of mostly recognized mushrooms and then categorized their kinds. Al-Mejibli and Abd (2017) created a smartphone application called the Mushroom Diagnosis Assistance System to ensure safety during mushroom collection. This study relies only on the most recognized features of mushrooms for type identification. A comparison was conducted between decision tree and Naïve Bayes classifiers, demonstrating that the decision tree outperformed in terms of error metrics, accurately categorized data, and misclassified samples. In addition to identifying mushroom infections, Chowdhury and Ojha (2017) recognized a method using data mining approaches, namely Naïve Bayes, RIDOR, and SMO algorithms. They concentrated on identifying the most prevalent symptoms of mushrooms to uncover mushroom illnesses. The comparison of the aforementioned algorithms revealed that Naïve Bayes exhibited superior accuracy in categorization.

Used several machine-learning approaches to categorize mushroom fungi using photos of mushrooms. Their research indicates that, while they extracted features from photos both with and without experiences, superior results were achieved solely with background images when using KNN algorithms in conjunction with Eigen feature extraction and the actual dimensions of the mushroom. They achieved 94% accuracy in feature extraction from photos with true dimensions and 87% accuracy alone in feature extraction from images. However, other researchers, said in their study that eliminating the backdrop from photographs may enhance accuracy. They devised a classification system centered on a feature-based computer. Concentrated only on identifying mushroom toxicity using visual characteristics of mushrooms. The cap form and color of mushrooms served as the features for this study, using logistic regression, support vector machine, and multigrained cascade forest as the three pattern recognition algorithms for feature extraction. The multigrained cascade forest classifier achieved a classification accuracy of 98%, surpassing the performance of the other two models.

The perception of all mushrooms not included in one's diet as toxic reflects that knowledge and interest are contingent upon consumption traditions (Lampman 2007). Only a limited number of instances exist when a species not previously used within the familial nucleus has been included in the diet, indicating a tendency to retard the unfamiliar due to potential intoxication concerns (Ellen 2008). Meanwhile, mushroom pickers and dealers were more willing to acquire knowledge about species beyond their traditional heritage. The method by which individuals distinguish nonedible mushrooms during collecting reflects their expertise: species resembling food varieties are deemed more significant and valued. The general identification criteria correspond with those documented in ethnomycological studies performed in the highlands of Mexico (Zent and Zent 2016). Similar accuracy in species identification has been documented for the Kurya tribe in Tanzania and the Karbi culture in India, which utilize traditional indicators linked to distinct characteristics associated with the mushrooms native to their regions, such as odor, hue, dimensions, habitat, and substrate (Ruan-Soto 2020). In contrast, civilizations from the Northwestern Andes in Colombia use vibrant colors and pungent or bitter flavors as signs of danger. Certain parameters used by cultures parallel those utilized in classical mycology (Prila and Teron 2018).

Local names provide a collective cognitive framework, whereby each ethnotaxon is defined by discernible traits informed by the unique knowledge of each culture (Cano-Estrada and Romero-Bautista 2016). Ramírez-Terrazo et al. (2021) demonstrated that local nomenclature for nonedible mushrooms in their research reflects a substantial body of knowledge on the topic, accounting for 60% of all conventional names of inedible mushrooms now acknowledged in Latin America. On a geographic basis, their results reflect more than ten times the number of local names previously documented in the area. Research from several global regions indicates that nonedible mushrooms are often referred to in broad terms, without precise nomenclature, and are only categorized as nonedible (Ammirati et al. 1985). Several studies indicate one or two local designations for nonedible ethnotaxa (Arora 2008). In Masovia, Poland, the average number of recorded inedible or toxic fungal species was 1.7, with a high of six (Pacheco-Cobos et al. 2010). The names arise from the distinction between nonedible and edible mushrooms, with each ethnotaxon defined by certain traits that facilitate their identification.

5.6 Toxic Compounds in Inedible Mushrooms

Several classes of mushrooms are classified as poisonous and have toxic compounds. The consumption of harmful mushrooms typically occurs inadvertently, often resulting from the misidentification of species (Mueller et al. 2007; Guillen and Baeza 2014). This unintentional occurrence frequently arises from a lack of comprehensive data regarding poisonous mushrooms, including details on their potential toxicity. It is essential to delineate the toxicological characteristics of mushroom species before their use by people (Kosentka et al. 2013).

A variety of very strong poisons have been found in many mushrooms. The ingestion of even a small quantity may pose a risk to human safety. Mushroom toxidromes are categorized based on the specific toxins and their clinical manifestations. Mushroom toxins are classified into seven primary categories: coprine, ibotenic acid, gyromitrin (monomethylhydrazine), psilocybin, muscarine, orellanus (Cortinarius species), and amatoxins (cyclopeptides). This text discusses these toxins, which include their origins, molecular characteristics, clinical effects, mechanisms, and probable toxicity (Kim et al. 2017). Tables 5.3 and 5.4 represent molecular mechanisms, clinical symptoms, toxicity locations, and possible toxic effects of diverse mushroom species.

Table 5.3 Clinical symptoms, toxicity locations, and possible toxic effects of diverse mushroom species

Toxic syndrome	Toxins	Sites of action	Mushroom species	Mortality rate	References
Acute gastrointestinal illness absent liver damage	Gastrointestinal I irritants	Gastrointestinal Tract	*Chlorophyllum molybdites*, *Clitocybe nebularis*, *Omphalotus illudens*	Rare	Brent and Palmer (2007)
Hallucinogenic	Psilocybin, psilocin	CNS (hallucinogenic effects)	*Psilocybe cubensis*, *P. mexicana*, *Conocybe cyanopus, G. aeruginosa*	Rare	Rumack and Spoerke (1994), Goldfrank (2011)
CNS stimulation and inhibition (Stupor, coma, psychosis, agitation, and hallucinations)	Ibotenic acid and muscimol	CNS (depressant and excitatory effects)	*Amanita muscaria*, *A. pantherina*, *A. gemmata*	Rare	Berne et al. (2005)
Cholinergic hyperactivity (emesis, diarrhea, bradycardia, bronchorrhea, lacrimation, bronchospasm, sialorrhea)	Muscarine	The autonomic nerve system, which includes muscarinic responses	*C. dealbata*, *C. illudens*, *I. fastigiata*, *B. calopus*	Rare	Bédry and Saviuc (2002)
Flushing, cephalalgia, tachycardia, thoracic discomfort, anxiety	Coprine	–	*Coprinus atramentarius*	Rare	Bédry and Saviuc et al. (2002)
Gastroenteritis and subsequent kidney failure	Allenic norleucine	Kidney, gastrointestinal tract	*Amanita smithiana*	Rare	Rumack and Spoerke (1994), Goldfrank (2011)

(continued)

Table 5.3 (continued)

Toxic syndrome	Toxins	Sites of action	Mushroom species	Mortality rate	References
Hepatic intoxication	Phallotoxins, amatoxins,	Gastrointestinal tract, liver, kidney	*Amanita phalloides, A. virosa, A. verna, A. bisporigera, Galerina autumnalis, G. marginata, G. venenata, Lepiota helveola*	5 ~ 30%	Dehay et al. (2009)
Seizures, delayed gastroenteritis, and hepatic damage	Gyromitrin	Gastrointestinal tract, CNS, liver and blood	*Gyromitra esculenta, G. infula, Sarcosphaera coronaria, Chrysina macropus*	0 ~ 10%	Saviuc et al. (2002)
Postponed nephric failure, cellular	Cortinarin, orellanine, orellinine	Renal tissues, gastrointestinal tract	*C. speciosissinus, Mycena pura, O. orarius Cortinarius orellanus*	20%	Bedry et al. (2001)
Delayed rhabdomyolysis	Unknown	Muscle	*Tricholoma equestre*	Rare	Mount et al. (2002), Zuzek et al. (2006)
Delayed encephalopathy in people with kidney damage	Unknown	Encephalopathy	*Pleurocybella porrigens*	27%	Aygul et al. (2010)
Abdominal pain, diarrhea, and intense sweating	–	–	*Clitocybe rivulosa*	–	Curtis et al. (1979), Ennecker-Jans et al. (2007)

5.6.1 *Amatoxins*

Fungi containing amatoxins are included in the families of Cortinariaceae (genus Galerina), Amanitaceae (genus Amanita), and Agaricaceae (genus Lepiota). Several species of Amanita, such as *Amanita virosa, A. bisporigera, A. tenuifolia, A. verna, A. ocreata, A. hygroscopica*, and *A. phalloides*, have high concentrations of amatoxins. Amatoxins are bicyclic octapeptides that serve as very effective thermoplastic

Table 5.4 Molecular properties and mechanism of toxicity of different toxins found in various mushroom species

Name of toxin	Source	Molecular characteristics	Mechanism of action (Toxicity)
Amatoxin	Different species of *Amanita*	Aegerolysin proteins are 16-kDa acidic proteins. It has 137 amino acids. Thirteen positively and sixteen negatively charged residues	Temporary elevation in arterial pressure followed by a gradual decline to mid-circulatory pressure, accompanied by bradycardia, myocardial ischemia, and ventricular extrasystoles. The hyperkalaemia resulting from haemolytic action presumably plays a substantial role in its toxicity
Ostreolysin	*Pleurotus ostreatus*	Bicyclic octapeptide with thermodynamic stability. 9 amatoxins have been reported, with α-amanitine being the most potent	RNA polymerase II is inhibited, resulting in the cessation of DNA transcription, which leads to protein production and cellular necrosis
Phallotoxin	*Amanita phalloides*	Peptides have a bicyclic structure and transannular thioether linkage	The toxin specifically binds to F-actin in liver cells, hence inhibiting the depolymerization of F-actin into G-actin
Agaritin	*Agaricus bisporus*	b-N-(g-L(+) glutamyl)-4-hydroxymethyl) phenylhydrazine, which is an L-glutamic acid	The toxin specifically binds to F-actin in liver cells
Orellanine	*Pleurotus ostreatus*	It is a bipyridine oxide that is heat-stable. When orellanine is deoxidized, it becomes nontoxic orellanine, which has chemical similarities with the pyridine herbicides diquat and paraquat	In vitro *research shows that orellanine creates oxygen radicals at the target region* via *iron redox cycling and/or activation. Further cellular results reveal that certain toxin metabolites may inhibit protein synthesis*
Acromelic acid	*Clitocybe acromelalga*	It is a member of the kainoid family, a group of nonproteinogenic pyrrolidine dicarboxylic acids	Acromelic acid A demonstrates neuroexcitatory properties, binds to glutamate receptors, resembles glutamic acid, induces specific behavioral alterations, and causes targeted damage to interneurons in the lower spinal cord when evaluated in an animal model
Gyromitrin	*Gyromitra esculenta*	Gyromitrin is a very unstable, volatile liquid that may exist free or bound to glucosides. It oxidizes at ambient temperature	Hydrazines operate as convulsants and interact with pyridoxal phosphate to produce a hydrazine, leading to reduced activity of glutamic acid decarboxylase and decreased synthesis of GABA

(continued)

Table 5.4 (continued)

Name of toxin	Source	Molecular characteristics	Mechanism of action (Toxicity)
Ibotenic acid	*A. muscaria* and *A. pantherina*	Is an α-amino-3-hydroxy-5--isoxazole acetic acid	A stimulant of NMDA receptors. The acidic nature of the isoxazole moiety resembles that of glutamic acid and replicates its effects in animals
Muscarine	*C. serussata, C. dealbata, A. Muscaria, C. phyllophilla, C. rivulosa*	It is tetrahydro-4-hydroxy-*N,N,N*-5-tetramethyl-2-furanmethanaminium	This chemical resembles GABA and mimics its central nervous system inhibitory effect
Muscimol	*Amanita muscaria* and *A. Pantherine*	It is a decarboxylated product of ibotenic acid	It interacts with the same receptors as acetylcholine and has a similar structure. It causes parasympathomimetic symptoms because cholinesterase does not hydrolyze it
Psilocybin and psilocin		4-phosphoryloxy-N, N-dimethyltryptamine, a tyramine-type component	Alkaline phosphatase and nonspecific esterases break the phosphoric esters group, showing that psilocybin is a prodrug and its hydroxyl metabolite, psilocin, is active. Psilocybin activates serotonin 2A

The molecular methodologies were derived from publications by Gonmori et al. (2011), Ward et al. (2013), Karimi and Razavi (2015), Smith and Davis (2016), and Zuliani et al. (2016)

toxins. α-Amanitine is an exceedingly potent amatoxin and is one of the nine recognized amatoxins. Symptoms associated with amatoxin poisoning include euphoria, depression, motor ataxia, dizziness, gastrointestinal disturbances, somnolence, muscle twitching, and changes in mental and emotional state. Its action is via the suppression of DNA-dependent RNA polymerase II (Asakawa et al. 1992; Carhart-Harris and Goodwin 2017). The decrease in mRNA levels consequently lowers protein synthesis, ultimately resulting in necrosis or death of the affected cells (Diaz 2006; Fu et al. 2017). Figure 5.1 shows the chemical structure of amatoxins (National Center for Biotechnology Information 2024a, b, c, d, e, f, g). Symptomatic therapy is mostly recommended for this inebriation, including the administration of benzodiazepines or phenobarbitone during seizures. Detoxification is essential at the start of treatment for persons poisoned by amatoxin, requiring careful supervision. Liver transplantation might be necessary in cases of serious damage (Eyad Sameh et al. 2019).

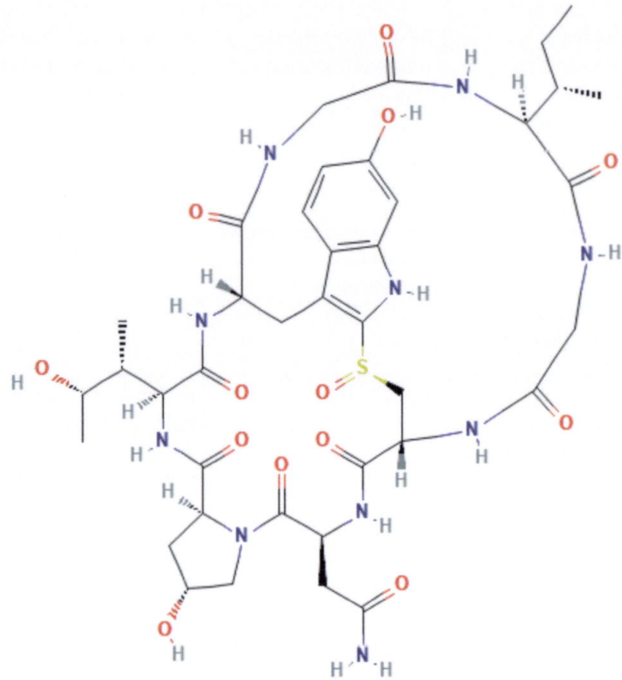

Fig. 5.1 Chemical structure of amatoxins (National Center for Biotechnology Information 2024a, b, c, d, e, f, g)

5.6.2 *Gyromitrin*

It is a volatile liquid that is oxidizable at ambient temperature, rendering it an unstable compound, and is found in some species of Gyromitra. The consumption of Gyromitra and the inhalation of fumes released during its preparation might result in toxicity (Berch et al. 2017; Kotowski et al. 2019). The predominant symptoms of this intoxication are nystagmus, ataxia, tiredness, tremor, vertigo, diarrhea, and vomiting. Moderate hemolysis and hepatic disorders may occur in some patients. Hydrazines, with convulsant, cytotoxic, and irritating characteristics, may be generated by the hydrolysis of gyromitrin in the stomach upon intake. Figure 5.2 shows the chemical structure of gyromitrin (National Center for Biotechnology Information 2024a, b, c, d, e, f, g). Gyromitrin induces carcinogenesis as an undesirable consequence. Symptoms of intoxication start 8–12 h postingestion. The antidote is pyridoxine (Li et al. 2022).

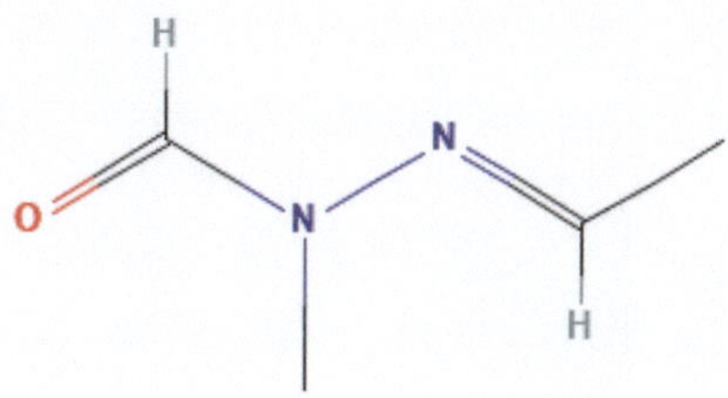

Fig. 5.2 Chemical structure of gyromitrin (National Center for Biotechnology Information 2024a, b, c, d, e, f, g)

5.6.3 Coprine

Mushrooms that contain coprine may induce organ toxicity when consumed with alcohol. This component accumulates in the bloodstream due to alcohol metabolism when alcohol is consumed before, concurrently with, or after mushroom consumption. Copine is present in *Coprinus atramentarius* mushrooms. The antidote remains unidentified. Certain Clitocybe species induce muscarinic syndrome (Lawton 2013; Liu et al. 2015). Muscarine is prevalent in *C. candicans, Clitocybe phyllophila, Clitocybe dealbata,* and *Clitocybe rivulose*. It is also found in *A. muscaria* and the Inocybe genus. Muscarinic syndrome symptoms include elevated pulse rate, cephalalgia, nausea, emesis, vertigo, and tachypnea. Patients negatively impacted may exhibit hypersecretion, miosis, and GI disturbances. Severe instances of poisoning with this toxin may result in bradycardia and potential collapse (Kurdil et al. 2016) (Fig. 5.3).

5.6.4 Muscimol

Muscarinic poisoning resulting from mushroom intake manifests as sweating, lacrimation, sialorrhea, emesis, diarrhea, hypotension, and dyspnea, attributable to its impact on the parasympathetic nervous system. The majority of patients recuperate within 1 day/night, except in instances when cardiac arrest occurs and resuscitation is unsuccessful. Symptomatic therapy is advised to manage this condition (Felix et al. 2004).

Atropine is advised to counteract the muscarinic effects, as this poisonous fungus also contains muscimol. Muscimol affects the central nervous system and induces paralysis 20–22 min after ingestion. It functions as a pesticide and is capable of exterminating flies. The mushrooms that contain muscimol include *Tricholoma muscarium* and *Amanita muscaria*. The remedy is physostigmine (Fig. 5.4).

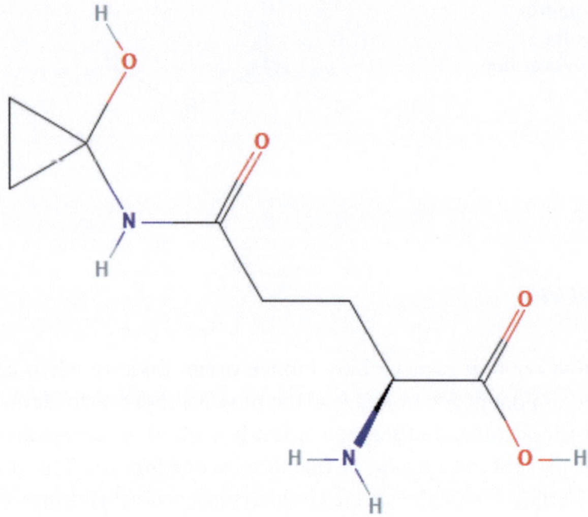

Fig. 5.3 The chemical structure of coprine (National Center for Biotechnology Information 2024a, b, c, d, e, f, g)

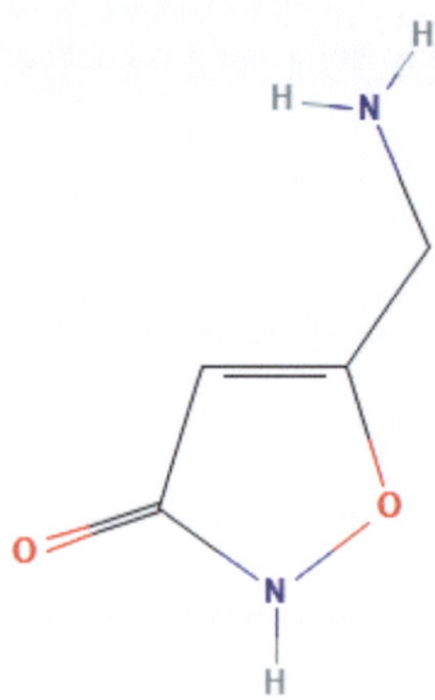

Fig. 5.4 Chemical structure of muscimol (National Center for Biotechnology Information 2024a, b, c, d, e, f, g)

5.6.5 Psilocybin

Psilocybin is prevalent in the genus Psilocybe. Psilocybin is typically present in *P. baeocystis, P. semilanceata, P. mexicana, P. cubensis,* and *P. bohemica. The symptoms of psilocybin poisoning manifest 30 min postingestion of the*

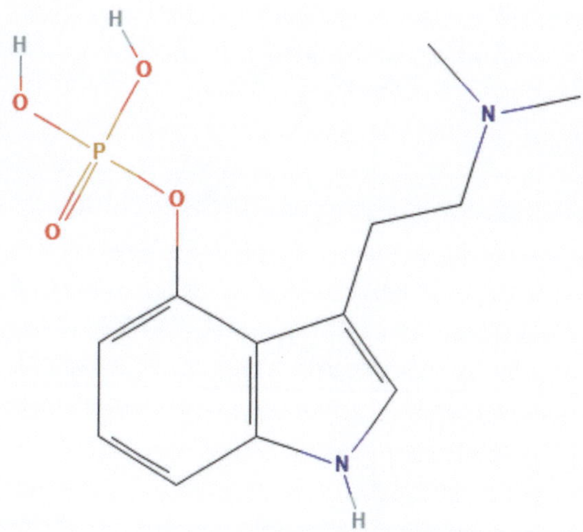

Fig. 5.5 Chemical structure of psilocybin (National Center for Biotechnology Information 2024a, b, c, d, e, f, g)

abovementioned mushrooms. Commonly reported symptoms of this intoxication include disorientation, motor incoordination, mydriasis, visual disturbances, anxiety, asthenia, vertigo, tachycardia, and hypertension, as well as nausea (Shao et al. 2016) (Fig. 5.5).

Issues related to psilocybin toxicity fully resolve between 4 and 12 h postingestion. Typically, hospitalization is unnecessary, and myocardial infarction in adults after psilocybin mushroom intoxication is infrequent (Hashimoto and Asakawa 1998; He et al. 2022).

Children may have a coma, hyperthermia, and convulsions. Psilocin mushrooms containing indole derivatives are often blue in hue. Indicators of psilocin intoxication include contorted extremities, tongue-related anxiety, impaired understanding, vivid visual hallucinations, and auditory hallucinations (Iwafuchi et al. 2003). The mushrooms that contain psilocin include *Panaeolus subbalteatus, Panaeolus sphinctrinus*, and *Panaeolus papilionaceus*. The antidote is chlorpromazine.

5.6.6 Orellanine

Orellanine is a thermally stable bipyridine N-oxide. The cyclopeptide orellanine is found in *Cortinarius speciosissimus* and *Cortinarius orellanus*, rendering these two mushrooms nephrotoxic. The quinone family of chemicals may accumulate in renal tissue due to orellanine oxidation, and the covalent interaction of these quinones with biological structures may lead to cellular damage. Symptoms of orellanine

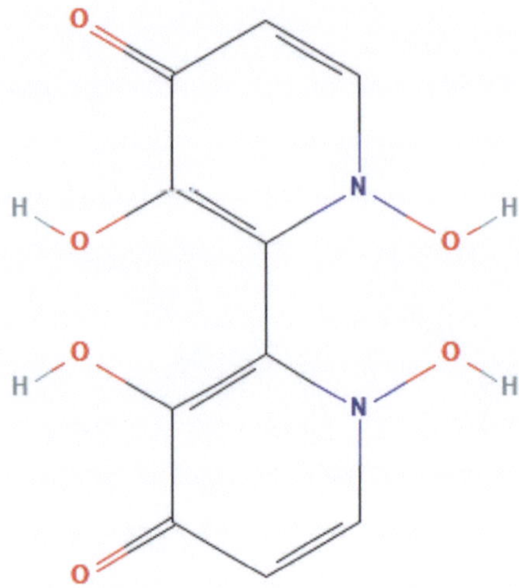

Fig. 5.6 Chemical structure of orellanine (National Center for Biotechnology Information 2024a, b, c, d, e, f, g)

poisoning may manifest 2–20 days postingestion. Nausea, stomach discomfort, and emesis are the first manifestations of poisoning (Hashimoto et al. 1987; Khani et al. 2017). Chills, severe thirst, polyuria, or oliguria, and likely anuria ensue. It is generally advised to cope with it via hemodialysis until there are improvements in renal functioning. Figure 5.6 shows the chemical structure of orellanine (National Center for Biotechnology Information 2024a, b, c, d, e, f, g). In vitro, findings demonstrate that orellanine produces radicals containing oxygen at the target location via redox cycling. Advanced knowledge of cellular processes has also indicated that a metabolite of orellanine may limit the production of proteins (Govorushko 2012).

5.6.7 Other Toxic Compounds

Lampteromyces japonicus has laterals. Lampterol poisoning causes emesis, diarrhea, and stomach pain. Mushroom fasciculol poisoning causes emesis, diarrhea, and neuromuscular paralysis (Blinova and Nedbaesvkaya 1995; Cai et al. 2016). The mushrooms that contain fasciculol include *Hypholoma fasciculare* and *Naematoloma fasciculare*. Upon intake, choline from toxic mushrooms is metabolized into acetylcholine inside the organism (Garcia et al. 2015; Ginterova et al. 2014). Choline poisoning symptoms include elevated blood pressure, reduced breathing, pupil constriction, enhanced blood flow, and heightened gastrointestinal activity. The mushrooms that contain choline include *Lactarius chrysorrheus* and *Rhodophyllus rhodopolius*.

Symptoms of poisoning from glycyrrhizin-containing mushrooms manifest 4–5 days after ingestion and last for over 1 month (Chen et al. 2009; Comelli et al. 2013). A prime instance of a glycine-containing toxic fungus causes intense discomfort in the hands and feet, resulting from peripheral neuropathy, which persists for over a month but is not life threatening (Jo et al. 2014).

5.7 Mortality Resulting from Poisonous Mushroom Ingestion

The prevalence of mushroom poisoning significantly varies globally, influenced by local customs, lifestyle, dietary variables, climate, and the availability of wild mushrooms. Table 5.5 presents data on morbidity and death across various geographies. Countries whose culinary traditions emphasize wild mushrooms have elevated incidences of mushroom poisoning. In Eastern Europe, around 150 kinds of mushrooms are collected (Musselius and Ryk 2002). The zenith of poisoning incidents occurred during and just after the Second World War, while mushrooms constituted a primary nutritional source (Grossman and Malbin 1954).

Numerous instances of mushroom poisoning on a larger scale have occurred in the past. In 1952, in the Poznan area of Poland, a simultaneous poisoning incident occurred, affecting 102 individuals, resulting in 11 fatalities. The poisoning is believed to have been caused by the fool's webcap (*C. orellanus*), which is sometimes confused with the edible Goliath webcap (*Cortinarius praestans*) and gypsy mushroom (*Cortinarius caperatus*) (Shao et al. 2016). Between 1952 and 1964, 144 instances of poisoning were documented after the ingestion of this fungus, resulting in 25 fatalities (Sun et al. 2018).

Between June and November 1981, 160 instances of mushroom poisoning were reported in northern Italy, resulting in four fatalities attributed to the consumption of *Amanita phalloides* (Fantozzi et al. 1986). Guangdong Province reported six incidences of mushroom poisoning in March 2000, resulting in 33 affected individuals and 20 fatalities; the toxic mushroom responsible for the poisoning was the Guangzhou destroying angel (*Amanita exitialis*) (Dinis-Oliveira et al. 2016). An epidemic occurred from April 28 to May 28, 2018, in 13 western and northwestern regions of Iran. Reports indicated that 1247 individuals were poisoned from consuming wild toxic mushrooms, including the common ink cap (*Coprinopsis atramentaria*), sulfur tuft (*Hypholoma fasciculare*), and deadly dapperling (*Lepiota brunneioncarnata*), resulting in 19 fatalities (Soltaninejad 2018); nonetheless, authorities attributed the poisoning to *Amanita virosa*. Mushroom poisoning is a significant issue in several Asian nations. China has the highest fatality rates, with the greatest number of fatalities happening in the provinces of Yunnan and Guizhou (Sun et al. 2018). In Nepal, the yearly fatalities resulting from the consumption of toxic mushrooms range from 15 to 20 individuals (Adhikari et al. 2005). Moreover, considerable death rates were documented in Iran (Soltaninejad 2018), Türkiye (Yardan et al. 2010), and Thailand (Parnmen et al. 2016).

Table 5.5 Reported fatalities and morbidity from various areas and nations

Location (country)	Average annual poisonings	Average number of fatalities/years	Time period	References
Bulgaria (Varna region)	120 (2014–2015)	3.9	2005–2015	Sun et al. (2018)
Belarus	60.9	1.1	1990–2015	Xiang et al. (2018)
China	326.4	70.6	2003–2015	Mikhasiova and Pleshkunova (2016)
Czechia (Prague Toxicological centre)	58.8	8.7	2004–2013	Marinov et al. (2018)
China (Guizhou	138.6	20.3	1985–2000	Soltaninejad (2018)
China (Yunnan)	660	24.5	2002–2006	Li et al. (2003)
China	145	8	1990	Jo et al. (2014)
France	5000–10,000[a]	–	1990	Pelclová and Rakovcová (1993)
Iran (Razi Hospital of Rasht)	1043	15	2018	Badsar et al. (2013)
Iran	14.6		2006–2010	Cervellin et al. (2018)
Italy	20		1995–2015	Soltaninejad (2018)
Japan	190	10	2001–2010	Yamaura (2013) Pérez-Moreno and Ferrera-Cerrato (1995)
Japan	184.2	20	1988–1997	Ishihara and Yamaura (1992)
Japan (Nagano Prefecture)	364.1	24	1959–1988	Yamaura et al. (1997), Gonmori and Yoshioka (2005)
Mexico	–	15	2002–20,065	Adhikari (2004)
Nepal	–	15–20	–	Beug (2007)
Russia	1128.3	66.5	1931	Sanz et al. (1989)
Poland			1995–2007	Schmutz et al. (2018)
Spain	10.3	1.2	1985–1990	Govorushko (2012), Boa (2004)
Switzerland	347.4[a]	–	1965–2015	Schenk-Jaeger et al. (2016)
Switzerland	375.9[a]	–	1995–2010	Schenk-Jaeger et al. (2012)
Taiwan	4.4	–	1985–1990	Yang et al. (1995), Yardan et al. (2010)
Turkey	50.8		2000–2005	Parnmen et al. (2016)
Ukraine	40.5	5.0	2005–2014	Romaniuk et al. (2012), Mengs et al. (2012)
USA	7976	25	1900–1995	Broussard et al. (2001), Kurdil et al. (2016)
USA	8214	–	2001–2011	Kintziger et al. (2011)

(continued)

Table 5.5 (continued)

Location (country)	Average annual poisonings	Average number of fatalities/years	Time period	References
USA (from 37 regional poison centers)	3419.5	–	1987–1988	Trestrail and Lampe (1990), Trestrail (1991)
USA (Florida)	85.6	0	2003–2005	Horowitz (2015)
USA and Europe	–	100–200	–	Pinson and Bradley (1996), Smith and Davis (2016)
Western Europe	–	50–90	–	Dadpour et al. (2017)
Zambia	55	5	1980–1981	Tagwireyi et al. (2016), Kavalci et al. (2010)

Between 1959 and 1988 in Japan, 2096 occurrences were documented, resulting in the poisoning of 10,924 individuals, of whom 72 succumbed. On average, there were 70 poisoning instances involving 365 victims (6.2 patients in each case), with a mortality rate of 2.4 individuals per year, and a progressive decline in the number of victims annually (Beug 2008). Poisoning mostly occurs in the northeastern region of Japan, namely in the prefectures of Hokkaido, Nagano, Niigata, Iwate, and Fukushima, with a peak incidence during September and October. The species accountable for the greatest incidence of poisonings are burnt knight (*Tricholoma ustale*) and wood pinkgel (*Rhodophyllus rhodopolius*) (Ishihara and Yamaura 1992).

Mexico possesses the greatest fatality rates attributed to mushrooms in North America. Fatalities in the United States and Canada are few (Leathem and Dorran 2007; Fleury et al. 2008; Berch et al. 2017), partly attributable to superior medical treatment. In southern Mexico, mortality rates from consuming "death caps" and "damaging angels" exceed 50%. In the United States and Canada, individuals who experience mushroom poisoning and seek immediate medical attention have a greater than 90% likelihood of survival (Méndez-Navarro et al. 2011).

Mushroom-related fatalities are also infrequent in Australia. During the period of 1999–2012, public hospitals in Sydney and Canberra documented four fatalities (Roberts et al. 2013). Poisoning due to mushrooms is also prevalent in other continents, with confirmed fatalities in South America (Meijer et al. 2007) and Africa (Tagwireyi et al. 2016); however, there is a lack of appropriate quantitative information from these regions. Establishing a dependable average rate of yearly morbidity and death worldwide is challenging. Frequently, disparate information on the same country is disseminated by several sources. Examinations of yearly deaths from mushroom poisoning in the US range from 20 individuals (Broussard et al. 2001) to 100–200 (Mengs et al. 2012). Brandenburg and Ward (2018) revealed more trustworthy statistics, noting 133,700 mushroom exposure instances (7428 per year) from 1999 to 2016. Almost 704 exposures (39 annually) led to significant damage.

Conversely, there are instances of insufficient death registration. According to official data from the Russian Health Ministry, an average of 26 fatalities occur each year due to mushroom intoxication (Public Health Service in Russia 2017). These data are significantly underestimated. Initially, a review of news Internet sites from 1995 to 2007 indicated that the death rate in Russia was 66 individuals per year (Govorushko 2012). The fatality rate from mushroom poisonings in Russia is consistently double that of Ukraine, where 31 fatalities were recorded between 2005 and 2014 (Kurdil et al. 2016). Asian countries reportedly have the greatest mortality rates due to mushroom poisoning. The United States, Türkiye, Nepal, Iran, Belarus, and Poland have elevated incidences of fatalities due to poisoning from mushrooms. The global average yearly death rate from the intake of toxic mushrooms is estimated to be between 200 and 250 individuals. Table 5.5 shows the disease and death rates due to mushroom poisoning.

Ramírez-Terrazo et al. (2021) reported that members of both examined societies correlated the ingestion of hazardous mushrooms with mortality; however, they also acknowledged that many species induce simply pain or drunkenness. Certain studies indicate that identifying undesirable symptoms is an element of conventional mycological understanding, particularly the recognition of various forms of mycetism (Lin et al. 2015a, b). Nonetheless, certain exemptions exist. The Mayan population of Chiapas delineates a typical progression: initial psychoactive intoxication, succeeded by gastrointestinal complications, tissue, and organ impairment, culminating in mortality (Duvic et al. 2003; West et al. 2009). Their findings contradict this idea since participants in their research said that various mushroom intoxications are the same, given that each mushroom possesses its distinct toxin. Thus, the induced symptoms are particular and not universally lethal. Numerous varieties of native medicines exist, demonstrating the depth of indigenous information about nonedible fungi. In Northeast India, the Karbis use 12 distinct conventional remedies for mycetisms, including maize, citrus, rice, tamarind, earth, and human excrement (Murphy 2012). Numerous studies in Central and Southern Mexico have shown that the most common traditional treatments with laxative properties are cooking oil, mezcal, lemon juice, crushed garlic, and salted water (Matsura et al. 2016). This indicates a certain level of convergence of local knowledge across various places.

5.8 Clinical Classifications of Mushroom Toxicity

Six clinical classifications of toxic conditions of mushrooms exist (Gonmori et al. 2011; Puschner 2012): (1) cytotoxic mushroom toxicity; (2) mushroom neurotoxicity; (3) myotoxic mushroom toxicity; (4) metabolic toxicity from mushrooms; (5) gastrointestinal toxicity caused by mushrooms; and (6) diversified harmful effects from mushrooms. Many are also categorized into subcategories (Wong and Ng 2006).

5.8.1 *Cytotoxic Mycological Toxicity*

This extensive category encompasses just those forms of intoxication that specifically impact an organ, resulting in either first hepatotoxicity or beginning nephrotoxicity (Marinov et al. 2018). It is categorized into three classes of mushrooms, resulting in (A1) primary toxicity of the liver, (B1) primary toxicity of the renal, or (C1) delayed primary toxicity of the renal. The A1 subgroup comprises mushrooms that possess amatoxins. They induce potentially fatal hepatotoxicity, accompanied by a distinct array of clinical manifestations (Dinis-Oliveira et al. 2016). Numerous fungus species have sufficient amatoxins to provide significant toxicity. Many Amanita species (e.g., *Amanita phalloides*) and Galerina and Lepiota species are included.

This particular type of mushroom poisoning has been the subject of several studies (Vargas et al. 2011; Ward et al. 2013; Yilmaz et al. 2015; Varvenne et al. 2015; Zhou et al. 2017). Subgroup B1 includes mushrooms that result in acute renal damage and contain aminohexadienoic acid (AHDA), including *Amanita pseudoporphyria* and *A. smithiana* (Wong and Ng 2006; Mengs et al. 2012). The toxicity associated with this group of mushrooms was elucidated by Apperley et al. (2013) and Karlson-Stiber and Persson (2003). Subgroup C1 includes orellanine-containing fungi, especially certain Cortinarius species, that cause delayed renal failure. Hedman et al. (2017) and Anantharam et al. (2016) reported poisoning from subgroup C1 mushrooms.

5.8.2 *Neurotoxic Mycological Toxicity*

This extensive category includes traditional forms of fungal toxicity that induce early neurotoxic effects. It is categorized into four subgroups: (A2) hallucinogenic mushrooms, (B2) mushrooms causing autonomic toxicity, (C2) mushrooms causing central nervous system toxicity, and (D2) morel neurologic syndrome. Subgroup A2 comprises fungi, which mostly induce hallucinations or similar symptoms. This group of fungi includes gymnopilus, psilocybin, psilocin, and related poisons. The wide variety of mushrooms involved includes Psilocybe, Conocybe, Copelandia, Gymnopilus, Panaeolus, Pluteus, and maybe Stropharia. Dydak et al. (2015) demonstrated this kind of poisoning. Fungi that directly affect vegetative processes, such as internal organs, glands that secrete both internally and externally, and blood and lymphatic arteries, are included in subgroup B2 (Saviuc et al. 2010). Muscarine and related toxins cause this; this subgroup contains certain species of Inocybe, Clitocybe, Mycena, and Boletus rubinus W.G. (Michelot and Melendez-Howell 2003).

Subgroup 2D encompasses a poorly characterized illness characterized by neurological and gastrointestinal symptoms after the consumption of certain morel mushrooms. No particular poison was administered to this subgroup, and the nature of this kind of "intoxication" remains ambiguous (Kosentka et al. 2013; Ginterova et al. 2014). Most cases were linked to the ingestion of Morchella spp. (Mikaszewska-Sokolewicz et al. 2016).

5.8.3 Mushroom Causing Myotoxicity

Intoxication with rhabdomyolysis (acute skeletal muscle necrosis) as the main symptom falls under this group. It falls under one of two mushroom subgroups, which result in (A3) myotoxicity with swift onset and (B3) myotoxicity with delayed onset (Klimaszyk and Rzymski 2018). Fungi that cause fast-onset myotoxicity, which may lead to partial or complete destruction of the red bone marrow's myeloid tissue cells, are included in subgroup 3A. Cycloprop-2-ene carboxylic acid is the causal toxin that causes poisoning when mushrooms of this subgroup are consumed (Yoshikawa et al. 2002). In these cases, Nisekurohatsu (Russula subnigricans) is one of the Russula species that are consumed (Lin et al. 2015a, b ; Cho and Han 2016; Matsura et al. 2016). Fungi that cause delayed onset myotoxicity are included in subgroup 3B. Although current research suggests that saponaceolide B and M may be the major poisons, the toxins that cause the poisoning from these mushrooms are still not fully known (Clericuzio et al. 2018; Gozzini et al. 2018).

5.8.4 Gastrointestinal Mycological Noxiousness

The following category comprises fungal species that generate a kind of "poisoning" not classified under the preceding five categories. These mushrooms cause gastrointestinal issues instead of particular toxicity, and from a practical therapeutic perspective, they are categorized under this categorization system. Mushroom intoxications that induce gastroenteritis often present moderate and brief gastrointestinal symptoms (McPartland et al. 1997; Marinov et al. 2018; Leudang et al. 2017; Brandenburg and Ward 2018; Sohn 2015).

5.8.5 Metabolic Toxicity Associated with Mushroom Poisoning

This large group causes a variety of toxic disorders and clinical symptoms; thus, it should be categorized by different clinical presentations rather than homogeneous clinical findings. Subgroup 4A fungi suppress GABA synthesis, causing metabolic problems in numerous biological systems (Wong and Ng 2006; Sohn 2015). Some Gyromitra mushrooms contain gyromitrin. Fungi that elicit alcohol-like disulfiram-like effects are included in subgroup 4B. Coprine-containing mushrooms like Coprinus species elicit this response (Saviuc and Flesch 2003; Wong and Ng 2006; Sohn 2015). Fungi in subgroup 4C induce neurological symptoms and may harm other organs. Polyporic acid-containing mushrooms like *Hapalopilus rutilans* are examples (Kraft et al. 1998; Diaz 2005a, b; Villa et al. 2013). Subgroup 4D includes fungi that induce bone marrow failure and palm, sole, and face lamellar desquamation. Trichothecenes-containing mushrooms like *Podostroma cornu-damae* exemplify it. This mushroom toxicity has been studied by Saikawa et al. (2001), Gonmori et al. (2011), and Park et al. (2016). *Trogia venenata,* a Yunnan mushroom, causes fast hypoglycemia in subgroup 4E. The unusual amino acids 2R-amino-5-hexynoic acid and 2R-amino-4S-hydroxy-5-hexynoic acid cause it, according to Shi et al. (2012). Seven French subgroup 4F poisoning victims ate Satan's bolete (*Rubroboletus satanas*) and developed hyperproinsulinemia (Kretz et al. 1991; Schenk-Jaeger et al. 2012; Patocka 2018).

Yoon et al. (2011) described an uncommon poisoning in subgroup 4G induced by the continued use of a *Ganoderma neojaponicum* herbal treatment for several days. Only two instances have been documented in Japan.

5.8.6 Various Adverse Responses Associated with Mushrooms

Subgroup A6 includes species that cause acute dermatitis after being eaten. This concerns the Shiitake mushroom (*Lentinula edodes*). The molecular pathway in this process remains unclear; however, it seems to be nonallergic and might be linked to the thermolabile polysaccharide lentinan present in these species of mushrooms (Boels et al. 2014; Mossong et al. 2015; Saupe et al. 2016). Mushrooms associated with a syndrome resembling erythromelalgia are classified within subgroup 6B. Observations were conducted following the ingestion of the paralysis funnel (*Clitocybe amoenolens*) and *C. acromelalga* (also called *Paralepistopsis acromelalga*). Taguchi et al. (2009), Miyazaki et al. (2013), and Yoshioka et al. (2017) suggested that acromeliac acid is the implicated toxin. Following repeated exposure, mushrooms associated with autoimmune hemolytic anemia are categorized within subgroup 6C. This pertains to brown roll-rim mushrooms, specifically Paxillus species, including poison pax (*Paxillus involutus*), as noted by Chen et al. (2014a, b) and Brzostowski et al. (2009, 2011). Hydrogen cyanide (HCN) poisoning resulting from the consumption of mushrooms with elevated HCN levels is believed to be the cause of subgroup 6D poisoning, characterized by encephalopathy syndrome, as reported in Japan. The mushrooms linked to this type of poisoning are *Pleurotus eringii*, *Grifola frondosa*, and *Pleurocybella porrigens* (Gonmori et al. 2011; Yamada et al. 2012; Nomura et al. 2017).

5.9 Nutritional and Medicinal Significance of Inedible Mushrooms

Before the advent of modern synthetic pharmaceuticals, ailments were addressed using natural resources, including flora, fauna, minerals, and fungi (Vunduk and Biketova 2022). One of the first evidence of mushrooms utilized for medical purposes was discovered in a frozen bag by Ötzi, the Iceman. One of the two mushrooms, *Fomitopsis betulina*, found in the frozen mummy's bag, was believed to possess therapeutic properties. It is hypothesized to serve as a treatment for intestinal parasites, although recent studies have demonstrated its health advantages extend beyond antiparasitic properties (Vunduk et al. 2015). Besides edible mushrooms, some inedible varieties exhibit bitter and pungent flavors, particularly those in the Polyporaceae family, which have been utilized as pharmacological substances in China since ancient times (Quang et al. 2006). Applying natural compounds derived from fungi in treating infectious and neoplastic disorders is a fundamental aspect of contemporary medicine. Bioactive compounds have been extracted from both edible and inedible plants, including those from the Paxillaceae, Polyporaceae, Thelephoraceae, Xylariaceae, and Scutigeraceae families (Quang et al. 2006). The documented bioactivities of mushrooms encompass antiatherogenic, anti-inflammatory, antitumor, immunosuppressive, antifungal, antiallergic, antibacterial, hypoglycemic, cytostatic, antioxidant, and antiviral, as well as hepatoprotective properties (Wasser and Weis 1999; Lindequist et al. 2005; Reis et al. 2011).

The nutrient and antioxidant profile of biomolecules in inedible mushroom species was nearly identical to that of edible mushrooms (Heleno et al. 2010; Grangeia et al. 2011). Inedible mushrooms contain significant quantities of energetic biomolecules, such as the monosaccharide alcohol derivative mannitol, as well as the fatty acids, oleic, linoleic, and palmitic acids, as well as the disaccharide trehalose (Reis et al. 2011). Additionally, they possess a substantial quantity of polyunsaturated fatty acids (PUFAs) and monounsaturated fatty acids (MUFAs) (Grangeia et al. 2011; Reis et al. 2011). Figure 5.7 illustrates the quantities of MUFAs and PUFAs in several inedible mushroom species, as documented by Reis et al. (2011).

Numerous chemical compounds of inedible mushrooms have been gathered globally in recent years. Certain isolated metabolites exhibited significant pharmacological properties, including bitter drimane sesquiterpenoids and cryptoporic acids derived from *Cryptoporus volvatus*, which demonstrated the capacity to inhibit superoxide anion radical release, skin, and colon cancer progression in murine models, as well as antihuman immunodeficiency virus (HIV)-1 activity. Inedible mushrooms possess many flavonoids, ascorbic acid, phenolics, tocopherols, and carotenoids. *Fomitopsis pinicola* exhibited a significantly elevated concentration of phenolics and robust antioxidant capabilities, particularly in reducing

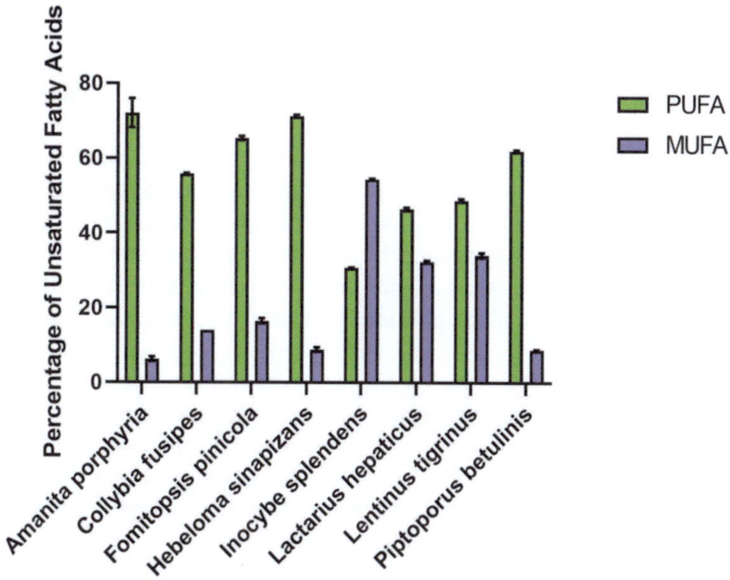

Fig. 5.7 Quantities of MUFAs and PUFAs in several inedible mushroom species

power. It may have uses in the protection from the disorders associated with free radicals, which lead to inflammation and apoptotic cell death (Reis et al. 2011; Ghafoor et al. 2024; Salar et al. 2024).

Alongside antioxidants and nutrients, the poisons found in mushrooms are equally significant. These toxins may serve as instruments for study in cell biology, developmental biology, and structural biology (He et al. 2022). Amanitin selectively binds to eukaryotic RNA polymerase II, suppressing transcription at nanomolar doses (Brueckner and Cramer 2008; Samwer et al. 2013). Additionally, psilocybin, a hallucinogenic compound derived from mushrooms, is extensively utilized as a psychedelic in human research, particularly in neurobiological studies (Hasler et al. 2004; Johnson et al. 2008; Carhart-Harris and Goodwin 2017). It functions as an agonist that targets serotonin-2A receptors, facilitating the comprehension of the connection between serotonin and psychotic states generated by drugs or disorders (Geyer and Vollenweider 2008; Petri et al. 2014). Moreover, psilocybin may effectively treat treatment-resistant severe depression as well as obsessive–compulsive disorder (Moreno et al. 2006; Carhart-Harris et al. 2016; Carhart-Harris and Goodwin 2017). Consequently, inedible mushrooms provide significant therapeutic value, serving as a viable source of antioxidants and minerals that can be isolated. Furthermore, their toxins can be employed in medical therapies such as cancer treatment.

5.10 Prevention of Mushroom Poisoning

This guideline serves as a fundamental set of instructions for managing situations of unknown mushroom consumption and is not exhaustive. Exceptions to the outlined rules may exist, and users should consult unique practice instructions, alternative treatment manuals, or more comprehensive literature to specifically handle detected mushroom poisoning cases. Antidotal therapy is limited for fungal intoxications; hence, the primary treatments for poisoning are symptomatic, decontamination, supportive care, clinical monitoring, as well as diligent nursing care. Comprehensive treatment of the poisoned patient: (1) stabilization and thorough patient assessment.

Implement standard clinical interventions to avert further decline in the patient's condition. Fortunately, fungal poisoning is not a common life-threatening condition; if airway, breathing, or perfusion are impaired, use appropriate resuscitative techniques. In instances of fatal poisoning, the first assessment of the patient must be conducted concurrently with resuscitation. Following the stabilization of vital signs, a precise history aids in directing subsequent care.

5.10.1 Standard Inquiries Should Encompass

(1) Which fungus or mushrooms did you want to collect? (2) What kind of tree was it situated near? (3) Was it thriving in the forest, meadow, or the garden? (4) On which substrate was the fungus growing? (5) Did the fungus emit an odor? What was its size and color? Did it possess gills or a "sponge" texture? (6) How many species of fungi were ingested? (7) How many individuals consumed the fungus? (8) Was the fungus consumed throughout many meals (please specify the precise ingestion periods)? (9) What was the duration between exposure and the onset of symptoms? (10) In what manner was the fungus prepared (raw or cooked)? (11) In what manner was the fungus preserved from the time of collection till preparation? (12) How was the preprepared fungus kept before ingestion? (13) Was any alcohol ingested with the meal? (14) Are all individuals who ingested the fungus unwell? (15) Are there additional unwell individuals in the group who did not ingest the fungi? (16) Has the patient previously consumed this fungus? (17) Were the mushrooms purchased, and if yes, from where?

Should the mushrooms be accessible and the diagnosis remains ambiguous, seek analysis from a mycologist; nonetheless, initiate treatment promptly if severe poisoning is suspected. In some locations, the "newspaper" test (which requires the newspaper to contain lignin) is employed: compress the mushroom against the newspaper, allow it to air dry, and thereafter pour a single drop of hydrochloric acid (HCl). The chapter might display a blue or bluish-green tint, signifying the existence of amatoxins. The test is not wholly conclusive.

5.10.2 Identification of Fungi

Precise fungus identification (genus and species) facilitates proper treatment guidance. In managing a patient poisoned by a fungus, it is primarily more critical to ascertain what the fungus is not rather than identifying it. Only a limited number of fungus species possess toxins (e.g., Amantidin, orellanine, gyromitrin) that need therapy beyond symptomatic relief. A qualified mycologist should be contacted to exclude the more hazardous fungus species from consideration.

5.10.3 Decrease in Absorption

The primary method for reducing absorption is the use of activated charcoal. Due to the rapid absorption of toxins, it should be taken before the induction of gastric lavage or emesis with a safe dose (adults: 50–100 g and children: 10–25 g). Generally, emesis or lavage eliminates only around one-third of ingested substances; hence, an adequate dosage of activated charcoal should be more effective. Conduct a sample for microscopic study of gastrointestinal contents before the administration of charcoal, if feasible.

5.10.4 Augmentation of Elimination

Dialysis and hemoperfusion may be warranted in a limited number of fungal intoxications (e.g., cyclopeptides, orellanine), but their efficacy is contingent upon early intervention in the poisoning process.

5.10.5 Specific Antifungal Antidotes

Limited antidotes exist for fungal intoxication. Numerous proposed antidotes have not shown efficacy in controlled clinical trials but continue to have widespread advocates.

5.10.6 Principal Fungal Disorders and Their Management

Symptomatic therapy for the patient should start before the diagnosis of the specific fungus. Understanding particular disorders associated with fungal toxicity is essential. It is necessary to recognize that numerous factors may contribute to symptoms that are NOT directly associated with an exogenous fungal toxin: panic responses (anxiety over potential ingestion of a highly toxic fungus), digestive difficulties with fungi, bacterial contamination (such as Salmonella, Staphylococcus, or, in severe cases, botulism), consumption of raw fungi (many edible species can induce gastrointestinal symptoms if improperly prepared), individualized allergic reactions, or, infrequently, pesticide residues. It is essential to note that when addressing fungal toxins, a longer period (often exceeding 6–12 h) between consumption and the manifestation of symptoms correlates with the severity of the fungal toxin (e.g., orellanine, amatoxin, gyromitrin). Prolonged risk of acute poisoning.

5.10.7 Ongoing Care

Following diagnosis and early treatment, sufficient observation time must be instituted, particularly in cases of severe poisoning. Psychiatric assessment and counseling are warranted in instances of deliberate poisoning (e.g., suicide, drug misuse). Education is advised in cases of accidental fungal poisoning to prevent further poisoning occurrences. Preventive and instructional resources are accessible at many poison centers.

5.11 Recommendations for Risk Assessment

5.11.1 Accidental Pediatric Ingestion History

Typically under 4 years of age. Normally, it is a minor consumption of a "nibble." A significant historical clue is that the patient was not administered the mushrooms. Initiation of symptoms: no emesis—minimal to negligible risk emesis occurring within 6 h postingestion—minimal risk vomiting commences after 6 h—increased risk.

Suggestion
The majority of these individuals may be safely released from the emergency room. If the patient exhibits delayed gastrointestinal symptoms (more than 6 h postingestion), evaluating AST/ALT levels and identifying the mushroom involved may be prudent.

Asymptomatic
Evaluate the dosage of the medication administered to the patient.

Evaluate baseline hepatic enzymes.

Reassess liver enzymes in 12–24 h; this may be done outpatient. Advise the patient to return if symptoms arise, including vomiting or stomach discomfort.

Contemplate transmitting digital images of mushrooms for mycological identification.

Symptomatic

- Intravenous fluids and antiemetic medicines.
- Evaluate the dosage of the medication. Evaluate baseline hepatic enzymes.
- Contemplate transmitting digital photographs of mushrooms for mycological identification. Conduct serial liver enzyme assessments at 12-h intervals.
- If these parameters stay normal and the patient shows symptomatic improvement, they may be discharged with instructions to return if symptoms worsen or new symptoms arise. Reassess liver enzymes in 24 h. If liver enzymes are high at any point, intake should be managed immediately, such as in hepatotoxic mushroom poisoning.

5.12 Conclusion

Taken together, inedible mushrooms can be fatal if consumed accidentally without properly identifying them. They contain some hazardous toxicants that can damage the different organs of the body, leading to mortality. However, certain beneficial compounds are present in them, including vitamins, unsaturated fatty acids, antioxidants, and flavonoids. These compounds provide significant medicinal importance to inedible mushrooms. The compounds from inedible mushrooms are used in different research fields, including structural, developmental, and cell biology. However, there is a dire need for further research on inedible and poisonous mushrooms so that people can identify them easily and avoid eating them. Moreover, advanced research on the applications of bioactive substances from these mushrooms can also be of significant importance in enhancing our lifestyle.

References

Adhikari MK. Mushroom poisoning and its state in Nepal. Plant Resour Bull. 2004;25:56–8.

Adhikari MK, Devkota S, Tiwari RD. Ethnomycolgical knowledge on uses of wild mushrooms in western and Central Nepal. Our Nat. 2005;3:13–9. https://nepjol.info/index.php/ON/article/viewFile/329/321

Agnihotri S, Si A. Solega ethno-ornithology. J Ethnobiol. 2012;32:185–211. https://doi.org/10.2993/0278-0771-32.2.185.

Al-Mejibli, Abd DH. Mushroom diagnosis assistance system based on machine learning by using mobile devices. In: Intisar Shadeed AlMejibli University of Information Technology and Communications Dhafar Hamed Abd Al-Maaref University College, vol. 9, no. 2; 2017. p. 103–13.

Ammirati JF, Traquair JA, Horgen PA. Poisonous Mushrooms of the Northern United States and Canada. Minneapolis: University of Minnesota Press; 1985.

Anantharam P, Shao D, Imerman PM, Burrough E, Schrunk D, Sedkhuu T, Tang SS, Rumbeiha W. Improved tissue-based analytical test methods for orellanine, a biomarker of Cortinarius mushroom intoxication. Toxins. 2016;8(5):158. https://doi.org/10.3390/toxins8050158.

Apperley S, Kroeger P, Kirchmair M, Kiaii M, Holmes DT, Garber I. Laboratory confirmation of Amanita smithiana mushroom poisoning. Clin Toxicol. 2013;51(4):249–51. https://doi.org/10.3109/15563650.2013.778995.

Arora D. Mushrooms demystified. California, USA: Ten Speed Press; 1996.

Arora D. California porcini: three new taxa, observations on their harvest, and the tragedy of no commons. Econ Bot. 2008;62(3):356–75. https://doi.org/10.1007/s12231-008-9050-7.

Asakawa Y, Hashimoto T, Mizuno Y, Tori M, Fukazawa Y. Cryptoporic acids AG, drimane-type sesquiterpenoid ethers of isocitric acid from the fungus Cryptoporus volvatus. Phytochemistry. 1992;31(2):579–92.

Aygul N, Duzenli MA, Ozdemir K, Altunkeser BB. A case report of unusual complication of Amanita phalloides poisoning: development of cardiogenic shock and its successful treatment with intra-aortic balloon counterpulsation. Toxicon. 2010;55:630–2.

Badsar A, Taramsari MR, Alireza AM, Rad MR, Chatrnour G, Jahromi SK. Mushroom poisoning in the southwest region of the Caspian Sea, Iran: a retrospective study. Iran J Toxicol. 2013;7:798–803.

Barman B, Warjri S, Lynrah KG, Phukan P, Mitchell ST. Amanita nephrotoxic syndrome: presumptive first case report on the Indian subcontinent. Indian J Nephrol. 2018;28(2):170–2. https://doi.org/10.4103/ijn.IJN_353_16.

Basumatary M, Gogoi M. Uses of wild edible macrofungi by Bado community of Kokrojha district, Assam, India. Trop Plant Res. 2016;3:176–81.

Bédry R, Saviuc P. Severe mushroom poisoning excluding Amanita phalloides syndrome. Reánimation. 2002;11:524–32.

Bedry R, Baudrimont I, Deffieux G, Creppy EE, Pomies JP, Ragnaud JM, Dupon M, Neau D, Gabinski C, De Witte S, et al. Wild-mushroom intoxication as a cause of rhabdomyolysis. N Engl J Med. 2001;345:798–802.

Berch SM, Kroeger P, Finston T. The death cap mushroom (*Amanita phalloides*) moves to a native tree in Victoria, British Columbia. Botany. 2017;95(4):435–40. https://doi.org/10.1139/cjb-2016-0183.

Berne S, Sepci K, Anderluh G, Turk T, Macek P, Poklar UN. Effect of pH on the pore forming activity and conformational stability of ostreolysin, a lipid raft-binding protein from the edible mushroom Pleurotus ostreatus. Biochemistry. 2005;44:11137–47.

Beug MW. NAMA toxicology committee report for 2006: recent mushroom poisonings in North America. 2007. http://www.namyco.org/docs/2006ToxReport.pdf.

Beug MW. Toxicology: reflections on mushroom poisoning in north America. Fungi. 2008;1–2:42–4.

Blinova L, Nedbaesvkaya N. Leningrad NPP radioecological monitoring. NIMB. 1995;2–3:24–30. (In Russian)

Boa E. Wild edible fungi: a global overview of their use and importance to people. United Nations: Food & Agriculture Org; 2004.

Boels D, Landreau A, Bruneau C, Garnier R, Pulce C, Labadie M, de Haro L, Harry P. Shiitake dermatitis recorded by French Poison Control Centers – new case series with clinical observations. Clin Toxicol. 2014;52(6):625–8. https://doi.org/10.3109/15563650.2014.923905.

Boyden S. Evolution and health. Ecologist. 1973;3(8):304–9.

Brandenburg WE, Ward KJ. Mushroom poisoning epidemiology in the United States. Mycologia. 2018;110(4):637–41. https://doi.org/10.1080/00275514.2018.1479561.

Brent J, Palmer RB. Mushrooms. In: Shannon MW, Borron SW, Burns MJ, editors. Haddad and Winchester's clinical management of poisoning and drug overdose. 4th ed. Philadelphia: Saunders Elsevier; 2007. p. 455–72.

Bresinsky A, Besl H. A colour atlas of poisonous fungi: a handbook for pharmacists, doctors, and biologists. London: Wolfe Publishing Ltd; 1990.

Briggler M. A guide to Missouri's edible and poisonous mushrooms. Missouri Department of Conservation; 2018.

Broussard CN, Aggarwal A, Lacey SR, Post AB, Gramlich T, Henderson JM, Younossi ZM. Mushroom poisoning—from diarrhea to liver transplantation. Am J Gastroenterol. 2001;96(11):3195–8.

Brueckner F, Cramer P. Structural basis of transcription inhibition by a-amanitin and implications for RNA polymerase II translocation. Nat Struct Mol Biol. 2008;15(8):811e818.

Brzostowski A, Bielawski L, Orlikowska A, Plichta S, Falandysz J. Instrumental analysis of metals profile in poison Pax (Paxillus involutus) collected at two sites in bory tucholskie. Chem Anal. 2009;54(6):1297–308.

Brzostowski A, Falandysz J, Jarzynska G, Zhang D. Bioconcentration potential of metallic elements by Poison Pax (Paxillus involutus) mushroom. J Environ Sci Health A Toxic Hazard Subst Environ Eng. 2011;46(4):378–93. https://doi.org/10.1080/10934529.2011.542387.

Cai Q, Cui Y-Y, Yang ZL. Lethal Amanita species in China. Mycologia. 2016;108(5):993–1009.

Cano-Estrada A, Romero-Bautista L. Valor económico, nutricionaly medicinal de hongos comestibles silvestres Economic, nutritional and medicinal value of edible wild mushrooms. Rev Chil Nutr. 2016;43:75–80. https://doi.org/10.4067/S0717-75182016000100011.

Carhart-Harris RL, Goodwin GM. The therapeutic potential of psychedelic drugs: past, present, and future. Neuropsychopharmacology. 2017;42(11):2105e2113.

Carhart-Harris RL, Bolstridge M, Rucker J, Day CM, Erritzoe D, Kaelen M, Bloomfield M, Rickard JA, Forbes B, Feilding A. Psilocybin with psychological support for treatment-resistant depression: an open-label feasibility study. Lancet Psychiatry. 2016;3(7):619e627.

Cary E, editor. Dio Cassius: roman history. London: Heinemann; 1925.

Cassidy N, Duggan E, Tracey JA. Mushroom poisoning in Ireland: the collaboration between the national poisons information centre and expert mycologists. Clin Toxicol. 2011;49(3):171–6. https://doi.org/10.3109/15563650.2011.560854.

Cervellin G, Comelli I, Rastelli G, Sanchis-Gomar F, Negri F, De Luca C, Lippi G. Epidemiology and clinics of mushroom poisoning in Northern Italy: a 21-year retrospective analysis. Hum Exp Toxicol. 2018;37(7):697–703. https://doi.org/10.1177/0960327117730882.

Chang ST, Miles PG. Mushroom biology: a new discipline. Mycologist. 1992;6:64–5.

Chaudhary S, Chaurasia RK, Patel S, Agrawal KK, Aswani R, Jaiswal NK. Clinical profile and outcome of patients presenting with mushroom poisoning in a tertiary care center of eastern Nepal. JNMA J Nepal Med Assoc. 2013;52(8):543–8.

Chen X-H, Zhou H-B, Qiu G-Z. Analysis of several heavy metals in wild edible mushrooms from regions of China. Bull Environ Contam Toxicol. 2009;83:280–5.

Chen Z, Zhang P, Zhang Z. Investigation and analysis of 102 mushroom poisoning cases in Southern China from 1994 to 2012. Fungal Divers. 2014a;64:123–31. https://doi.org/10.1007/s13225-013-0260-7.

Chen Z, Zhang P, Zhan Z. Investigation and analysis of 102 mushroom poisoning cases in Southern China from 1994 to 2012. Fungal Divers. 2014b;64(1):123–31.

Cho JT, Han JH. A case of mushroom poisoning with Russula subnigricans: development of rhabdomyolysis, acute kidney injury, cardiogenic shock, and death. J Korean Med Sci. 2016;31(7):1164–7. https://doi.org/10.3346/jkms.2016.31.7.1164.

Chowdhury D, Ojha S. An empirical study on mushroom disease diagnosis: a data mining approach. Int Res J Eng Technol (IRJET). 2017;4(1):529–34.

Clericuzio M, Mellerio GG, Finzi PV, Vidari G. Secondary metabolites isolated from tricholoma species (basidiomycota, tricholomatacee): a review. Nat Prod Commun. 2018;13(9):1213–24.

Comelli I, Lippi G, De Blasio A, Cervellin G. Accidental mushroom poisoning mimicking stroke. A case report and literature review. Acta Biomed. 2013;84(3):229–33.

Curtis DR, Lodge D, McLennan H. The excitation and depression of spinal neurones by ibotenic acid. J Physiol. 1979;291:19–28.

Dadpour B, Tajoddini S, Rajabi M, Afshari R. Mushroom poisoning in the northeast of Iran, a retrospective 6-yearepidemiologic study. Emergency. 2017;5(1):e23.

Das K, Lamo A, Paul D, Jha LK. Ethnomycological knowledge on wild edible mushroom of Khasi Tribes of Meghalaya, North-Eastern India. Indian J Nat Prod Resour. 2014;2:3433–43.
Dehay MH, de Sainte MF, Assez N, Dherbecourt V, Goldstein P. Syndrome muscarinique par ingestion de champignon: à propos de deux cas dont un mortel. Eur J Emerg. 2009;22:18–23.
Diaz JH. Syndromic diagnosis and management of confirmed mushroom poisonings. Crit Care Med. 2005a;33:427–36.
Diaz JH. Syndromic diagnosis and management of confirmed mushroom poisonings. Crit Care Med. 2005b;33:427–36.
Diaz JH. The evolving environmental epidemiology of mushroom poisoning. Am J Epidemiol. 2006;163(suppl_11):1. https://doi.org/10.1093/aje/163.suppl_11.S122-d.
Diaz JH. Mistaken mushroom poisonings. Wilderness Environ Med. 2016a;27(2):330–5.
Diaz JH. Mistaken mushroom poisonings. Wilderness Environ Med. 2016b;27:330–5.
Dinis-Oliveira RJ, Soares M, Rocha-Pereira C, Carvalho F. Human and experimental toxicology of orellanine. Hum Exp Toxicol. 2016;35(9):1016–29. https://doi.org/10.1177/0960327115613845.
Duvic C, Hertig A, Herody M, Dot JM, Didelot F, Giudicelli CP, et al. Acute renal failure following ingestion of Cortinarius orellanus in 12 patients. Initial presentation and progress over a period of 13 years. Presse Med. 2003;32:249–53.
Dydak K, Sliwinska-Mosson M, Milnerowicz H. Psilocybin – public available psychodysleptic. Postep Hig Med Dosw. 2015;69:986–95. (In Polish)
Eden PT, editor. Seneca: Apocologyntosis. Cambridge: Cambridge University Press; 1990.
Ellen R. Ethnomycology among the Nuaulu of the Moluccas: putting Berlin's "general principles" of ethnobiological classification to the test. Econ Bot. 2008;62:483–96. https://doi.org/10.1007/s12231 008 9036-5.
Ennecker-Jans AS, Van Daele PL, Blonk MI, Varin DS, Van Laar JA. Amatoxin poisoning due to soup from personally picked deathcap mushrooms (Amanita phalloides). Ned Tijdschr Geneeskd. 2007;151:764–8.
Eren SH, Demirel Y, Serdal U, Can A, Korkmaz I, Güven FMK. Mushroom poısonıng: retrospectıve analysıs of 294 cases. Clinics. 2010a;65(5):491–6. (Eren et al. 2010)
Eren SH, Demirel Y, Ugurlu S, Korkmaz I, Aktas C, Güven FM. Mushroom poisoning: retrospective analysis of 294 cases. Clinics. 2010b;65:491–6.
Erenler AK, Dogan T, Kocak C, Ece Y. Investigation of toxic effects of mushroom poisoning on the cardiovascular system. Basic Clin Pharmacol. 2016;119(3):317–21.
Eyad Sameh A, Khaled AM, Mohamad M, Mohannad G, Bassem SA, Samy S. Prediction of whether mushroom is edible or poisonous using back-propagation neural network. Int J Acad Appl Res (IJAAR). 2019;3(2):1–8.
Fantozzi R, Ledda F, Caramelli L, Moroni F, Blandina P, Masini E, Botti P, Peruzzi S, Zorn M, Mannaioni PF. Clinical findings and follow-up evaluation of an outbreak of mushroom poisoning survey of Amanita phalloides poisoning. Klin Wochenschr. 1986;64:38–43.
Feldman LH. (transl). Josephus: Antiquities. Cambridge, MA: Harvard University Press; 1965.
Felix H, Grimberg U, Benz MA, Huber T, Vollenweider FX. Acute psychological and physiological effects of psilocybin in healthy humans: a double-blind, placebo-controlled dose-effect study. Psychopharmacology. 2004;172(2):145e156.
Filippova IA. Mushrooms-healers. Moscow: Tsentrpoligraf; 2018. (In Russian).
Fleury MD, Stratton J, Tinga C, Charron DF, Aramini J. A descriptive analysis of hospitalization due to acute gastrointestinal illness in Canada, 1995–2004. Can J Public Health. 2008;99:489–93.
Ford RI. History of ethnobiology. In: Anderson E, Pearsall D, Eugene Hunn NT, editors. Ethnobiology. Wiley-Blackwell; 2011. p. 420.
Fu X, Fu B, He Z, Gong M, Li Z, Chen Z. Acute renal failure caused by Amanita oberwinklerana poisoning. Mycoscience. 2017;58:21–7.
Garcia J, Costa VM, Carvalho A, Baptista P, de Pinho PG, Bastos MD, Carvalho F. Amanita phalloides poisoning: mechanisms of toxicity and treatment. Food Chem Toxicol. 2015;86:41–55.

Gawlikowski T, Romek M, Satora L. Edible mushroom-related poisoning: a study on circumstances of mushroom collection, transport, and storage. Hum Exp Toxicol. 2015;34(7):718–24. https://doi.org/10.1177/0960327114557901.

Geyer MA, Vollenweider FX. Serotonin research: contributions to understanding psychoses. Trends Pharmacol Sci. 2008;29(9):445e453.

Ghafoor N, Mehar T, Batool M, Salar MZ, Ahmed MZ, Atique U. Attenuative effects of poncirin against polyethylene microplastics-prompted hepatotoxicity in rats. J King Saud Univ Sci. 2024;36(10):103475.

Ginterova P, Sokolova B, Sevcik J, Maier V. Determination of mushroom toxins ibotenic acid, muscimol and muscarine by capillary electrophoresis coupled with electrospray tandem mass spectrometry. Talanta. 2014;125:242–7. https://doi.org/10.1016/j.talanta.2014.03.019.

Goldfrank LR. Mushrooms. In: Nelson LS, Lewin NA, Howland MA, Hoffman RS, Goldfrank LR, Flomenbaum NE, editors. Goldfrank's toxicologic emergencies. 9th ed. New York: McGraw-Hill; 2011. p. 1522.

Gonmori K, Yoshioka N. Mushroom toxins. In: Suzuki O, Watanabe K, editors. Drugs and poisons in humans: a handbook of practical analysis. Berlin: Springer; 2005. p. 469–80.

Gonmori K, Fujita H, Yokoyama K, Watanabe K, Suzuki O. Mushroom toxins: a forensic toxicological review. Forensic Toxicol. 2011;29(2):85–94. https://doi.org/10.1007/s11419-011-0115-4.

Govorushko SM. Natural processes and human impacts: interaction between humanity and the environment. Dordrecht: Springer; 2012.

Govorushko S, Rezaee R, Dumanov J, Tsatsakis A. Poisoning associated with the use of mushrooms: a review of the global pattern and main characteristics. Food Chem Toxicol. 2019a;128:267–79.

Govorushko S. et al. Food and chemical toxicology 128. 2019b. 267–279 276 821250-overview. Accessed 3 Aug 2018.

Gozzini D, Mellerio GG, Gilardoni G, Clericuzio M, Vidari G. New terpenoids from Tricholoma saponaceum. Nat Prod Commun. 2018;13(9):1097–100.

Graeme KA. Mycetism: a review of the recent literature. J Med Toxicol. 2014;10:173–89.

Grangeia C, Heleno SA, Barros L, Martins A, Ferreira ICFR. Effects of trophism on nutritional and nutraceutical potential of wild edible mushrooms. Food Res Int. 2011;44:1029–35.

Grossman CM, Malbin B. Mushroom poisoning: a review of the literature and report of two cases caused by a previously undescribed species. Ann Intern Med. 1954;40(2):249–59. https://doi.org/10.7326/0003-4819-40-2-249.

Guillen J, Baeza A. Radioactivity in mushrooms: a health hazard? Food Chem. 2014;154:14–25. https://doi.org/10.1016/j.foodchem.2013.12.083.

Guzmán G. The hallucinogenic mushrooms: diversity, traditions, use and abuse with special reference to the genus Psilocybe. In: Fungi from different environments; 2009. p. 256–77.

Hajicek-Dobberstein S. Soma siddhas and alchemical enlightenment: psychedelic mushrooms in Buddhist tradition. J Ethnopharmacol. 1995;48(2):99–118.

Hall IR, Buchanan PK, Cole AL, Yun WS. Stephenson Edible and poisonous mushrooms of the world, vol. 103. Timber Press Portland; 2003.

Halpern GM. Medicinal mushrooms. Prog Nutr. 2010;12:29–36.

Hashimoto T, Asakawa Y. Biologically active substances of Japanese inedible mushrooms. Heterocycles. 1998;2(47):1067–110.

Hashimoto T, Tori M, Mizuno Y, Asakawa Y. Cryptoporic acids A and B, novel bitter drimane sesquiterpenoid ethers of isocitric acid, from the fungus Cryptoporus volvatus. Tetrahedron Lett. 1987;28(50):6303–4.

Hawksworth DL. Mushrooms: the extent of the unexplored potential. Int J Med Mushrooms. 2001;3:333–7.

He MQ, Wang MQ, Chen ZH, Deng WQ, Li TH, Vizzini A, Jeewon R, Hyde KD, Zhao RL. Potential benefits and harms: a review of poisonous mushrooms in the world. Fungal Biol Rev. 2022;42:56–68.

Hedman H, Holmdahl J, Mölne J, Ebefors K, Haraldsson B, Nyström J. Long-term clinical outcome for patients poisoned by the fungal nephrotoxin orellanine. BMC Nephrol. 2017;18:121. https://doi.org/10.1186/s12882-017-0533-6.

Heleno SA, Barros L, Sousa MJ, Martins A, Ferreira ICFR. Tocopherols composition of Portuguese wild mushrooms with antioxidant capacity. Food Chem. 2010;119:1443–50.

Holsen DS, Aarebrot S. Poisonous mushrooms, mushroom poisons and mushroom poisoning. A review. Tidsskr Nor Laegeforen. 1997;117(23):3385–8.

Horowitz BZ. Mushroom toxicity. 2015. http://emedicine.medscape.com/article/

Hyde KD, AlHatmi MSA, Andersen B, Boekhout T, Buzina W, Dawson TL, Eastwood DC, Jones EBG, de Hoog S, Kang Y, Longcore JE, McKenzie EHC, Meis JF, PinsonGadais L, Rathnayaka AR, RichardForget F, Stadler M, Theelen B, Thongbai B, Tsui CKM. The world's ten most feared fungi. Fungal Divers. 2018;93(1):161–94.

Image by 1benjones – Own work, CC BY-SA 3.0, https://commons.wikimedia.org/w/index.php?curid=20691162.

Image by Alan Rockefeller – Own work, CC BY-SA 4.0, https://commons.wikimedia.org/w/index.php?curid=95423433.

Image by Bernypisa – Own work, CC BY-SA 3.0, https://commons.wikimedia.org/w/index.php?curid=4821886.

Image by Dominicus Johannes Bergsma – Own work, CC BY-SA 4.0, https://commons.wikimedia.org/w/index.php?curid=84242752.

Image by Drew Henderson (Hendre17), CC By SA 3.0, https://mushroomobserver.org/images/1672908?q=1vQ9C.

Image by Hankwang, CC BY 1.0, https://commons.wikimedia.org/w/index.php?curid=507182.

Image by Irene Andersson (irenea), CC By SA 3.0, https://mushroomobserver.org/24743?q=1xRmb.

Image by Irma Zettur, licensed under http://creativecommons.org/licenses/by-nc/4.0/.

Image by John Carl Jacobs (JCJacobs) at Mushroom Observer, CC BY-SA 3.0, https://commons.wikimedia.org/w/index.php?curid=20555377.

Image by Jörg Hempel, CC BY-SA 2.0 de, https://commons.wikimedia.org/w/index.php?curid=4727638.

Image by Monserrat (MonserratM), CC By SA 4.0, https://mushroomobserver.org/images/1724764?q=1oaSB.

Image by purplecloud, CC By SA 3.0, https://mushroomobserver.org/images/1714444?q=1oaSB.

Image by Ron Pastorino (Ronpast), CC By SA 3.0, https://mushroomobserver.org/images/1726691?q=1oaSB.

Image by Strobilomyces – Self-photographed, CC BY-SA 3.0, https://commons.wikimedia.org/w/index.php?curid=1074447.

Image by Strobilomyces – Self-photographed, CC BY-SA 3.0, https://commons.wikimedia.org/w/index.php?curid=1235165.

Ishihara Y, Yamaura Y. Descriptive epidemiology of mushroom poisoning in Japan. Nihon Eiseigaku Zasshi. 1992;46(6):1071–8. (In Japanese)

Iwafuchi YA, Morita T, Kobayashi H, Kesuga K, Ito K, Nakagawa O, et al. Delayed onset acute renal failure associated with Amanita pseudoporphyria Hongo ingestion. Int Med. 2003;42(1):78–81.

Jo W-S, Hossain MA, Park S-C. Toxicological profiles of poisonous, edible, and medicinal mushrooms. Mycobiology. 2014;42(3):215–20.

Johnson MW, Richards WA, Griffiths RR. Human hallucinogen research: guidelines for safety. J Psychopharmacol. 2008;22(6):603e620.

Karaseva EI, Butvilovsky VE. Poisonous mushrooms and plants. Minsk: BSMU; 2014. ISBN 978-985-528-835-8 (In Russian)

Karimi G, Razavi BM. Poisonous mushrooms. In: Gopalakrishnakone P, editor. Clinical toxinology in Asia Pacific and Africa. Dordrecht: Springer Science+Business Media; 2015. p. 587–608.

Karlson-Stiber C, Persson H. Cytotoxic fungi—an overview. Toxicon. 2003;42:339–49.

Kavalci C, Kavalci G, Guzel A, Inal M, Durukan P, Karasalihoglu S, Cevik Y, Sut N, Sayhan MB. Poisoning from wild mushroom in the Trakya region: an analysis of seven years. Hong Kong J. Emerg Med. 2010;17(4):341–6.

Khani R, Moudi M, Khojeh V. Contamination level, distribution and health risk assessment of heavy and toxic metallic and metalloid elements in a cultivated mushroom Pleurotus florida (Mont.) Singer. Environ Sci Pollut Res. 2017;24:4699–708.

Kieslichova E, Frankova S, Protus M, Merta D, Uchytilova E, Fronek J, Sperl J. Acute liver failure due to Amanita phalloides poisoning: therapeutic approach and outcome. Transplant Proc. 2018;50(1):192–7. https://doi.org/10.1016/j.transproceed.2017.11.032.

Kim T, Lee D, Lee JH, Lee Y-S, Oh BJ, Lim KS, Kim WY. Predictors of poor outcomes in patients with wild mushroom-induced acute liver injury. World J Gastroenterol. 2017;23(7):1262–7. https://doi.org/10.3748/wjg.v23.i7.1262.

Kintziger KW, Mulay P, Watkins S, Schauben J, Weisman R, Lewis-Younger C, Blackmore C. Wild mushroom exposures in Florida, 2003–2007. Pub Health Rep. 2011;126(6):844–52. https://doi.org/10.1177/003335491112600610.

Klimaszyk P, Rzymski P. The yellow knight fights back: toxicological, epidemiological, and survey studies defend edibility of Tricholoma equestre. Toxins. 2018;10:468–9. https://doi.org/10.3390/toxins10110468.

Kosentka P, Sprague SL, Ryberg M, Gartz J, May AL, Campagna SR, Matheny PB. Evolution of the toxins muscarine and psilocybin in a family of mushroom-forming fungi. PLoS One. 2013;8(5):e64646. https://doi.org/10.1371/journal.pone.0064646.

Kotowski MA, Pietras M, Luczaj L. Extreme levels of mycophilia documented in Mazovia, a region of Poland. J Ethnobiol Etnomed. 2019;15:12. https://doi.org/10.1186/s13002-0190291-6.

Kraft J, Bauer S, Keilhoff G, Miersch J, Wend D, Riemann D, Hirschelmann R, Holzhausen HJ, Langner J. Biological effects of the dihydroorotate dehydrogenase inhibitor polyporic add, a toxic constituent of the mushroom Hapalopilus rutilans, in rats and humans. Arch Toxicol. 1998;72(11):711–21. https://doi.org/10.1007/s002040050565.

Krakowiak A, Piekarska-Wijatkowska A, Kobza-Sindlewska K, Rogaczewska A, Politanski P, Hydzik P, Szkolnicka B, Klopotowski T, Picheta S, Porebska B. Poisoning deaths in Poland: types and frequencies reported in lodz, Krakow, sosnowiec, gdansk, wroclaw and poznan during 2009–2013. Int J Occup Med Environ Health. 2017;30(6):897–908. https://doi.org/10.13075/ijomeh.1896.01063.

Krenova M, Pelclova D, Navratil T. Survey of Amanita phalloides poisoning: clinical findings and follow-up evaluation. Hum Exp Toxicol. 2007;26(12):955–61. https://doi.org/10.1177/0960327107085832.

Kretz O, Creppy EE, Dirheimer G. Characterization of bolesatine, a toxic protein from the mushroom Boletus satanas Lenz and its effects on kidney cells. Toxicology. 1991;66(2):213–24. https://doi.org/10.1016/0300-483X(91)90220-U.

Kurdil NV, Padalka VM, Ivashchenko OV, Lutsenko OH, Bohomol AH, Struk VF, Andriushchenko VV. Toxic syndromes in acute poisonings by conditionally edible and poisonous mushrooms. Emerg Med. 2016;2(73):111–9. (In Ukrainian)

Lampman AM. General principles of ethnomycological classification among Tseltales in Tenejapa Chiapas. J Ethnobiol. 2007;27:11–27. https://doi.org/10.2993/0278-0771(2007)27[11:GPOECA]2.0.C.O;2.

Latha SS, Naveen S, Pradeep CK, Sivaraj C, Dinesh MG, Anilakumar KR. Toxicity assessment of wild mushrooms from the western ghats, India: an in vitro and sub-acute in vivo study. Front Pharmacol. 2018;9:90. https://doi.org/10.3389/fphar.2018.00090.

Lawton LD. Accidental poisoning by death cap mushrooms: be careful what you eat. Wilderness Environ Med. 2013;11:168–70. https://doi.org/10.1016/j.wem.2012.11.001.

Leathem AM, Dorran TJ. Poisoning due to raw Gyromitra esculenta (false morels) west of the Rockies. Can J Emerg Med. 2007;9(2):127–30.

Leudang S, Sikaphan S, Parnmen S, Nantachaiphong N, Polputpisatkul D, Ramchiun S, Teeyapant P. DNA-based identification of gastrointestinal irritant mushrooms in the genus chlorophyllum:

a food poisoning case in Thailand. J Health Res. 2017;31(1):41–9. https://doi.org/10.14456/jhr.2017.6.
Li X-Y, Tao R-G, Zhao S-W. Analysis of mushroom poisoning in Yunnan, 1985–2000. Chin J Food Hygiene. 2003;15(1):49–51.
Li HJ, Zhang H, Zhang Y, Zhou J, Yin Y, He Q, Jiang S, Ma P, Zhang Y, Yuan Y, Lang N. Mushroom poisoning Outbreaksd China, 2021. China CDC Wkly. 2022;4(3):35e40.
Lin S, Mu M, Yang F, Yang C. Russula subnigricans poisoning: from gastrointestinal symptoms to rhabdomyolysis. Wilderness Environ Med. 2015a;26(3):380–3.
Lin SD, Mu MY, Yang FW, Yang CF. Russula subnigricans poisoning: from gastrointestinal symptoms to rhabdomyolysis. Wilderness Environ Med. 2015b;26(3):380–3. https://doi.org/10.1016/j.wem.2015.03.027.
Lindequist U, Niedermeyer THJ, Julich W-D. The Pharmacological potential of mushrooms eCAM. 2005;2:285–99.
Liu B, Huang Q, Cai H, Guo X, Wang T, Gui M. Study of heavy metal concentrations in wild edible mushrooms in Yunnan Province, China. Food Chem. 2015;188:294–300.
Logemann H, Argueta J, Guzmán G, Montoya Bello L, Bandala Munoz VM, De Leon Chocooj R. A deadly poisoning by mushrooms in Guatemala. Rev Mex Micol. 1987;3:211–6. (In Spanish)
Mao XL. Mao Poisonous mushrooms and their toxins in China. Mycosystema. 2006;3:345–63.
Marinov P, Bonchev G, Ivanov D, Zlateva S, Dimitrova T, Georgiev K. Mushrooms intoxications. J IMAB. 2018;24(1):1887–90. https://doi.org/10.5272/jimab.2018241.1887.
Matsura M, Kato S, Saikawa Y, Nakata M, Hashimoto K. Identification of cyclopropylacetyl-(R)-carnitine, a unique chemical marker of the fatally toxic mushroomnRussula subnigricans. Chem Pharm Bull (Tokyo). 2016;64(6):602–8.
McPartland JM, Vilgalys RJ, Cubeta MA. Mushroom poisoning. Am Fam Physician. 1997;55(5):1797–806.
Meijer AA, Amazonas MAL, Rubio GBG, Curial RM. Incidences of poisonings due to *Chlorophyllum molybdites* in the state of Parana, Brazil. Braz Arch Biol Technol. 2007;50(3):118–24.
Méndez-Navarro J, Ortiz-Olvera NX, Villegas-Ríos M, Méndez-Tovar LJ, Andersson KL, Moreno-Alcantar R, Gallardo-Cabrera VE, Félix S, Galván C, Vargas G, Gómez LM, Dehesa-Violante M. Hepatotoxicity from ingestion of wild mushrooms of the genus Amanita section Phalloideae collected in Mexico City: two case reports. Ann Hepatol. 2011;10(4):568–74.
Mengs U, Pohl RT, Mitchell T. Legalon® SIL: the antidote of choice in patients with acute hepatotoxicity from amatoxin poisoning. Curr Pharmaceut Biotechnol. 2012;13:1964–70.
Michelot D, Melendez-Howell LM. Amanita muscaria: chemistry, biology, toxicology, and ethnomycology. Mycol Res. 2003;107:131–46. https://doi.org/10.1017/S0953756203007305.
Mikaszewska-Sokolewicz MA, Pankowska S, Janiak M, Pruszczyk P, Lazowski T, Jankowski K. Coma in the course of severe poisoning after consumption of red fly agaric (Amanita muscaria). Acta Biochim Pol. 2016;63(1):181–2.
Mikhasiova EI, Pleshkunova IV. The laboratory diagnosis of the acute poisonings with fungi of the genus *Amanita*. In: Proceedings of the 70th international; 2016.
Miles PG, Chang ST. Mushroom biology: concise basics and current developments. Singapore: World Scientific; 1997.
Miyazaki S, Minami T, Mizuma H, Kanazawa M, Doi H, Matsumura S, Lu JS, Onoe H, Furuta K, Suzuki M. The action site of the synthetic kainoid (2S,3R,4R)-3-carboxymethyl-4-(4-methylphenylthio) pyrrolidine-2-carboxylic acid (PSPA-4), an analogue of Japanese mushroom poison acromelic acid, for allodynia (tactile pain). Eur J Pharmacol. 2013;710(1–3):120–7. https://doi.org/10.1016/j.ejphar.2012.10.023.
Moreno FA, Wiegand CB, Taitano EK, Delgado PL. Safety, tolerability, and efficacy of psilocybin in 9 patients with obsessive-compulsive disorder. J Clin Psychiatry. 2006;67(11):1735e1740.
Mossong J, Decruyenaere F, Moris G, Ragimbeau C, Olinger CM, Johler S, Perrin M, Hau P, Weicherding P. Investigation of a staphylococcal food poisoning outbreak combining case-

control, traditional typing and whole genome sequencing methods, Luxembourg, June 2014. Euro Surveill. 2015;20(45):4–10. https://doi.org/10.2807/1560-7917.ES.2015.20.45.30059.
Mottershead J, editor. Suetonius: Claudius. Bristol: Classical Press; 1969.
Motyka M, Marcinkowski JT. Why did they use psychoactive drugs? Part I. Role and importance of psychoactive substances in culture: from prehistory to early twentieth century. Probl Hig Epidemiol. 2014;95(2):223–33.
Mount P, Harris G, Sinclair R, Finlay M, Becker GJ. Acute renal failure following ingestion of wild mushrooms. Int Med J. 2002;32:187–90.
Mueller GM, Schmit JP, Leacock PR, Buyck B, Cifuentes J, Desjardin DE, Halling RE, Hjortstam T, Iturriage KH. Larsson Global diversity and distribution of macrofungi. Biodivers Conserv. 2007;16(1):37.
Murphy CM. Writing an effective review article. J Med Toxicol. 2012;8(2):89–90.
Musselius SG, Ryk AA. Mushroom Poisoning. Moscow: Medicine; 2002. (In Russian)
National Center for Biotechnology Information. PubChem Compound Summary for CID 125121811, Amatoxin. 2024a. Retrieved December 13, 2024 from https://pubchem.ncbi.nlm.nih.gov/compound/Amatoxin.
National Center for Biotechnology Information. PubChem Compound Summary for CID 9548611, Gyromitrin. 2024b. Retrieved December 13, 2024 from https://pubchem.ncbi.nlm.nih.gov/compound/Gyromitrin.
National Center for Biotechnology Information. PubChem Compound Summary for CID 108079, Coprine. 2024c. Retrieved December 13, 2024 from https://pubchem.ncbi.nlm.nih.gov/compound/Coprine.
National Center for Biotechnology Information. PubChem Compound Summary for CID 4266, Muscimol. 2024d. Retrieved December 13, 2024 from https://pubchem.ncbi.nlm.nih.gov/compound/Muscimol.
National Center for Biotechnology Information. PubChem Compound Summary for CID 10624, Psilocybin. 2024e. Retrieved December 13, 2024 from https://pubchem.ncbi.nlm.nih.gov/compound/Psilocybin.
National Center for Biotechnology Information. PubChem Compound Summary for CID 89579, Orellanine. 2024f. Retrieved December 13, 2024 from https://pubchem.ncbi.nlm.nih.gov/compound/Orellanine.
National Center for Biotechnology Information. PubChem Compound Summary for CID 128265, Fasciculol E. 2024g. Retrieved December 13, 2024 from https://pubchem.ncbi.nlm.nih.gov/compound/Fasciculol-E.
Nicholson FB, Korman MG. Death from *Amanita* poisoning. Aust NZ Med. 1997;27(4):448–9. https://doi.org/10.1111/j.1445-5994.1997.tb02212.x.
Nieminen P, Mustonen AM. Toxic potential of traditionally consumed mushroom species—a controversial continuum with many unanswered questions toxins. Toxins (Basel). 2020;12(10):639.
Niksic M, Klaus A, Argyropoulos D. Safety of foods based on mushrooms. In: Regulating safety of traditional and ethnic foods; 2015. p. 421.
Nomura C, Masayama A, Yamaguchi M, Sakuma D, Kajimura K. PCR-based method for the detection of toxic mushrooms causing food-poisoning incidents. Shokuhin Eiseigaku Zasshi. 2017;58(3):132–42. https://doi.org/10.3358/shokueishi.58.132.
Pacheco-Cobos L, Rosetti M, Cuatianquiz C, Hudson R. Sex differences in mushroom gathering: men expend more energy to obtain equivalent benefits. Evol Hum Behav. 2010;31:289–97. https://doi.org/10.1016/j.evolhumbehav.2009.12.008.
Pajoumand A, Shadnia S, Efricheh H, Mandegary A, Hassanian-Moghadam H, Abdollahi M. A retrospective study of mushroom poisoning in Iran. Hum Exp Toxicol. 2005;24:609–13.
Palapala VA, Aimi T, Inatomi S, Morinaga T. ITS-PCR- RFLP method for the distinguishing commercial cultivars of edible mushroom, Flammulina velutipes. J Food Sci. 2002;67:2486–90.
Park JS, Min JH, Kim H, Lee SW, Kang JH, An JY. Four cases of successful treatment after Podostroma cornu-damae intoxication. Hong Kong J Emerg Med. 2016;23(1):55–9.

Parnmen S, Sikaphan S, Leudang S, Boonpratuang T, Rangsiruji A, Naksuwankul K. Molecular identification of poisonous mushrooms using nuclear ITS region and peptide toxins: a retrospective study on fatal cases in Thailand. J Toxicol Sci. 2016;41(1):65–76. https://doi.org/10.2131/jts.41.65.

Patocka J. Bolesatine, a toxic protein from the mushroom rubroboletus satanas. Milit Med Sci Lett. 2018;87(1):14–20.

Pelclová D, Rakovcová H. Amanita phalloides poisoning queries recorded at the toxicology information center in Prague. Cas Lek Cesk. 1993,132(15):470–2.

Pérez-Moreno JS, Ferrera-Cerrato R. A review of mushroom poisoning in Mexico. Food Addit Contam A. 1995;12(3):355–60.

Petri G, Expert P, Turkheimer F, Carhart-Harris R, Nutt D, Hellyer PJ, Vaccarino F. Homological scaffolds of brain functional networks. J R Soc Interface. 2014;11(101):20140873.

Pinson CW, Bradley AL. A primer for clinicians on mushroom poisoning in the west. West J Med. 1996;165(5):318–9.

Precup L, Dumitra S, Precup I, Morgovan S, Lazăr D. Aspects of mushroom intoxication in children. J Pediatrului XV. 2012;57–58:43–4.

Prila, Teron RH. Eschewing poisons: an ingenious wisdom of foraging macrofungi by Karbi ethnic group in North East India. Curr Sci. 2018;115:1328–33. https://www.researchgate.net/figure/Map-of-Karbi-Anglongdistrict-Assam-North-East-India_fig1_328163140.

Public Health Service in Russia. In: Oxenoit GK (editor) Statistical Compendium. 2017, the Russian public health service was characterized by its constitutional guarantee of free. healthcare, but also by underfunding, staffing shortages, and uneven distribution of resources, particularly in rural areas.

Puschner B. Mushroom toxins. In: Gupta RC, editor. Veterinary toxicology: basic and clinical principles. San Diego: Elsevier/Academic; 2012. p. 1140–51. https://doi.org/10.1016/B978-0-12-385926-6.00111-3.

Quang DN, Hashimoto T, Asakawa Y. Inedible mushrooms: A good source of biologically active substances. Chem Record. 2006;6:79–99.

Rajarathnam S, Shashirekha MN. Mushrooms and truffles. In: Caballero B, editor. Encyclopedia of food sciences and nutrition. 2nd ed; 2003. p. 4048–54. https://doi.org/10.1177/0960327110361758.

Ramírez-Terrazo A, Adriana Montoya E, Garibay-Orijel R, Caballero-Nieto J, Kong-Luz A, Méndez-Espinoza C. Breaking the paradigms of residual categories and neglectable importance of non-used resources: the "vital" traditional knowledge of non-edible mushrooms and their substantive cultural significance. J Ethnobiol Ethnomed. 2021;17:1–18.

Reis FS, Pereira E, Barros L, Sousa MJ, Martins A, Ferreira IC. Biomolecule profiles in inedible wild mushrooms with antioxidant value. Molecules. 2011;16(6):4328–38.

Roberts DM, Hall MJ, Falkland MM, Strasser SI, Buckley NA. Amanita phalloides poisoning and treatment with silibinin in the Australian capital territory and new South Wales. Med J Aust. 2013;198:43–7. https://doi.org/10.5694/mja12.11180.

Romaniuk BP, Dubovaya GA, Fastova ON, Dubovaya JM, Ruban AI, Makaruk VA. Clinical and morphological characteristics in poisoning by death cup mushroom (*Amanita phalloides*) in the Luhansk region. Prob Ecol Med Genet Clin Immunol. 2012;109(1):21–8. (In Ukranian). Rosstat, Moscow

Romano GM, Iannone L, Novas MV, Carmaran C, Romero AI, Lopez SE, Lechner BE. Toxic fungi in buenos aires city and surroundings. Medicina (B Aires). 2013;73(5):406–10. (in Spanish)

Ruan-Soto F. Highly cultural significant edible and toxic mushrooms among the Tseltal from the Highlands of Chiapas, Mexico. Ethnobiol Conserv. 2020;32:1–20. https://ethnobioconservation.com/index.php/ebc/article/view/414

Ruiz-Gonzalez LE, Vazquez-Zea JA, Vega-Villasante F, Guzman-Davalos L, Guerrero-Galvan SR. Evaluation of the toxicity of Basidiomycota fungi on the hatching of Artemia franciscana cysts. Rev Iberoam De Micol. 2017;34(4):220–4. https://doi.org/10.1016/j.riam.2017.03.007. (In Spanish).

Rumack BH, Spoerke DG. Handbook of mushroom poisoning: diagnosis and treatment. Boca Raton: CRC Press; 1994. p. 330.
Saikawa Y, Okamoto H, Inui T, Makabe M, Okuno T, Suda T, Hashimoto K, Nakata M. Toxic principles of a poisonous mushroom Podostroma cornudamae. Tetrahedron. 2001;57(39):8277–81. https://doi.org/10.1016/S0040-4020(01)00824-9.
Salar MZ, Bibi Z, Hassan T, Hussain A, Ijaz MU. Cellular crossroads: oxidative stress, inflammation and apoptosis in molecular biology. In: Ijaz MU, Ehsan N, Imran M, Yousaf S, editors. Zoology: advancements and research trends. Lahore: FahumSci; 2024. p. 37–43. https://doi.org/10.61748/Zool.2024/06.
Samorini G. New data on the ethnomycology of psychoactive mushrooms. Int J Med Mushrooms. 2001;3(2–3):257–78.
Samwer M, Dehne HJ, Spira F, Kollmar M, Gerlich DW, Urlaub H, Görlich D. The nuclear F-actin interactome of Xenopus oocytes reveals an actin-bundling kinesin that is essential for meiotic cytokinesis. EMBO J. 2013;32(13):1886e1902.
Sanz P, Reig R, Piqueras J, Marti G, Corbella J. Fatal mushroom poisoning in Barcelona, 1986–1988. Mycopathologia. 1989;108:207–9.
Saupe A, Slimmer S, Anderson D, Carlton K, Smith K. Shiitake dermatitis – a case of food poisoning from a common edible mushroom. Clin Toxicol. 2016;54(8):671–2.
Saviuc P, Flesch F. Acute higher fungi mushroom poisoning and its treatment. Presse Med. 2003;32(30):1427–35.
Saviuc PF, Danel VC, Moreau PA, Claustre AM, Ducluzeau R, Carpentier PH. Acute erythermalgia: look for mushroons. Rev Med Interne. 2002;23:394–9.
Saviuc P, Harry P, Pulce C, Garnier R, Cochet A. Can morels (Morchella sp.) induce a toxic neurological syndrome? Clin Toxicol. 2010;48(4):365–72. https://doi.org/10.3109/15563651003698034.
Schenk-Jaeger KM, Rauber-Lüthy C, Bodmer M, Kupferschmidt H, Kullak-Ublick GA, Ceschi A. Mushroom poisoning: a study on circumstances of exposure and patterns of toxicity. Eur J Intern Med. 2012;23(4):e85–91. https://doi.org/10.1016/j.ejim.2012.03.014.
Schenk-Jaeger KM, Egli S, Hanimann D, Senn-Irlet B, Kupferschmidt H, Büntgen U. Introducing mushroom fruiting patterns from the Swiss national poisons information centre. PLoS One. 2016;11(9):e0162314. https://doi.org/10.1371/journal.pone.0162314.
Schmutz M, Carron P-N, Yersin B, Trueb L. Mushroom poisoning: a retrospective study concerning 11-years of admissions in a Swiss Emergency Department. Intern Emerg Med. 2018;13:59–67. https://doi.org/10.1007/s11739-016-1585-5.
Shao DH, Tang SS, Healy RA, Imerman PM, Schrunk DE, Rumbeiha WK. A novel orellanine containing mushroom Cortinarius armillatus. Toxicon. 2016;114:65–74. https://doi.org/10.1016/j.toxicon.2016.02.010.
Shepherd CJ, Totterdell CJ. Mushrooms and toadstools of Australia. Inkata Press; 1988.
Shi GQ, Huang WL, Zhang J, Zhao H, Shen T, Fontaine RE, Yang L, Zhao S, Lu BL, Wang YB. Clusters of sudden unexplained death associated with the mushroom, Trogia venenata, in rural yunnan province, China. PLoS One. 2012;7(5):e35894. https://doi.org/10.1371/journal.pone.0035894.
Smith MR, Davis RL. Mycetismus: a review. Gastroenterol Rep. 2016;4(2):107–12. https://doi.org/10.1093/gastro/gov062.
Sohn CH. Type and treatment of toxic mushroom poisoning in Korea. Korean Med Assoc. 2015;58(9):818–24. https://doi.org/10.5124/jkma.2015.58.9.818.
Soltaninejad K. Outbreak of mushroom poisoning in Iran: April–May, 2018. Int J Occup Environ Med. 2018;9:152–6. https://doi.org/10.15171/ijoem.2018.1380.
Sönmez BM, Yılmaz F, Karakılıç ME, Durdu T, Şahin T, Cebiçci H, Beydilli İ, Yücel M. Demographic and clinical characteristics of mushroom poisonings presenting to emergency department. J Clin Anal Med. 2015;6(4):469–72. https://doi.org/10.4328/JCAM.2176.
Suliga EM, Yarovova AV. Mushrooms as a source of danger to human health. In: Kashitsyna LV, editor. Current problems of physical culture and life safety. Saratov: Publishing House "Saratovsky istochnik; 2017. p. 124–7. (In Russian).

Sun J, Zhang J, Wang YZ. Multivariate characterization of elements accumulated in Wolfiporia extensa mushroom from Yunnan province of China. J Environ Sci Health B. 2017;52(3):206–13. https://doi.org/10.1080/03601234.2017.1261552.

Sun J, Li H-J, Zhang H-S, Zhang Y-Z, Xie J-W, Ma P-B, Guo C, Sun CY. Investigating and analyzing three cohorts of mushroom poisoning caused by Amanita exitialis in Yunnan, China. Hum Exp Toxicol. 2018;37(7):665–78.

Taguchi T, Tomotoshi K, Mizumura K. Excitatory actions of mushroom poison (acromelic acid) on unmyelinated muscular afferents in the rat. Neurosci Lett. 2009;456(2):69–73. https://doi.org/10.1016/j.neulet.2009.03.102.

Tagwireyi D, Chingombe P, Khoza S, Maredza M. Pattern and epidemiology of poisoning in the East African region: a literature review. J Toxicol. 2016; https://doi.org/10.1155/2016/8789624.

Taofiq O, Gonzalez-Paramas AM, Martins A, Barreiro MF, Ferreira ICFR. Mushrooms extracts and compounds in cosmetics, cosmeceuticals and nutricosmetics–a review. Ind Crop Prod. 2016;90:38–48.

Trestrail JH. Mushroom poisoning in the United States—an analysis of 1989 United Poison centers data. Clin Toxicol. 1991;29:459–65.

Trestrail JH, Lampe KF. Mushroom toxicology resources utilized by certified regional poison centers in the United States. Clin Toxicol. 1990;28(2):169–76.

Ukwuru MU, Muritala A, Eze LU. Edible and non-edible wild mushrooms: nutrition, toxicity and strategies for recognition. J Clin Nutr Metab. 2018;2:2.

VanderMolen KM, Little JG, Sica VP, El-Elimat T, Raja HA, Oberlies NH, Baker TR, Mahony C. Safety assessment of mushrooms in dietary supplements by combining analytical data with in silico toxicology evaluation. Food Chem Toxicol. 2017;103:133–47.

Vargas N, Bernal A, Sarria V, et al. Amatoxin and phallotoxin composition in species of the genus *Amanita* in Colombia: a taxonomic perspective. Toxicon. 2011;58:583–90. https://doi.org/10.1016/j.toxicon.2011.09.005.

Varvenne D, Retornaz K, Metge P, De Haro L, Minodier P. Amatoxin-containing mushroom (Lepiota brunneoincarnata) familial poisoning. Pediatr Emerg Care. 2015;31(4):277–8. https://doi.org/10.1097/PEC.0000000000000399.

Vetter J, Vetter S. Death cap poisoning and the animals. Literature review. MAGY ALLATORVOSOK. 2014;136(12):750–8. (In Hungarian)

Villa AF, Saviuc P, Langrand J, Favre G, Chataignerl D, Garnier R. Tender nesting polypore (Hapalopilus rutilans) poisoning: report of two cases. Clin Toxicol. 2013;51(8):798–800. https://doi.org/10.3109/15563650.2013.827708.

Vishnevsky MV. Poisonous mushrooms of Russia. Prospect, Moscow (In Russian). In: Vo KT, Montgomery ME, Mitchell ST, Scheerlinck PH, Colby DK, Meier KH, Kim-Katz S, Anderson IB, Offerman SR, Olson KR, Smollin CG, editors. US department of health and human services/centers for disease control and prevention. MMWR (Morb. Mortal. Wkly. Rep.), vol. 66, No. 21; 2017. p. 549–53.

Vo KT, Montgomery ME, Mitchell ST, Scheerlinck PH, Colby DK, Meier KH, Kim-Katz S, Anderson IB, Offerman SR, Olson KR, Smollin CG. US department of health and human services/centers for disease control and prevention. MMWR (Morb Mortal Wkly Rep). 2017;66(21):549–53.

Vunduk J, Biketova AY. Chapter 9: The cultural use of mushrooms. In: Stojković D, Barros L, editors. Edible fungi: chemical composition, nutrition and health effects. The Royal Society of Chemistry; 2022. p. 327–58.

Vunduk J, Klaus A, Kozarski M, Petrovic P, Zizak Z, Nikšić M, Van Griensven LJLD. Did the Iceman know better? screening of the medicinal properties of the birch polypore medicinal mushroom, Piptoporus betulinus (Higher Basidiomycetes). Int J Med Mushrooms. 2015;17(12):1113.

Ward J, Kapadia K, Brush E, Salhanick SD. Amatoxin poisoning: case reports and review of current therapies. J Emerg Med. 2013;44(1):116–21. https://doi.org/10.1016/j.jemermed.2012.02.020.

Wasser SP, Weis AL. Therapeutic effects of substances occurring in higher Basidiomycetes mushrooms: a modern perspective. Crit Rev Immunol. 1999;19:65–96.
Wasson RG. The soma of the Rig Veda: what was it? J Am Orient Soc. 1971;91:169–87.
Wasson VP, Wasson RG. Mushrooms Russia and history by Valentina Pavlovna Wasson and R. Gordon Wasson. Pantheon Books; 1957.
West PL, Lindgren J, Horowitz BZ. Amanita smithiana mushroom ingestion: a case of delayed renal failure and literature review. J Med Toxicol. 2009;5(1):32–8.
Wexler P. History of toxicology and environmental health: toxicology in antiquity II. Academic; 2014.
Wong JH, Ng TB. Toxins from basidiomycete fungi (mushroom): amatoxins, phallotoxins, and virotoxins. In: Kastin AJ, editor. Handbook of biologically active peptides; 2006. p. 131–5. https://doi.org/10.1016/B978-012369442-3/50023-4.
Wu F, Zhou LW, Yang ZL, Bau T, Li TH, Dai YC. Resource diversity of Chinese macrofungi: edible, medicinal and poisonous species. Fungal Divers. 2019;98:1–76.
Xiang H, Zhou Y, Zhou C, Lei S, Yu H, Wang Y, Zhu S. Investigation and analysis of *Galerina sulciceps* poisoning in a canteen. Clin Toxicol. 2018;56(5):365–9. https://doi.org/10.1080/15563650.2017.1388386.
Yamada M, Tokumitsu N, Saikawa Y, Nakata M, Asano J, Miyairi K, Okuno T, Konno K, Hashimoto K. Molybdophyllysin, a toxic metalloendopeptidase from the tropical toadstool, Chlorophyllum molybdites. Bioorg Med Chem. 2012;15(22):6583–8. https://doi.org/10.1016/j.bmc.2012.09.036.
Yamaura Y. Recent trends of mushroom poisoning in Japan. Chudoku Kenkyu. 2013;26:39–43. (In Japanese)
Yamaura Y, Nakamura K, Ishihara Y. Descriptive epidemiology of mushroom poisoning in nagano prefecture. Nihon Eiseigaku Zasshi. 1997;38(2):110–5. (In Japanese)
Yang CC, Tsai WJ, Chuo HT, Shiau FS, Deng JF. Mushroom poisoning in Taiwan. J Nat Toxins. 1995;4(2):185–94.
Yang W-S, Lin C-H, Huang J-W, Fang C-C. Acute renal failure caused by mushroom poisoning. J Formos Med Assoc. 2006;105(3):263–7.
Yardan T, Baydin A, Eden AO, Akdemir HU, Aygun D, Acar E, Arslan B. Wild mushroom poisonings in the middle black sea region in Turkey: analyses of 6 years. Hum Exp Toxicol. 2010;29(9):767–71. https://doi.org/10.1177/0960327110361758.
Yilmaz I, Ermis F, Akata I, Kaya E. A case study: what doses of Amanita phalloides and amatoxins are lethal to humans? Wilderness Environ Med. 2015;26:491–6.
Yoon Y-H, Choi S-H, Cho H-J, Moon S-W, Kim J-Y, Lee S. Reversible pancytopenia following the consumption of decoction of Ganoderma neojaponicum Imazeki. Clin Toxicol (Phila). 2011;49(2):115–7. https://doi.org/10.3109/15563650.2011.553834.
Yoshikawa K, Kuroboshi M, Arihara S, Miura N, Tujimura N, Sakamoto K. New triterpenoids from Tricholoma saponaceum. Chem Pharm Bull. 2002;50(1):1603–6. https://doi.org/10.1248/cpb.50.1603.
Yoshioka N, Ouchi H, Kan T, Yoshida M, Nomura M. Rapid LC-MS determination of acromelic acids a and B, toxic constituents of the mushroom Paralepistopsis acromelalga. Shokuhin Eiseigaku Zasshi. 2017;58(5):241–5. https://doi.org/10.3358/shokueishi.58.241.
Yu XZ, Lu ZQ, Hong GL, Sun CY, Chen XR, Li HJ. Consensus of clinical experts on the diagnosis and treatment of mushroom poisoning in China. J Clin Emerg. 2019;20(8):583–98.
Zahan N, Hasan MZ, Malek MA, Reya SS. A deep learning-based approach for edible, inedible and poisonous mushroom classification. In: 2021 international conference on information and communication technology for sustainable development (ICICT4SD). IEEE; 2021. p. 440–4.
Zamudio F, Hilgert NI. Multi-dimensionality and variability in folk classification of stingless bees (Apidae: Meliponini). J Ethnobiol Ethnomed. 2015;11(1):41. https://doi.org/10.1186/s13002-015-0029-z.
Zent E, Zent S. Ebojto: Plantas Trepadoras entre los jotï, Guayana Venezolana. Etnobiología. 2016;14:5–38. https://cutt.ly/GgjLSnp

Zhang J-J, Li Y, Zhou T, Xu D-P, Zhang P, Li S, Li H-B. Bioactivities and health benefits of mushrooms mainly from China. Molecules. 2016;21(7):938.

Zhou ZY, Shi GQ, Fontaine R, Wei K, Feng T, Wang F, Wang GQ, Qu Y, Li ZH, Dong ZJ, Zhu HJ, Yang ZL, Zeng G, Liu JK. Evidence for the natural toxins from the mushroom Trogia venenata as a cause of sudden unexpected death in Yunnan Province, China. Angew Chem Int Ed Engl. 2012;51(10):2368–70. https://doi.org/10.1002/anie.201106502.

Zhou J, Yuan Y, Lang N, Yin Y, Sun CY. Analysis of hazard in mushroom poisoning incidents in China mainland. Chin J Emerg Med. 2016;25(6):724e728.

Zhou Q, Tang S-S, He ZM, Luo T, Chen Z-H, Zhang P. Amatoxin and phallotoxin concentrations in Amanita fuliginea: influence of tissues, developmental stages and collection sites. Mycoscience. 2017;58(4):267–73. https://doi.org/10.1016/j.myc.2017.03.003.

Zuliani AM, Kabar I, Mitchell T, Heinzow HS. Acute liver failure after ingestion of death cap mushrooms. DMW (Dtsch Med Wochenschr). 2016;141(13):940–2. https://doi.org/10.1055/s-0042-103987.

Zuzek MC, Macek P, Sepci K, Cestnik V, Frangez R. Toxic and lethal effects of ostreolysin, a cytolytic protein from edible oyster mushroom (Pleurotus ostreatus), in rodents. Toxicon. 2006;48:264–71.

Chapter 6
Mushroom Cultivation: An Integral Part of Circular Agriculture

Amjad Abbas, Muhammad Usman, Shehnaz Jamil, Ikhlas Shafique, Saima Yousaf, Amer Habib, and Muhammad Amjad Ali

Abstract Circular agriculture is the most efficient use of all waste resources to get a useful product in any agricultural production system. It is the key point in modern farming for long-term sustainability. The main target of recycling agriculture is to protect the environment and enhance food production. In this context, mushrooms are very important as they convert organic waste streams from the food industry or agricultural processing plants into human food, which is rich in protein and dietary fibers. Different substrates materials like cereal straw, coffee grounds, by-products of conventional farming, and organic waste from the sugar, oil, or rice industry are preferred wastes to produce different types of edible mushrooms. This chapter emphasizes the versatility of species such as Oyster and Shiitake mushrooms, which can grow on diverse substrates and offer high nutritional and medicinal value. Additionally, the chapter discusses the environmental, economic, and social benefits of mushroom cultivation, including waste reduction, soil improvement, carbon sequestration, job creation, and community engagement. Different challenges, like contaminations during spawn cultivation, market access, low levels of production technologies, and food habits of the local population, are also discussed in this chapter. However, the need for technological advancements, supportive policies, and public awareness to further integrate mushroom cultivation into sustainable food systems is very important, and there is a dire need to convert useless waste into nutritious food.

Keywords Circular agriculture · Sustainability · Shiitake mushrooms · Spawn Cultivation

A. Abbas · M. Usman · S. Jamil · I. Shafique · A. Habib · M. A. Ali (✉)
Department of Plant Pathology, University of Agriculture, Faisalabad, Pakistan
e-mail: amjad.ali@uaf.edu.pk

S. Yousaf
Department of Agronomy, University of Agriculture, Faisalabad, Pakistan

I. U. Haq et al. (eds.), *Mushrooms*, https://doi.org/10.1007/978-981-95-2539-3_6

6.1 Mushrooms as Decomposer

Many fungi, including mushrooms, play a significant role as decomposers in an ecosystem. They produce simpler compounds by breaking down complex organic matter waste, which also helps in nutrient recycling in the soil. This process is crucial for supporting plant growth and maintaining soil nutritional value and health (Libretexts. 2024).

Mushrooms serve as nature's recyclers by breaking down all types of organic matter like, dead animals and plants. Through the production and secretion of enzymes, they decompose complex organic matter into simpler compounds, making these available to other organisms and enriching the soil with vital nutrients produced. This soil type helps enhance plant growth and sustain the ecosystem (Pearson 2024).

Harmful dyes and substances are broken down or removed by mushrooms, either through secretory enzymes or by absorption and adsorption. Mushrooms behave as hyperaccumulators by degrading cellulose and lignin, thereby enhancing soil conditions and nutrients, and they play a vital role in bioremediation (Barkat Md Gulzar et al. 2020).

Mushrooms release different enzymes like lignin peroxidases, laccase, and manganese peroxidase, which are effective in the decomposition process. These enzymes break down the pollutants and complex compounds, and the nutrients produced during the breakdown of these pollutants and compounds return to the soil and enhance plant growth. Mushrooms' utilization of agricultural waste as a substrate reduces waste management challenges, improves soil fertility and structure, promotes biodiversity, and supports sustainable agriculture practices (Adenipekun and Lawal 2012).

Mushrooms help maintain environmental sustainability by recycling and reducing natural waste, improving soil structure, and enhancing the nutrient cycle. Their ability to decompose organic waste helps circulate the nutrients in the ecosystem (Pearson 2024).

6.2 Cultivation on Organic Waste Stream

Organic waste streams rich in nutrients, such as food scraps, agricultural residues, and industrial by-products, can be effectively used as substrates for cultivating mushrooms in different regions. These practices enhance sustainable agricultural food sources while reducing environmental pollutants. For example, agricultural waste like sawdust and straw have been proven effective for the cultivation and growth of mushrooms. The transformation of organic waste into valuable products promotes recycling and contributes to the circular economy (Pawar and Umesh 2021).

Several key steps exist to create a sterile and nutrient-rich substrate by preparing organic waste to cultivate mushrooms. The first step is selecting appropriate waste materials, such as sawdust, straw, and coffee grounds. After choosing the material, it is mixed with water to provide moisture up to 50–70%, and nutrients such as magnesium, nitrogen, and calcium are supplied. To ensure a contamination-free substrate, pasteurization or sterilization is a good choice (Sayner 2024).

Many mushrooms, including oyster, shiitake, and button mushrooms, are commonly cultivated on organic waste. Oyster mushrooms can be grown on many substrates and produce high yields, making them popular worldwide. Composted manure is used to cultivate button mushrooms, while hardwood sawdust is used to cultivate shiitake mushrooms. These mushrooms not only provide nutrition by serving as food but also help reduce environmental waste and promote environmental sustainability (Pawar and Umesh 2021).

6.3 Environmental Benefits

Mushroom cultivation significantly contributes to soil health by converting complex organic waste into nutrient-dense compost. The spent mushroom substrate (SMS) serves as a valuable soil amendment, enriching its nutritional profile, improving soil structure, and boosting microbial activity. Using bio-based substrates, mushroom cultivation enhances biodiversity and diverse species interactions. This approach helps support ecosystem balance and strengthen biodiversity (Genetics 2023). Cultivating mushrooms on agricultural waste is a cost-effective method that yields high returns. This method is not only economically sustainable, with low input costs and higher returns (Gupta et al. 2022), but also enhances food security by turning waste materials into a source of nutritious food. This sustainable method helps combat food shortages while improving overall dietary nutrition (Pawar and Umesh 2021).

6.3.1 Mushroom Cultivation Is an Alternative Way to Burn the Agricultural Waste

Burning agricultural waste is a serious environmental problem since it contributes to global warming and causes health problems. In comparison, using this waste for mushroom cultivation converts plant-based lignocelluloses into soil rich in organic materials. The key species of mushrooms used in sustainable mushroom farming can turn organic waste, such as agricultural organic matter, into nutritious, high-protein mushrooms. Cellulose, hemicellulose, and lignin, all of which are prevalent in agricultural wastes like wheat straw, ultimately provide mushrooms with their nourishment. The most common agricultural wastes utilized for mushroom production include sawdust, banana leaves, tea leaves, cotton straw, rice bran, molasses, wheat, paddy, and rice straw (Peng 2008). Rice, wheat straw, and cotton waste are commonly used for mushroom production. However, straw becomes filled with other microorganisms, and removing them is crucial because the mushrooms' mycelium cannot develop in their presence, particularly if they are to be grown inside. Sawdust must be supplemented with a nitrogen source, such as urea, bran, horse dung, and sunflower seed (Wu et al. 2020). The following legume crop wastes were useful in cultivation: soybean, mash clover, kulthi, and a mixture of soybean, mash, and kulthi (Ivarsson et al. 2021). Using straw from two plant sources is more productive in mushroom cultivation than single straw-based substrate. It was shown that the combination of gram pod waste and wheat straw (1:1) increased mushroom yields by 5.26% compared to wheat straw alone. While the growth of mushrooms produced a poor yield in sugarcane bagasse, it was similar in paddy straw and wheat straw. Appropriate concentrations of lignin, hemicellulose, and cellulose are necessary to achieve this selectively high yield (Dehariya and Vyas 2013).

6.3.2 Circular Economic Values

Several elements, such as initial investment, operating costs, revenue generation, and prospective profitability, must be considered to evaluate the economic viability of recycling substrates for mushroom cultivation (Iram 2022).

6.3.3 Initial Investment

The establishment of a mushroom cultivation facility involves several critical investments. One of the primary expenses is the cost of acquiring or leasing land, along with constructing or renovating a tailored building to meet the specific requirements of mushroom farming. Additionally, significant investment is required to procure equipment such as composting machines, pasteurization and sterilization units, and advanced climate control systems to ensure optimal growing conditions. Another crucial component is the initial procurement of substrate materials, which serve as the general medium for mushrooms. These substrates may include various organic materials, such as straw, corn stover, coffee grounds, and manure, all of which provide the necessary nutrients to support healthy mushroom growth (Grimm and Wösten 2018).

6.3.4 Operational Costs

A mushroom cultivation facility needs constant investment in several important areas. The salaries of employees in charge of several production stages, such as substrate preparation, inoculation, growth, and harvesting, constitute a significant portion of labor expenditures. Another important cost is utilities, which include water, power, and heating or cooling systems necessary to maintain the exact environmental conditions required for the best potential mushroom development. Regular maintenance of buildings and equipment is also essential to guarantee uninterrupted operation and avoid expensive malfunctions. Furthermore, to maintain output and satisfy market expectations, supplies including premium spawn, new substrates, and other necessary resources must be continuously acquired (Grimm and Wösten 2018).

6.3.5 Revenue Generation

In the mushroom industry, there are several ways to generate income. Selling freshly grown mushrooms to different establishments, such as neighborhood markets, eateries, and supermarkets, is the main source of revenue. Value-added products, such as powders, dietary supplements, and mushroom extracts, provide a profitable alternative to fresh produce, meeting the increasing need for healthier options. Moreover, recycling agricultural waste offers an extra source of income because compost and other organic materials can be sold as by-products, enhancing sustainability and profitability (Cunha Zied et al. 2020).

6.3.6 Potential Profitability

Several critical performance indicators assess a mushroom farming company's financial sustainability. The break-even study helps predict when a business will start turning a profit by identifying when total revenue equals total costs. The return on investment (ROI), which compares the net profit with the initial capital invested, is another crucial indicator of total profitability. Furthermore, the profit margin is a vital sign of sound financial standing. It shows the portion of income remaining after all costs are deducted, offering information about the operation's overall effectiveness and profitability (Grimm and Wösten 2018; Iram 2022).

6.3.7 Vertical Farming

Certain farming crops can be cultivated in indoor environments having precise temperature, nutrients and light. In vertical farming, tall multi-story structures are made in which mushrooms are stacked in layers (Gupta et al. 2024). As a trade term in modern agriculture, vertical farming aims to minimize the use of land. In that way, areas with limited soil could be used at their maximum. Introducing this advanced method into mushroom production can reduce the use of space and bacterial contamination rates. Mushroom cultivation by vertical farming requires less land area

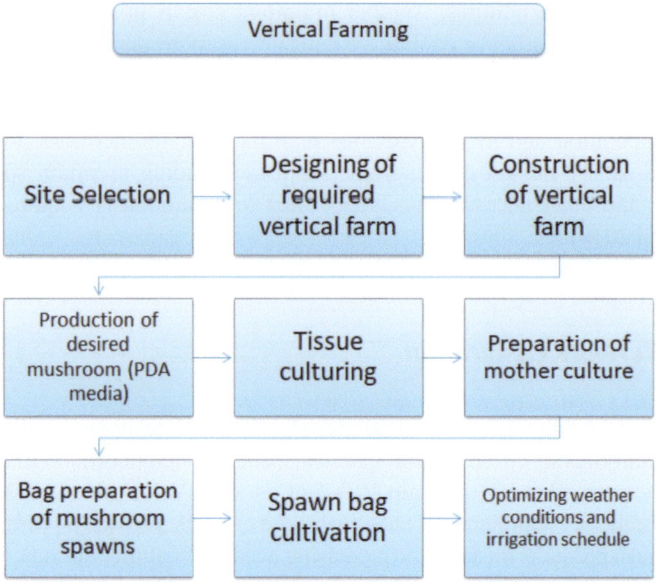

Fig. 6.1 Detailed general layout of vertical farming for mushroom cultivation

and water. Mushrooms are rich in nutrients and dietary fibers, with low carbohydrates and calories but high in protein, so mushrooms have become a priority in list of functional foods, with a dire need to enhance their cultivation given limited land resources. This method can be used to get maximum growth and yield by maintaining optimal growing conditions like humidity and temperature (Al Mamun et al. 2021; Gupta et al. 2024). The general layout for conducting vertical farming practices to cultivate mushrooms is given in Fig. 6.1.

6.3.8 Automated Farming

Automated farming systems are designed to reduce manual labor and optimize growing conditions for mushroom cultivation. To create a controlled and efficient growing environment, automated mushroom farming uses advanced technologies like machine learning (ML), Internet of things (IoT), and renewable energy sources. These systems monitor and adjust environmental parameters like light exposure, humidity, temperature, and CO_2 levels in real time to ensure optimal conditions for mushroom growth (Rukhiran et al. 2023). Key components of automated farming are shown in Fig. 6.2.

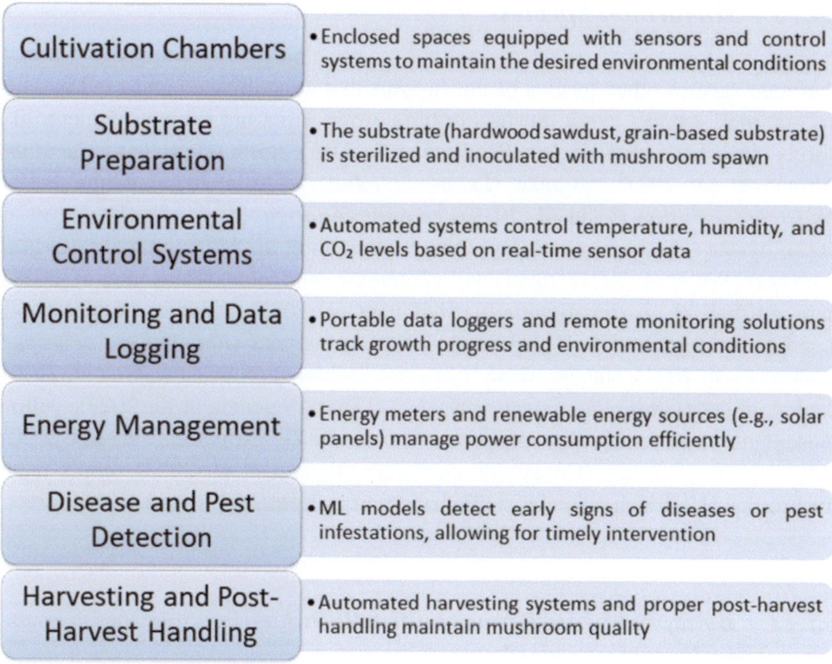

Fig. 6.2 General overview of automated farming for mushroom cultivation

6.3.9 Bioreactors

In mushroom cultivation bioreactors are meant to be an advanced system of cultivation to control growing conditions and enhance productivity. Bioreactors are specially designed vessels that provide favorable conditions for the growth of mushroom fruiting bodies and mycelium. In bioreactors, for liquid-state cultivation of mushroom mycelia, submerged culture bioreactors are used while solid-state bioreactors are used for cultivation of solid-state substrates like grain or sawdust (Berovic et al. 2022). Bioreactors provide optimal growing conditions (temperature, humidity, CO_2 levels, and oxygen supply). After that substrates are prepared and sterilized (to avoid any contamination) and then mushroom spawn is inoculated. Sensors monitor the environmental parameters in real time, and the automation system manages or adjusts conditions based on sensor data. Bioreactors are used for continuous large-scale production of mushrooms, providing consistency in quality and purity (reducing contamination risks) (Tang et al. 2007; Berovic et al. 2022).

6.3.10 Mushroom Species

There are various other species of mushrooms that are cultivated and used in daily diet routines. Almost every mushroom type grows in its unique environment, like morels (*Morchella angusticeps*) (Richard et al. 2015), reishi (*Ganoderma lucidum*) (Ahmad et al. 2021), shiitake (*Lentinula edodes*) (Atila 2021), white button (*Agaricus bisporus*) (Li et al. 2019), cremini (*Agaricus bisporus*) (Davila et al. 2022), oyster (*Pleurotus ostreatus*) (Nongthombam et al. 2021), maitake (*Grifola frondosa*) (Wu et al. 2021), turkey tail (*Trametes versicolor*) (Benson et al. 2019), giant puffball (*Calvatia gigantea*) (Shukla et al. 2023), chicken of the woods (*Laetiporus sulphureus*) (Wang et al. 2015), enokitake (*Flammulina velutipes*) (Sharma et al. 2021), shaggy mane (*Coprinus comatus*) (Nowakowski et al. 2020), black trumpet (*Craterellus cornucopioides*) (Ozuna-Valencia et al. 2023), yellow houseplant mushroom (*Leucocoprinus birnbaumii*) (Adikaram et al. 2020), Cordyceps mushrooms (*Cordyceps militaris*) (Nguyet et al. 2023), lion's mane mushrooms (*Hericium erinaceus*) (Tachabenjarong et al. 2022), and many more.

6.3.11 Benefits of Mushroom Cultivation in Recycle Agriculture

Mushroom cultivation is important in recycling agriculture because it decomposes organic waste that is harmful for the environment. Mushrooms require fewer resources for their cultivation as compared to all other crops that are being cultivated. They also use less space as compared to other crops and provide high yield (Genetics 2023).

Mushroom industries first use horse manure collected from racetrack and boarding stables as a substrate, breaking its compound down into simpler one and increasing its value. The utilization of these manures increases as the mushroom industry expands, and these stables and manures eventually become insufficient for mushroom cultivation. The increase in demand promote farmers to use more consistent materials like second-grade hay. This practice helps the environment by preventing the landfilling and becomes a source of income for farmers (Beyer 2023). Mushroom industries in the world use many kinds of waste for cultivation and use them as different kind of nutrient source that are available in the waste.

Agricultural residues and organic waste like sawdust, coffee grounds, straw, corn cobs, etc., are used in mushroom farming. The utilization of these waste and by-products reduces their quantity in the environment and also helps in the reduction of their effect on the environment (Beyer 2023). Mushrooms consume the by-products from industries and other residues like husks and straw as a source of medium (substrate) for the growth, which reduces the waste in the environment and increases the value of discarded material (Pawar and Umesh 2021).

Mushrooms can contribute to the sequestration of carbon from harmful residues and prevent the emission of greenhouse gas into the environment, which also contributes to the mitigation of climate change. Mushroom cultivation uses less amount of water compared to other crops, which is very important for water-scarce regions (Jasinska 2023).

Mushroom industries and their cultivation play an important role in the management of poultry industry manure, which affects the environment due its odor and pest issues, and the release of harmful gasses. Mushroom industries also use this poultry manure as a substrate, which provide benefits to both industries. Furthermore, the use of broken leaves collected from the field and roadsides and paper waste also prevents landfilling. Using it for making compost enhances soil structure and also provides environmental benefits (Beyer 2023).

Spent mushroom substrate (SMS) is a nutrient-rich residue from mushroom growth. This compost is made from organic elements, including straw, sawdust, and agricultural waste, that are used to cultivate mushrooms. Crucial nutrients like potassium, phosphorus, and nitrogen are released as mushrooms break down the substrate. SMS is a great soil supplement that increases soil fertility, encourages plant development, and lessens the need for artificial fertilizers. Studies show that SMS promotes beneficial microbial activity in the soil, improves water retention, and strengthens soil structure (Mwangi et al. 2024).

The ability of mushrooms to absorb and store carbon dioxide (CO_2) from the environment makes them essential for carbon sequestration. Particularly effective in this process are mycorrhizal fungi, which develop symbiotic associations with plant roots. By transferring carbon from plants to the soil, these fungi help the earth retain carbon as organic matter. According to studies, mycorrhizal fungi can store up to 36% of the world's yearly emissions from fossil fuels, underscoring their significance in the fight against climate change. Mushrooms improve soil health and reduce greenhouse gas emissions by increasing soil carbon storage (Sheffield 2023).

The production of 1 kg of mushrooms uses up to 5 kg of organic waste. The decomposition of organic waste in landfills also occurs anaerobically, which produces methane gases, a harmful gas that causes global warming rather than carbon dioxide. The decomposition of this organic waste by mushrooms not only reduces the emission of methane but also makes this process aerobic. This process also helps lessen the environmental effect of these organic waste and contributes to the mitigation of climate change (Beyer 2023).

The cultivation of mushrooms has significant economic benefits, which converts the complex organic agricultural waste into useful simpler compounds. Farmers can reduce disposal expenses and generate new revenue streams by recycling waste materials like sawdust and straw as growth substrates. Furthermore, a waste management option that is environmentally benign is offered via the sale of premium organic compost made from spent mushroom substrate (SMS), a by-product of the growing process (Mwangi et al. 2024).

Mushroom farming is an interesting option for small-scale farmers and urban agriculture because it has inexpensive beginning costs and requires little space. By turning waste materials into useful goods, this approach promotes economic sustainability and lessens its negative effects on the environment, all of which contribute to a circular economy (Warrick 2024).

6.3.12 Potential Highlights for Generating Income from the Sale of Mushrooms and Value-Added Products

Mushrooms are cultivated globally. Mushroom production is well suited to exemplify the potential synergies of the circular economy through its recycling of organic matter and nutrients via the decomposition of organic waste, production of yields high-nutritional-value food, and generation of job opportunities (Dorr et al. 2021). Mushroom cultivation has the potential to be a highly profitable agricultural operation that addresses several current challenges, including resource usage, circular economy, job development, sustainability, unpredictable weather, and nutritional security in the country (Khan et al. 2024).

6.3.13 Job Creation Potential and Social Impact

Depending on the scale, mushroom farms can create various job roles, from labor-intensive positions (such as substrate preparation, harvesting) to more skilled roles in production management, marketing, and distribution. Direct employment opportunities in mushroom cultivation are substantial, particularly in small-scale and indoor environments. This approach makes it possible for people to operate in confined circumstances. As a sustainable agricultural technique, mushroom farming provides steady employment in cultivation, harvesting, processing, and distribution. Additionally, it supports local economies by creating employment in managing organic waste, such as agricultural by-products, which are commonly used in mushroom cultivation (Jayaraman et al. 2024).

Small-scale mushroom farms focusing on recycled materials may thrive through local and regional sales, contributing to the growth of local economies. One can start a business that grows edible mushrooms like oyster mushrooms, shiitake mushrooms, or other gourmet types. A regulated environment for cultivating mushrooms can be maintained indoors, in greenhouses, or with specialized mushroom bags. Whether done on a large or small scale, mushroom cultivation requires minimal upkeep and can be pursued part-time (Sharma et al. 2017). Raising mushroom production would make them readily accessible across the country, which would further aid in eliminating nutritional insecurity and boosting the incomes of small and marginal producers (Vaishnavi et al. 2022).

This practice can boost income stability for smallholder farmers and aid in poverty reduction (Shackleton et al. 2011). Moreover, it promotes sustainable small-scale farming by repurposing organic waste as a growing substrate, similar to fertilizer (Jasińska 2018; Khan et al. 2024). By selling fresh mushrooms and value-added products, farmers and entrepreneurs can generate additional income, which can be used to purchase other essential food items. Mushroom farming is a quick and environmentally beneficial option to generate revenue with little investment. It offers the ability to repurpose agricultural by-products, transforming waste into affluence. Mushroom farming can help rural people avoid poverty while also improving their incomes through the advancement of agri-entrepreneurship (Joseph et al. 2024).

The demand for mushrooms and mushroom-based products can support ancillary businesses, such as packaging companies, transportation services, and local retail shops. The development of value-added mushroom products, such as dried mushrooms, mushroom powders, and mushroom extracts could enhance this business. Health-conscious customers may find these products appealing due to their diverse culinary and nutritional uses (Dhivya and Monika 2024).

6.3.14 Community Impact and Empowerment

Mushroom farming that utilizes recycled materials can lead to community-based solutions for waste management and sustainable agriculture. These initiatives can empower local farmers to become active participants in the circular economy. Farmers can produce mushrooms more effectively by adopting innovative techniques and using the available space by growing them vertically in stacked layers as compared to traditional farming, where crops are cultivated horizontally in large fields (Rahmann et al. 2021). Local mushroom farming initiatives can foster community engagement, skill development, and entrepreneurship, contributing to overall food security in light of economic progression is listed in Fig. 6.3.

6.3.15 Local Food Production

Mushrooms have a relatively short growth cycle, allowing for multiple harvests throughout the year. This ensures a steady supply of fresh produce. The best strategy to lessen food shortages in Pakistan is to produce mushrooms because of their short growing cycle, low environmental impact, high nutritional value, and efficient land management. Because of initiatives by various governmental and nongovernmental groups, the production and cultivation of mushrooms in Pakistan have the potential to cause a significant rise in interest among farmers and consumers (Khan et al. 2024).

A diverse range of mushrooms can be grown, offering a wealth of nutritious choices for local diets. Incorporating edible mushrooms into our daily meals shows great potential as a preventative and therapeutic strategy against chronic diseases,

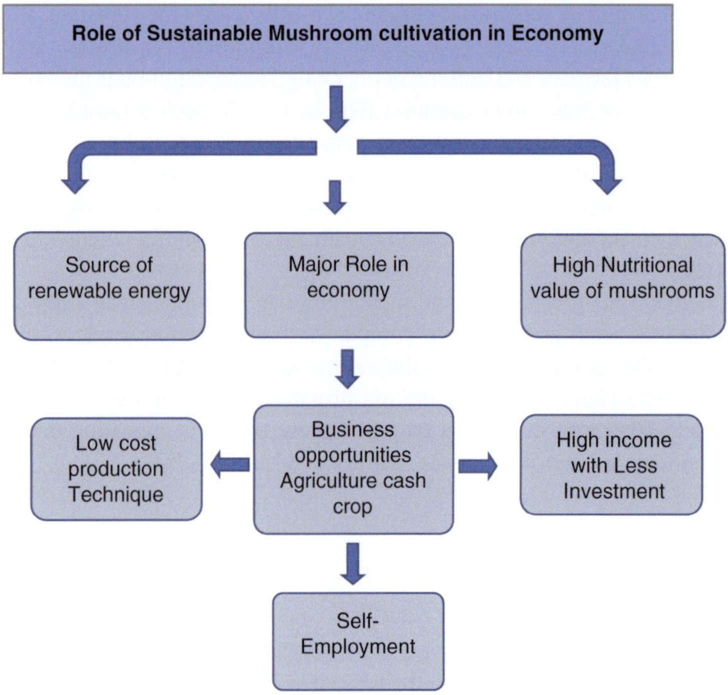

Fig. 6.3 Role of sustainable mushroom cultivation in the economy

thanks to their rich array of nutritional and functional properties (Qing et al. 2021; Losoya-Sifuentes et al. 2025).

Integrating mushroom cultivation with conventional farming methods can greatly enhance a circular economy. Mushrooms can grow on a variety of locally available plant materials and industrial by-products, such as paddy straw, wheat straw, soybean straw, cotton waste, coffee grounds, water hyacinth, tree sawdust, sugarcane bagasse, and wild grasses. These adaptable fungi employ their biochemical processes to transform waste products, which are rich in lignocellulosic content and possess an optimal carbon-to-nitrogen (C: N) ratio, into a highly nutritious food source. This approach requires minimal space while supplying all necessary nutrients (Singh 2011; Thakur 2020).

The primary problem of the twenty-first century will be providing sustainable food for the world's 8–11 billion inhabitants. Land use change, food waste, and the supply chain are only a few of the human activities directly related to agriculture that contribute to 21–37% of annual greenhouse gas emissions (Cowan et al. 2022).

Mushrooms not only recycle agricultural waste but also fill the nutritional gap of this large population. There is a 4.3% annual increase in mushroom cultivation every year. Recently, many people have been adopting mushroom cultivation to make their livelihoods better as well bridging the gap in the nutrition of the society (Vaishnavi et al. 2022).

Mushrooms are a rich source of protein, vitamins, minerals, and antioxidants. They can help address nutritional deficiencies and improve overall health. They are prized for their organoleptic properties, culinary versatility, and significant nutritional content. Serving as an excellent source of protein and essential amino acids, they also provide abundant dietary fiber while maintaining low-fat content (Reis et al. 2012).

Mushroom cultivation requires minimal initial investment and can be pursued on a small scale, making it accessible to low-income households and smallholder farmers. By selling fresh mushrooms and value-added products, farmers and entrepreneurs can generate additional income, which can be used to purchase other essential food items.

Mushrooms can be grown indoors and in controlled environments, making them less susceptible to the adverse effects of climate change compared to traditional crops. As food demand increases, climate models predict that conditions for crop production in much of the world will deteriorate due to increases in the frequency and intensity of extreme hydrometeorological events, such as extended periods of drought and flooding.

References

Adenipekun C, Lawal R. Uses of mushrooms in bioremediation: a review. Biotechnol Mol Biol Rev. 2012;7(3):62–8. https://doi.org/10.5897/BMBR12.006.

Adikaram NKB, Yakandawala DMD, Jayasinghe L. Leucocoprinus birnbaumii (Agaricales: Basidiomycota), an attractive yellow houseplant mushroom, revisited after 100 years. Ceylon J Sci. 2020;49(2):209–11.

Ahmad R, Riaz M, Khan A, Aljamea A, Algheryafi M, Sewaket D, Alqathama A. Ganoderma lucidum (Reishi) an edible mushroom; a comprehensive and critical review of its nutritional, cosmeceutical, mycochemical, pharmacological, clinical, and toxicological properties. Phytother Res. 2021;35(11):6030–62.

Al Mamun MR, Deb I, Hridoy T, Soeb MJA, Shammi S. Effects of different lighting conditions on growth, yield and nutrient content of white oyster Mushroom in vertical farm. Eur J Agric Food Sci. 2021;3(6):61–7.

Atila F. Cultivation and utilization of shiitake mushroom. In: Medicinal plants: domestication, biotechnology and regional importance; 2021. p. 383–413.

Barkat Md Gulzar A, Kumar Vandana U, Paul P, Mazumder PB. The role of mushrooms in biodegradation and decolorization of dyes. IntechOpen; 2020. https://doi.org/10.5772/intechopen.90737.

Benson KF, Stamets P, Davis R, Nally R, Taylor A, Slater S, Jensen GS. The mycelium of the Trametes versicolor (Turkey tail) mushroom and its fermented substrate each show potent and complementary immune activating properties in vitro. BMC Complement Altern Med. 2019;19(1):1–14.

Berovic M, Podgornik BB, Gregori A. Cultivation technologies for production of medicinal mushroom biomass. Int J Med Mushrooms. 2022;24(2):1.

Beyer DM. Impact of the Mushroom industry on the environment. Penn State Extension; 2023. https://extension.psu.edu/impact-of-the-mushroom-industry-on-the-environment?utm_source=chatgpt.com

Cowan N, Ferrier L, Spears B, Drewer J, Reay D, Skiba U. CEA systems: the means to achieve future food security and environmental sustainability? Front Sustain Food Syst. 2022;6:891256.

Cunha Zied D, Sánchez JE, Noble R, Pardo-Giménez A. Use of spent mushroom substrate in new mushroom crops to promote the transition towards a circular economy. Agronomy. 2020;10(9):1239.

Davila M, Muniz A, Du X. The impact of roasting and steaming on savory flavors contributed by amino acids, 5′-nucleotides, and volatiles in Agaricus bisporus mushrooms. Int J Gastron Food Sci. 2022;30:100590.

Dehariya P, Vyas D. Effect of different agro-waste substrates and their combinations on the yield and biological efficiency of Pleurotus sajor-caju. IOSR J Pharm Biol Sci. 2013;8(3):60–4.

Dorr E, Koegler M, Gabrielle B, Aubry C. Life cycle assessment of a circular, urban mushroom farm. J Clean Prod. 2021;288:125668.

Genetics FC. Environmental and ecological aspects of Mushroom cultivation. Full Canopy; 2023. https://fullcanopygenetics.com/environmental-and-ecological-aspects/

Grimm D, Wösten HA. Mushroom cultivation in the circular economy. Appl Microbiol Biotechnol. 2018;102:7795–803.

Gupta N, Mehta M, Singh K. Benefits, challenges and opportunities in mushroom production: a review. Pharma Innov J. 2022;SP-11(3):360–4.

Gupta R, Uniyal PL, Kumar Y. Chapter 1: Mushroom cultivation in relation to vertical farming: a booming entrepreneurship. In: Futuristic trends in biotechnology, vol. 3; 2024. e-ISBN: 978-93-6252-440-9.

Iram S. Economic analysis of production and Marketing of Button Mushroom in Central Kashmir, Doctoral dissertation, SKUAST Kashmir. 2022.

Ivarsson E, Grudén M, Södergren J, Hultberg M. Use of faba bean (Vicia faba L.) hulls as substrate for Pleurotus ostreatus–potential for combined mushroom and feed production. J Clean Prod. 2021;313:127969.

Jasińska A. Spent mushroom compost (SMC)–retrieved added value product closing loop in agricultural production. Acta Agraria Debreceniensis. 2018:185–202.

Jasinska A. Sustainability of mushroom cultivation systems. Horticulturae. 2023;9(11):1191. https://doi.org/10.3390/horticulturae9111191.

Jayaraman S, Yadav B, Dalal RC, Naorem A, Sinha NK, Rao CS, et al. Mushroom farming: a review focusing on soil health, nutritional security and environmental sustainability. Farming Syst. 2024;2(3):100098.

Joseph B, Krishnakumar NM, Ceasar SA, Devassy SM. Role of rural women in achieving United Nations sustainable development goals through Mushroom agri-entrepreneurship: Mushroom cultivation as a sustainable livelihood option for rural women. In: The role of female leaders in achieving the sustainable development goals. IGI Global; 2024. p. 214–31.

Khan A, Murad W, Ali S, Shah SS, Halim SA, Khalid A, et al. Contribution of mushroom farming to mitigating food scarcity: current status, challenges and potential future prospects in Pakistan. Heliyon. 2024;10:e40362.

Li B, Ding Y, Tang X, Wang G, Wu S, Li X, et al. Effect of L-arginine on maintaining storage quality of the white button mushroom (Agaricus bisporus). Food Bioprocess Technol. 2019;12:563–74.

Libretexts. 24.2A: fungi habitat, decomposition, and recycling. Biology LibreTexts; 2024. https://bio.libretexts.org/Bookshelves/Introductory_and_General_Biology/General_Biology_%28Boundless%29/24%3A_Fungi/24.02%3A_Ecology_of_Fungi/24.2A%3A_Fungi_Habitat_Decomposition_and_Recycling

Losoya-Sifuentes C, Cruz M, del Refugio Rocha-Pizaña M, Loredo-Treviño A, Belmares R. Edible mushrooms: a nutrient-rich ingredient for healthier food products–a review. Curr Nutr Rep. 2025;14(1):9.

lovemyindia425. Mushroom cultivation using organic waste: 3 types of mushrooms – guidebest compost techniques. Guidebest Compost Techniques; 2023. https://guidebest.in/mushroom-cultivation-using-organic-waste/

Mushroom Spawn FAQ. North Spore. Retrieved 24 from https://northspore.com/pages/spawn-faq
Mwangi RW, Mustafa M, Kappel N, Csambalik L, Szabó A. Practical applications of spent mushroom compost in cultivation and disease control of selected vegetables species. J Mater Cycles Waste Manag. 2024;26(4):1918–33. https://doi.org/10.1007/s10163-024-01969-9.
Nguyet LTB, Hung TV, Phu PQ, Cuong NM, Bui TT. A controlled environmental chamber for growing Cordyceps Militaris. In: The international conference on sustainable energy technologies. Singapore: Springer Nature Singapore; 2023. p. 53–60.
Nongthombam J, Kumar A, Ladli BGVVSN, Madhushekhar M, Patidar S. A review on study of growth and cultivation of oyster mushroom. Plant Cell Biotechnol Mol Biol. 2021;22(5–6):55–65.
Nowakowski P, Naliwajko SK, Markiewicz-Żukowska R, Borawska MH, Socha K. The two faces of Coprinus comatus—functional properties and potential hazards. Phytother Res. 2020;34(11):2932–44.
Ozuna-Valencia KH, Moreno-Robles AL, Rodríguez-Félix F, Moreno-Vásquez MJ, Barreras-Urbina CG, Madera-Santana TJ, et al. Black trumpet (Craterellus cornucopioides). In: Mushrooms. CRC Press; 2023. p. 131–52.
Pawar JK, Umesh. Mushroom cultivation on agricultural wastes for environmental sustainability and food security. Int J Bot Stud. 2021;6(3):279–82. https://www.botanyjournals.com/assets/archives/2021/vol6issue3/6-3-31-841.pdf
Pearson S. How mushrooms are decomposers (mushrooms unmasked). MushroomGoodCom; 2024. https://mushroomgood.com/how-mushrooms-are-decomposers/
Peng JT. Agro-waste for cultivation of edible mushrooms in Taiwan. Food and Fertilizer Technology Center; 2008.
Qing Z, Cheng J, Wang X, Tang D, Liu X, Zhu M. The effects of four edible mushrooms (Volvariella volvacea, Hypsizygus marmoreus, Pleurotus ostreatus and Agaricus bisporus) on physicochemical properties of beef paste. LWT. 2021;135:110063.
Rahmann G, Azim K, Brányiková I, Chander M, David W, Erisman JW, et al. Innovative, sustainable, and circular agricultural systems for the future. Org. Agric. 2021;11:1–7. https://doi.org/10.1007/s13165-021-00356-0.
Reis FS, Barros L, Martins A, Ferreira IC. Chemical composition and nutritional value of the most widely appreciated cultivated mushrooms: an inter-species comparative study. Food Chem Toxicol. 2012;50(2):191–7.
Richard F, Bellanger JM, Clowez P, Hansen K, O'Donnell K, Urban A, et al. True morels (Morchella, Pezizales) of Europe and North America: evolutionary relationships inferred from multilocus data and a unified taxonomy. Mycologia. 2015;107(2):359–82.
Rukhiran M, Sutanthavibul C, Boonsong S, Netinant P. IoT-based mushroom cultivation system with solar renewable energy integration: assessing the sustainable impact of the yield and quality. Sustainability. 2023;15(18):13968.
Sayner A. A complete guide to Mushroom substrates. GroCycle; 2024. https://grocycle.com/mushroom-substrate/
Shackleton S, Delang CO, Angelsen A. From subsistence to safety nets and cash income: exploring the diverse values of non-timber forest products for livelihoods and poverty alleviation. Berlin/Heidelberg: Springer; 2011. p. 55–81.
Sharma SK, Roy SS, Ansari MA, Ningombam A, Punitha P, Surjit T. Integration of mushroom cultivation in integrated farming system. In: Integrated farming system for sustainable hill agriculture: an option for climate smart agriculture & natural resource management; 2017. p. 161.
Sharma VP, Barh A, Bairwa RK, Annepu SK, Kumari B, Kamal S. Enoki mushroom (Flammulina velutipes (Curtis) singer) breeding. In: Advances in plant breeding strategies: vegetable crops: volume 10: leaves, Flowerheads, green pods, mushrooms and truffles; 2021. p. 423–41.
Sheffield TUO. Fungi stores a third of carbon from fossil fuel emissions and could be essential to reaching net zero. The University of Sheffield; 2023. https://www.sheffield.ac.uk/news/fungi-stores-third-carbon-fossil-fuel-emissions-and-could-be-essential-reaching-net-zero

Shukla PR, Skeg J, Buendia EC, Masson-Delmotte V, Pörtner HO, Roberts DC, …, Malley J. Climate change and land: an IPCC special report on climate change, desertification, land degradation, sustainable land management, food security, and greenhouse gas fluxes in terrestrial ecosystems; 2019. https://philpapers.org/rec/SHUCCA-2.

Sivrikaya H, Bacak L, Saraçbaşı A, Toroğlu I, Eroğlu H. Trace elements in Pleurotus sajor-caju cultivated on chemithermomechanical pulp for bio-bleaching. Food Chem. 2002;79(2):173–6.

Singh M, Vijay B, Kamal S, Wakchaure GC. Mushrooms Cultivation, Marketing and Consumption. New Delhi: Indian Council of Agricultural Research; 2011.

Tachabenjarong N, Rungsardthong V, Ruktanonchi U, Poodchakarn S, Thumthanaruk B, Vatanyoopaisarn S, et al. Bioactive compounds and antioxidant activity of lion's mane mushroom (Hericium erinaceus) from different growth periods. In: E3S web of conferences, vol. 355. EDP Sciences; 2022. p. 02016.

Tang YJ, Zhu LW, Li HM, Li DS. Submerged culture of mushrooms in bioreactors–challenges, current state-of-the-art, and future prospects. Food Technol Biotechnol. 2007;45(3):221–9.

Thakur MP. Advances in mushroom production: key to food, nutritional and employment security: a review. Indian Phytopathol. 2020;73:377–95.

Vaishnavi M, Sharma A, Tiwari J, Singh S, Sharma S. Production of edible mushrooms to meet the food security: a review. J Posit School Psychol. 2022:4316–25.

Wang J, Sun W, Luo H, He H, Deng WQ, Zou K, et al. Protective effect of eburicoic acid of the chicken of the woods mushroom, Laetiporus sulphureus (higher Basidiomycetes), against gastric ulcers in mice. Int J Med Mushrooms. 2015;17(7):619.

Warrick H. The economics of Mushroom farming: financial insights for new farmers. Mushroom Media Online; 2024. https://mushroommediaonline.com/economics-of-mushroom-farming/

Wu CY, Liang CH, Liang ZC. Evaluation of using spent mushroom sawdust wastes for cultivation of Auricularia polytricha. Agronomy. 2020;10(12):1892.

Wu JY, Siu KC, Geng P. Bioactive ingredients and medicinal values of Grifola frondosa (Maitake). Foods. 2021;10(1):95.

Chapter 7
Insect Pests and Diseases of Cultivated Mushrooms

Amer Habib, M. Talha Ashraf, M. Madni Afzal, Sadia Murad, M. Tahir Murad, Ishrat Fatima, Waqar Akram, and Hayssam Mohamed Ali

Abstract The natural resource known as mushrooms has the potential to be used as a functional food for both medical and nutritional purposes. Essential fatty acids, which are polyunsaturated fatty acids that are essential for optimal health and well-being, are abundant in them. Given their nutritional and medicinal importance, mushrooms are a significant vegetable that typically flourishes in forests. People without land can also grow it on a limited basis at home. Growing mushrooms is a profitable agricultural occupation. Sciarid flies are among the most challenging pests in mushroom cultivation due to their direct feeding damage, contamination of fruiting bodies, and ability to vector pathogens. Effective management requires an integrated approach combining environmental controls, biological methods, and chemical interventions. By understanding their biology and behavior, growers can implement measures to reduce infestations and minimize economic losses. Integrated pest management (IPM) for phorid flies offers a sustainable and effective approach for controlling these pests while minimizing environmental impact. By combining monitoring, sanitation, biological control, mechanical methods, and judicious chemical treatments, an IPM program can significantly reduce the prevalence of phorid flies in urban, agricultural, and food-handling environments. Successful management requires ongoing effort, careful monitoring, and a holistic approach that addresses the root causes of infestations.

A. Habib (✉) · M. T. Ashraf · M. M. Afzal · M. T. Murad · W. Akram
Department of Plant Pathology, Faculty of Agriculture, University of Agriculture, Faisalabad, Pakistan
e-mail: amer.habib@uaf.edu.pk

S. Murad
Department of Soil Science, Faculty of Agriculture and Environment Science, The Islamia University of Bahawalpur, Bahawalpur, Pakistan

I. Fatima
Department of Botany, University of Agriculture, Faisalabad, Pakistan

H. M. Ali
Department of Botany and Microbiology, College of Science, King Saud University, Riyadh, Saudi Arabia

I. U. Haq et al. (eds.), *Mushrooms*, https://doi.org/10.1007/978-981-95-2539-3_7

Keywords Mushrooms · Essential fatty acids · Sciarid flies · Integrated Pest Management (IPM) · Phorid flies

7.1 Introduction

7.1.1 *Importance of Cultivated Mushrooms in Agriculture and Economy*

Mushrooms are natural resources that can serve as functional foods for nutritional and medicinal purposes. They are a good source of essential fatty acids, which are polyunsaturated fatty acids vital for health and well-being (Adedokun et al. 2022). Because of their nutritive and medicinal value, mushrooms are an important vegetable that usually grows in the forest. They can also be cultivated domestically on a small scale by landless people. The cultivation of mushrooms is one of the lucrative agricultural jobs (Carvalho et al. 2024).

By 2050, agriculture will be an industry that requires precise information and specific protocols. The complete recycling of nutrients, water, and agro-waste will be a necessity rather than a luxury (Ghose and Mitra 2022). With the world's increasing population and its decrease in per capita arable land, along with rapid urbanization and industrialization, climate change, and a growing demand for quality and functional foods, it will be necessary to focus on secondary agriculture and novel crops such as mushrooms (Bell et al. 2022). Diversification in farming systems improves sustainability, and mushroom production contributes to this. The commercial production of edible mushrooms represents the unique exploitation of microbial technology for the bioconversion of agricultural, industrial, forestry, and household wastes into nutritious food (mushrooms) (Kosre et al. 2021). Currently, 14,000 mushroom species are known to exist. Of these, 31 genera are regarded as prime edible mushrooms. Only 200 of these are experimentally grown, and 100 are cultivated economically, with approximately 60 being commercially relevant and ~10 having reached industrial-scale production (Kaliyaperumal et al. 2018).

7.1.2 *Challenges Faced by Mushroom Growers*

1. *Temperature fluctuations*: Mushrooms require specific temperature ranges for optimal growth; even slight fluctuations can severely impact their development. Inconsistent temperatures can slow growth, encourage disease development, or reduce the overall yield.
2. *Humidity control*: Mushrooms thrive in high-humidity environments, but excessive moisture can also lead to mold or bacterial growth, damaging the entire

crop. Proper humidity control is crucial to maintaining balance and preventing costly losses.

3. *Ventilation issues*: Mushrooms need a well-ventilated growing environment. Poor ventilation can lead to higher levels of CO_2, which negatively affects mushroom quality and growth. Ensuring proper air circulation is essential for maintaining healthy crops.
4. *Energy consumption*: Running climate control systems for mushroom farming can be expensive, especially when using inefficient equipment. High-energy costs can eat into profits, particularly for smaller operations.

7.2 Insect Pests of Cultivated Mushrooms

7.2.1 Sciarid Flies (Fungus Gnats)

7.2.1.1 Identification and Life Cycle

Fungus gnats, belonging to the family *Sciaridae* in the order *Diptera*, are small flies commonly associated with damp environments and decaying organic material (Merritt et al. 2009). These pests are frequently found in greenhouses, indoor plants, and agricultural systems, where they can damage plant roots during their larval stage. Proper identification and understanding of their life cycle are essential for effective management.

Fungus gnats can be identified by their morphological and behavioral characteristics.

1. *Adults*:

    Size: Adult fungus gnats are small, typically 2–5 mm long.

    Color: They are dark gray to black, with a slender body.

    Wings: Adults possess *transparent or smoky gray wings*, often held flat over their abdomen. A *Y-shaped vein pattern* is a distinguishing feature.

    Antennae: Their long, thread-like antennae are longer than their head.

    Behavior: Adults are weak fliers and often hover around moist soil or plants. They are attracted to light and may congregate around windows or lamps.

2. *Larvae*:

    - The larval stage is white or translucent, with a distinct *black head capsule*.
    - They are 4–10 mm long and found in the topsoil, feeding on organic matter, fungi, algae, and occasionally plant roots.

3. *Eggs and Pupae*:

    Eggs: Tiny, oval-shaped, and difficult to observe without magnification, laid in clusters in moist soil.

    Pupae: Pupae are found in the soil, where larvae metamorphose into adults.

7.2.1.2 Damage Caused

Sciarid flies, commonly called fungus gnats, are significant pests in mushroom cultivation systems. These flies, primarily from the genera Bradysia and Lycoriella, cause direct and indirect damage to cultivated mushrooms. Their presence impacts the yield and quality of mushrooms and creates favorable conditions for secondary infections by pathogens. This note provides a detailed overview of the damage caused by sciarid flies in mushroom production, focusing on economic and biological impacts (Buxton and Mallet 1980).

Direct Damage

1. *Feeding Damage*:

   - Larvae of sciarid flies feed on the mycelium of mushrooms, resulting in significant disruption to the growth and development of the fungal network.
   - Damaged mycelium exhibits reduced vigor, leading to diminished mushroom production.
   - Feeding activity can also lead to dry, patchy areas in the compost or substrate where mycelium fails to colonize effectively (Jess and Kilpatrick 2000).

2. *Physical Contamination*:

   - Adult flies often contaminate mushroom-fruiting bodies by resting on their surfaces. This contamination reduces the marketability of the mushrooms.
   - The presence of dead adults in the growing media or harvested mushrooms further degrades product quality.

3. *Crop Yield Reduction*:

   - Heavy infestations result in lower yields due to the destruction of the mycelium and interference with normal mushroom growth cycles.
   - In severe cases, infestations can render entire batches of mushrooms unharvestable.

Indirect Damage

1. *Vectoring of Pathogens*:

   - Sciarid flies act as vectors for a variety of fungal and bacterial pathogens, including:
   - *Verticillium fungicola* (dry bubble disease)
   - *Mycogone perniciosa* (wet bubble disease)
   - *Pseudomonas tolaasii* (bacterial blotch)
   - These pathogens can spread rapidly through mushroom beds, causing widespread crop losses.

2. *Facilitation of Secondary Infections*:
   - Wounds caused by larval feeding create entry points for opportunistic pathogens, exacerbating the damage to mushroom crops.
   - The combination of physical damage and introduced pathogens creates a compounded negative effect on the health and productivity of the crop.
3. *Disruption of Environmental Conditions*:
   - Infestations increase the biological load in mushroom production facilities, leading to higher levels of ammonia and CO_2, which can negatively affect mushroom growth.
   - The presence of flies also attracts other pests or scavengers, compounding the problem.

7.2.1.3 Economic Impact

1. *Loss of Revenue*:
   - Yield reductions and quality degradation due to Sciarid fly infestations lead to significant economic losses for mushroom growers.
   - Due to contamination and reduced quality, infected batches may be rejected in the marketplace or sold at a lower price.
2. *Increased Management Costs*
   - Managing sciarid fly populations involves additional costs for implementing integrated pest management (IPM) strategies, such as:
   - Use biological control agents (e.g., predatory mites like *Hypoaspis miles* and nematodes like *Steinernema feltiae*).
   - Application of insecticides or other control measures.
   - Preventive measures, such as maintaining hygiene and modifying environmental conditions, add to the operational expenses.

7.2.1.4 Management Strategies

To mitigate the damage caused by sciarid flies in mushroom production, the following strategies are commonly employed:

1. *Environmental Management*
   - Maintaining optimal moisture levels in compost to discourage egg-laying by adult flies.
   - Ensuring proper ventilation to minimize humidity buildup and reduce fly activity.

2. *Biological Controls*

   - Introducing natural predators such as predatory mites (*Hypoaspis miles*) and parasitic nematodes (*Steinernema feltiae*), which target larvae in the substrate (Grewal et al. 2005; Harris and Oetting 1997).

3. *Physical Barriers*

   - Using insect-proof screens and sealing cracks to limit adult fly entry into growing rooms.
   - Installing sticky traps to monitor and reduce adult populations.

4. *Chemical Controls*

   Judicious use of approved insecticides, targeting both larvae and adults, while minimizing harm to the mushrooms and the environment (White 1986).

7.2.1.5 Conclusion

Sciarid flies are among the most challenging pests in mushroom cultivation due to their direct feeding damage, contamination of fruiting bodies, and ability to vector pathogens. Effective management requires an integrated approach combining environmental controls, biological methods, and chemical interventions. By understanding their biology and behavior, growers can implement measures to reduce infestations and minimize economic losses.

7.2.2 Phorid Flies (Humpbacked Flies)

7.2.2.1 Identification and Behavior

Due to their distinctive body shape, phorid flies, called humpbacked flies, belong to the family *Phoridae* in the order *Diptera*. These small flies are widely distributed across various habitats and are often associated with decaying organic matter. Phorid flies are known for their characteristic "humpbacked" appearance, erratic flight patterns, and diverse ecological roles, ranging from decomposers to parasites (Disney 2008). This note provides a detailed overview of their identification and behavior. Phorid flies can be identified by their unique morphological characteristics and distinguishing features:

1. *Physical Characteristics*

   Size: Small, typically ranging from 0.5 to 6 mm in length.

   Color: Most species are dull brown, black, or yellowish.

   Body shape: A distinctive *humpbacked thorax* gives them a unique, arched appearance, making them easily recognizable.

Wings: Phorid flies have *short and broad wings*, with reduced venation compared to other flies. Their wing veins may appear incomplete or faint.

Antennae: Short antennae with an arista (bristle-like structure) are present.

Legs: Long legs, particularly the hind legs, aid in their jumping and running behaviors.

2. *Larvae*

   - The larvae are legless, cylindrical, and taper at one end.
   - They are typically white or translucent and can be found in decaying organic matter, fungi, or carrion.

3. *Eggs and Pupae*

   - Eggs are tiny and laid in suitable substrates such as decaying material or animal remains.
   - Pupae are oval-shaped and develop in the same substrate as the larvae.

7.2.2.2 Behavior

Phorid flies exhibit various behaviors influenced by their ecological roles and environmental conditions.

1. *Flight Pattern*

   - Phorid flies are known for their *erratic and darting flight behavior*, differentiating them from other small flies, such as fruit flies.
   - They are weak fliers and often prefer to run or jump on surfaces rather than fly over long distances.

2. *Feeding Habits*

   Larvae

   - Larvae feed on decaying organic matter, including decomposing plant material, animal remains, and fungi.
   - In some species, larvae exhibit parasitic behavior, attacking live hosts such as ants, bees, or other insects.

   Adults

   - Adults feed on various substances, including nectar, decomposing organic matter, and fungal spores.

3. *Reproductive Behavior*

   Females lay eggs in decaying organic material or other substrates that provide sufficient food for the larvae. In parasitic species, females may lay eggs on or inside the body of a host insect (Fatchurochim et al. 1989).

4. *Ecological Roles*

   Decomposers: Many species play a crucial role in breaking down organic matter, aiding nutrient recycling.

Parasites: Certain species exhibit parasitic behavior, particularly targeting social insects like ants and bees. For instance, *Apocephalus borealis* lays its eggs inside bees, leading to the phenomenon known as zombie bees (Greenberg 1973).

Disease transmission: Phorid flies can mechanically transmit pathogens due to their association with decaying matter, animal waste, and human environments.

5. *Habitat and Activity*
   - Phorid flies thrive in moist, decaying organic environments such as compost piles, garbage bins, animal carcasses, and sewer systems.
   - They are also common in indoor environments, particularly where hygiene is poor, such as hospitals, kitchens, and food storage areas.
6. *Response to Environmental Stimuli*
   - Phorid flies are attracted to light and often gather near windows or light sources.
   - They are highly responsive to odors emitted from decaying organic matter, guiding their feeding and oviposition behaviors.

7.2.2.3 Economic and Medical Importance

Phorid flies are of concern in both agricultural and medical contexts:

1. *Pest Status*
   - In food production and storage facilities, phorid flies can become a nuisance due to their association with decomposing organic matter and their ability to contaminate food.
2. *Disease Transmission*
   - They are potential mechanical vectors of pathogens such as bacteria and fungi, posing risks in healthcare and food-handling environments.
3. *Impact on Livestock*

   Some parasitic species target livestock and poultry, causing stress and health issues.
4. *Infestation Symptoms*

   Phorid flies, also known as humpbacked flies, can infest a variety of environments, including homes, healthcare facilities, food storage areas, and agricultural setups. Recognizing the symptoms of their infestation is critical for timely management and prevention of further problems. Below are a phorid fly infestation's key signs and symptoms (Baker and Dunn 2011).

1. *Presence of Adult Flies*

   Appearance: Adult phorid flies are small (0.5–6 mm), with a distinct humpbacked thorax and dark brown, black, or yellowish coloration.

Erratic flight: Their characteristic darting and erratic flight patterns, as opposed to smooth flying, clearly indicate their presence.

Clustering near moist areas: Adults are often seen near sinks, drains, garbage bins, decomposing organic matter, or areas with leaks and standing water (Disney 2008).

2. *Larvae in Decaying Organic Matter*

   Larval activity: Larvae, which are white or translucent and taper at one end, can often be found feeding on decomposing organic material, including food waste, animal carcasses, and fungi.

   Breeding sites: Larvae are frequently found in moist, nutrient-rich areas such as:

- Compost piles.
- Clogged drains.
- Soil in overwatered potted plants.
- Animal waste or remains.

3. *Odors*

   Foul smell: Infested areas often have a strong, unpleasant odor caused by the decaying organic material that serves as breeding and feeding grounds for phorid flies.

- *Associated with breeding sites*: The odor is usually strongest near garbage bins, drains, or organic waste accumulation.

4. *Infestation in Unusual Locations*

   Phorid flies are known for exploiting a wide range of breeding materials and environments. Their presence in uncommon places, such as wall voids, crawl spaces, and decaying insulation, can indicate an infestation.

   Dead animals: Phorid flies can also signal the presence of dead rodents, birds, or other animals in hard-to-reach areas, such as wall cavities or attics.

5. *Contamination*

   Food contamination: Phorid flies may land on and contaminate food, food preparation areas, or storage facilities, introducing bacteria and pathogens.

   Visible adults on surfaces: Seeing adult flies resting on food, counters, or near drains is a strong sign of infestation.

6. *Wounds and Parasitism (Rare but Notable)*

   In rare cases, parasitic phorid fly species may infest living hosts, including wounds in humans or animals. This condition, known as *myiasis*, is typically associated with poor hygiene or neglected wounds.

7. *Increase in Fly Population*

   A sudden increase in small flies in the home or facility can signal an ongoing infestation. Adult phorid flies have a short lifecycle (approximately 20–28 days), meaning breeding populations can grow rapidly under favorable conditions.

8. *Attraction to Light*

   Adult phorid flies are drawn to light and often congregate around windows, light fixtures, or lamps, especially during the evening.

9. *Drains and Plumbing Issues*
   Phorid flies often breed in organic matter inside drains, pipes, or septic systems. If flies emerge from sinks, showers, or other drainage points, it could indicate a buildup of organic debris or clogs in the plumbing.

7.2.2.4 Integrated Pest Management (IPM)

Phorid, or humpbacked flies, are notorious pests in urban and agricultural environments. They often infest areas with decaying organic matter, dampness, and poor hygiene. These flies are a nuisance and pose significant risks by spreading pathogens and contaminating food products, medical environments, and crops. Effective control of phorid flies requires an integrated pest management (IPM) approach that combines multiple strategies, such as biological, chemical, cultural, and mechanical methods, to minimize their impact in a sustainable and environmentally friendly manner.

Components of Integrated Pest Management (IPM) for Phorid Flies
Integrated pest management is a holistic approach that aims to control pest populations through a combination of strategies while minimizing the use of chemical pesticides (Pfeiffer and Dove 2009). The goal of IPM for phorid flies is to reduce their populations to acceptable levels, prevent future infestations, and mitigate potential damage to crops, facilities, and public health. The following are key components of IPM for phorid flies:

1. *Monitoring and Detection*
   Monitoring and early detection are essential in identifying the presence of phorid flies and determining the extent of the infestation. Effective monitoring helps ensure that control measures are implemented promptly.
   Fly traps: Sticky and *pheromone traps* can effectively capture adult phorid flies. These traps provide valuable information about the density of the infestation and help track fly movement. They can be placed near potential breeding sites such as drains, trash areas, or compost piles.
   Inspection of breeding sites: Regular inspection of breeding sites is crucial. Phorid flies typically breed in *decaying organic matter, moist environments, and food waste*. Common breeding sites include *garbage bins, compost piles, drains, sewage systems*, and *decaying organic material* in homes, kitchens, and food processing areas.
   Larval sampling: Besides monitoring adults, sampling larvae from suspected breeding sites is important. Identifying larvae presence helps confirm that breeding is occurring and provides insights into the developmental stages, which are critical for effectively targeting control strategies effectively targeting control strategies.
2. *Sanitation and Cultural Control*
   Sanitation is one of the most important and effective components of IPM, as it helps eliminate breeding sites and prevent the attraction of phorid flies.

Remove decaying organic material: Phorid flies are attracted to decaying organic matter, so it is essential to *remove food scraps, decomposing vegetables, and animal waste* from kitchens, garbage bins, and food storage areas. Proper organic waste disposal in *sealed containers* is crucial to prevent fly attraction.

Cleaning and decontaminating breeding sites: *Drains, garbage disposals, and wastewater systems* are common breeding grounds for phorid flies. Regular cleaning and disinfection of these areas, using nontoxic cleaners or drain treatments, will reduce the organic buildup and remove breeding substrates. For drains, *drain cleaners* or *microbial agents* may help eliminate organic matter.

Moisture control: Reducing moisture levels in areas prone to dampness is key to limiting fly reproduction. Fixing leaks, improving ventilation, and ensuring no standing water sources (e.g., in sinks, bathrooms, and basements) will help make the environment less conducive to phorid fly survival.

3. *Biological Control*

   Biological control is environmentally friendly and sustainable for managing phorid fly populations. It involves using natural predators or microorganisms to reduce fly numbers without the need for harmful chemicals.

   Predators: Several natural predators can help control phorid fly populations. For example, *predatory mites* and *nematodes* can be introduced to target phorid fly larvae. These predators feed on the larvae, reducing their numbers and disrupting the fly lifecycle.

   Parasitic flies: In certain situations, specific parasitic fly species, such as those from the genus *Pseudacteon*, may be used to control populations of specific pests like fire ants. While *Pseudacteon* parasitic flies are not commonly used for phorid fly management, their potential application in biological control is an area of ongoing research.

4. *Mechanical Control*

   Mechanical control strategies focus on physically preventing phorid flies from accessing areas where they might breed or thrive.

   Fly screens: Installing *mesh screens* on windows and doors is an effective physical barrier that prevents adult phorid flies from entering indoor spaces.

   Seal cracks and gaps: Phorid flies can enter buildings through cracks and gaps in windows, doors, and walls. Sealing these openings can help minimize the risk of flies entering a structure from external sources.

   Traps: In addition to sticky traps, *electric fly zappers* or light traps can catch adult flies. These traps effectively reduce the adult fly population and provide additional monitoring insights.

5. *Chemical Control*

   While chemical control is typically used as a last resort in an IPM program, it may be necessary in severe infestations or when other methods have proven ineffective.

   Insecticides: Insecticides, particularly those designed for fly control, can be applied to *breeding sites* and *fly resting areas* to target adult flies and larvae. Using residual insecticides can help reduce fly populations for an extended period. However, insecticides should be applied judiciously to avoid harming beneficial insects or polluting the environment.

Insect growth regulators (IGRs): IGRs, which disrupt the development of insect larvae, can be used as an alternative to conventional insecticides. These chemicals prevent larvae from reaching maturity and reproducing, limiting the population.

7.2.2.5 Best Practices for IPM in Phorid Fly Management

1. *Comprehensive approach*: IPM for phorid flies should always combine multiple strategies tailored to the specific situation. A *one-size-fits-all approach* is less likely to succeed, as the fly's habits and breeding sites may differ depending on the environment.
2. *Regular monitoring and inspection*: Monitoring fly populations and inspecting potential breeding sites are key to successful management. Early detection ensures that control measures are implemented before an infestation becomes widespread.
3. *Education and training*: Educating employees or residents about sanitation and hygiene is essential to long-term control. Encouraging proper waste disposal, drain maintenance, and moisture management will help prevent future infestations.
4. *Evaluation and adjustment*: After implementing control measures, it is important to evaluate their effectiveness regularly and adjust tactics as needed. This might involve switching to different traps, using alternative biological control methods, or making environmental adjustments.

7.2.2.6 Conclusion

Integrated pest management (IPM) for phorid flies offers a sustainable and effective approach to controlling these pests while minimizing environmental impact. By combining monitoring, sanitation, biological control, mechanical methods, and judicious chemical treatments, an IPM program can significantly reduce the prevalence of Phorid flies in urban, agricultural, and food-handling environments. Successful management requires ongoing effort, careful monitoring, and a holistic approach to addressing infestations' root causes.

7.2.3 Mites

Mites are a significant pest in mushroom cultivation, with various species causing direct and indirect damage to the crop. The primary mite species that affect mushrooms include bulb mites (*Rhizoglyphus* spp.), straw mites (*Tarsonemus* spp.), and predatory mites that become pests in high populations. Their presence in mushroom farms can lead to substantial economic losses and reduced yield quality (Ekbom and Olsson 2021). Below is a detailed account of their impact:

7.2.3.1 Life Cycle of Bulb Mites

Bulb mites, belonging to the genus *Rhizoglyphus*, are common soil-dwelling pests that thrive in warm, moist environments. These mites are known for infesting various crops, including mushrooms. The life cycle of bulb mites is divided into the following stages:

1. *Egg*
   Female bulb mites lay their eggs in substrate or host material clusters. The eggs are oval, smooth, and translucent. A single female can lay up to 200 eggs over her lifespan, which hatch within 3–6 days under optimal conditions (20–25 °C).
2. *Larva*
   After hatching, the larva emerges as a six-legged, immobile stage that feeds sparingly on organic matter for 2–3 days before molting into the nymphal stage.
3. *Nymph*
   The nymph has eight legs and is more active than the larva. It undergoes two molting phases, the protonymph and deutonymph, during which it actively feeds on plant material, fungi, or other organic substrates. The nymphal stage lasts approximately 4–7 days.
4. *Adult*
   The adult mite is pear-shaped, creamy-white, or yellowish with brownish legs. It has a hard outer cuticle and specialized mouthparts for feeding on soft plant tissues and fungi. Adults live for 2–4 weeks and reproduce rapidly under favorable conditions. Under optimal conditions, the entire life cycle of bulb mites can be completed in as little as 2 weeks, leading to rapid population growth in infested areas.

7.2.3.2 Impact of Bulb Mites on Mushroom Cultivation

Bulb mites are significant pests in mushroom cultivation, primarily affecting species like *Agaricus bisporus*. They cause damage directly and indirectly, resulting in reduced yield and quality.

1. *Direct Damage*
   Feeding on mycelium: Bulb mites feed on the mycelium of mushrooms, destroying the fungal network. This disrupts nutrient uptake and the growth of fruiting bodies.
   Physical damage: Their feeding activity creates wounds on the mycelium and mushroom caps, leaving them susceptible to secondary infections.

2. *Indirect Damage*

   Transmission of pathogens: Bulb mites act as vectors for fungal and bacterial pathogens, such as *Verticillium fungicola*, which causes dry bubble disease in mushrooms.

   Contamination of substrate: Mites carry contaminants and spores of undesirable fungi, reducing the quality of the growing medium.
3. *Economic Losses*

   Mushroom cultivation is a high-value industry, and infestations can lead to significant economic losses due to decreased marketability and yield. Severely infested batches often need to be discarded entirely.

7.2.3.3 Control Measures

Managing bulb mite infestations in mushroom cultivation requires an integrated pest management (IPM) approach, including:

Sanitation: Ensuring clean substrates, tools, and growing environments to reduce initial mite populations.

Biological control: Introducing predatory mites, such as *Hypoaspis miles*, to control bulb mite populations (Zhang and Kells 2018).

Cultural practices: Maintaining optimal humidity and temperature levels to discourage mite proliferation.

Chemical control: Acaricides are applied as a last resort, ensuring they are safe for mushroom production and have minimal environmental impact (Sánchez et al. 2020).

7.3 Fungal Diseases

7.3.1 Trichoderma *spp. (Green Mold)*

7.3.1.1 Causal Organism

Trichoderma spp., a genus of saprophytic fungi belonging to the family *Hypocreaceae*, is a frequent contaminant in mushroom farming (Campbell and Anselme 2011).

7.3.1.2 Symptoms

1. Dense, greenish mold growth will be seen in the substrate or mushroom mycelium-affected areas.

2. Mushrooms may be stunted or malformed, and growth begins at the base of the fruiting body or substrate.
3. *Trichoderma* can outcompete the edible mushroom mycelium for nutrients by rapidly colonizing substrates.
4. *Trichoderma* exposure can cause serious damage to mushrooms, which can manifest as symptoms like decreased size and changed cap shapes.

7.3.1.3 Management

Prevention: Strict hygiene is essential to prevent *Trichoderma* infection. Growing environments, tools, and equipment should all be routinely sterilized.

Environmental control: Ensure adequate ventilation and less moisture in the growing environment. Static and humid air are ideal for *Trichoderma* growth.

Biological control: By acting as *Trichoderma's* competitors for resources, some helpful fungi, such as *Gliocladium* species, can be introduced.

Chemical control: Although fungicides can be used, their efficacy may be limited, and they may damage beneficial microorganisms in the growing environment.

Resistance strains: Although their availability may differ, certain commercial mushroom strains are more resistant to *Trichoderma*.

7.3.1.4 Disease Cycle

1. Conidia, or *Trichoderma* spores, are dispersed through the air or transported by water.
2. They adhere to the surface and germinate, infecting the substrate or mushroom mycelium.
3. *Trichoderma* produces more spores after establishing itself, which feeds the infection cycle.
4. The spread happens quickly in warm, humid settings with inadequate ventilation.

7.3.2 Verticillium spp. *(Dry Bubble)*

7.3.2.1 Causal Organism

1. *Verticillium* species, particularly *Verticillium fungicola*, is a soil-borne fungus belonging to the *Hypocreaceae* family.

7.3.2.2 Symptoms

1. Mushrooms produce low-quality fruits because of their dry, leathery, and malformed caps, which frequently stay closed.
2. Caps may look yellowish, and the edges may become deformed or dry.
3. The overall yield of mushrooms may also be decreased due to the infection.
4. The disease typically strikes mushrooms at various stages, especially in stressful environmental conditions (variations in humidity and temperature).

7.3.2.3 Management

Environmental control: To lower the risk of infection, maintain the ideal temperature and humidity levels, 16–18 °C and 80–90% humidity.

Sanitation: Sanitize substrates, instruments, and growing conditions. The disease may continue to exist in contaminated environments.

Resistant strains: Although not all strains of mushrooms possess this quality, use those resistant to *Verticillium*.

Crop rotation: Breaking the disease cycle can be achieved by letting substrates rest between crops.

Fungicides: Although they are less frequently used in the production of mushrooms, chemical controls may be used.

7.3.2.4 Disease Cycle

Spores produced by the fungus infect mushrooms while in the mycelium stage.

1. Once established, it releases toxins that result in aberrant mushroom growth, making it a "dry bubble."
2. Contaminated surfaces, water, or equipment can spread infection. Between cropping cycles, spores can live in the soil or on contaminated debris.

7.3.3 Mycogone spp. *(Wet Bubble)*

7.3.3.1 Causal Organism

Mycogone species, particularly the harmful fungus *Mycogone perniciosa*, belong to the *Clavicipitaceae* family (Jeffries and Hayward 2008).

7.3.3.2 Symptoms

1. *Mycogone*-infected mushrooms have wet, enlarged, and deformed caps.
2. Affected mushrooms may be reddish or darkly discolored and appear slimy and greasy.
3. The mushrooms may eventually deteriorate if the infected caps do not fully develop or open.
4. The disease is more frequently seen in the early fruiting stages when the mushrooms are most vulnerable.

7.3.3.3 Management

Environmental control: Maintain steady temperatures, lower humidity, and ensure adequate air circulation. Extremely moist environments are ideal for *mycogone* growth.

Sanitation: To stop additional contamination, remove contaminated mushrooms right away.

Substrate sterilization: Reducing the initial inoculum requires high levels of substrate sterilization.

Prevent overcrowding: By restricting airflow, dense plantings can aid in the spread of disease.

7.3.3.4 Disease Cycle

1. Spores of *mycogone* are dispersed by air, water, or contaminated surfaces.
2. The fungus causes swelling and aberrant growth in the fruiting bodies when it infects them early in development.
3. The cycle is perpetuated when pores are released from infected tissue and dispersed throughout the developing environment.

7.3.4 Cladobotryum spp. (Cobweb Disease)

7.3.4.1 Causal Organism

Cladobotryum spp., a fungus in the family *Cladobotryaceae*.

7.3.4.2 Symptoms

1. Cobweb-like, grayish-white growth emerged on substrate surfaces and around the bases of mushrooms.

2. This fungus spreads quickly throughout the growing area, and infected mushrooms may get engulfed by the webbing, resulting in damaged or deformed mushrooms.
3. *Cladobotryum*-infected mushrooms may grow off-color caps and have an unappealing appearance in general.

7.3.4.3 Management

Increase ventilation: Low airflow and high humidity are ideal conditions for cobweb disease. To avoid the illness, ensure adequate air circulation.
Sanitation: Frequently clean and sanitize the growing area and its equipment.
Chemical controls: Fungicides can be useful, but because they may affect beneficial fungi, they should be used cautiously.
Eliminate infected material: To prevent the spread of disease, infected mushrooms and substrates should be eliminated immediately.

7.3.4.4 Disease Cycle

1. *Clodobotryum* spores are airborne and readily dispersed via contaminated equipment or air.
2. After landing on a substrate or mushroom, the fungus produces mycelium that quickly covers the mushroom, giving it the distinctive "cobweb" look.
3. High humidity and inadequate ventilation, which encourage the growth of fungal mycelium, are the main causes of the disease cycle.

7.4 Bacterial Diseases

7.4.1 Pseudomonas tolaasii *(Brown Blotch)*

7.4.1.1 Causal Organism

Pseudomonas tolaasii, a Gram-negative bacterium in the family *Pseudomonadaceae.*

7.4.1.2 Symptoms

1. The mushroom's cap develops brown, asymmetrical lesions.
2. Initially appearing as wet patches, the lesions eventually turn dark brown.
3. Due to bacterial breakdown, the impacted caps may soften and smell bad. Although the infection can spread to other areas of the fruiting body, it usually begins on the mushroom cap.

7.4.1.3 Management

Sanitation: Tools and surfaces should be cleaned frequently to stop Pseudomonas from spreading.
Humidity control: Steer clear of excessive moisture buildup while maintaining humidity levels between 85% and 90%.
Eliminate infected mushrooms: Throw away any mushrooms exhibiting symptoms to stop additional contamination.
Water quality: Verify that the water used to grow mushrooms is uncontaminated.

7.4.1.4 Disease Cycle

1. The bacteria can spread through the air, contaminated tools, or contaminated water.
2. The bacteria enter the mushroom tissues through wounds or other natural openings to start the infection.
3. As they colonize the tissue, the bacteria produce the pigments that give it brown discoloration. Contaminated surfaces can harbor spores.

7.4.2 Pseudomonas gingeri *(Ginger Blotch)*

7.4.2.1 Causal Organism

Pseudomonas gingeri is a species closely related to *P. tolaasii* and is part of the *Pseudomonadaceae* family (Zhang and Yu 2010).

7.4.2.2 Symptoms

1. Lesions on the ginger-colored mushroom caps are typically irregular in shape and initially watery before drying out and hardening as the disease worsens.
2. Stunted growth is possible in mushrooms; the blotches are usually concentrated close to the cap's center.

7.4.2.3 Management

1. Handle with hygiene, regulate humidity, and eliminate contaminated mushrooms, just like *P. tolaasii*.
2. To avoid bacterial contamination, use sterile water when watering the mushrooms rather than wetting them directly.
3. Clean the growing room's surfaces and make sure there is adequate airflow.

7.4.2.4 Disease Cycle

1. *Pseudomonas gingeri* is spread by coming into contact with contaminated objects, water, and surfaces.
2. When the bacteria get inside the mushroom tissue, they produce lesions that are distinctively ginger in color.
3. The bacteria keep growing on the contaminated tissue, which causes it to spread even more.

7.5 Viral Diseases

7.5.1 Mushroom Virus X (MVX)

7.5.1.1 Causal Organism

1. Mushroom Virus X is a member of the *Alphaflexivirus* family.

7.5.1.2 Symptoms

1. Mushrooms may be discolored, have deformed caps, and grow slowly.
2. The most typical symptoms are a general decrease in size or a yellowish discoloration.
3. The caps of infected mushrooms may be closed or have irregular surfaces, and the mushrooms have irregular shapes.

7.5.1.3 Management

1. *Sanitation*: Prevent cross-contamination between batches to stop the spread of MVX.
2. *Quarantine*: Batches of contaminated mushrooms need to be removed and isolated immediately.
3. *Resistant strains*: MVX may be managed by using virus-resistant mushroom strains.

7.5.1.4 Disease Cycle

1. Spores, physical contact, or contaminated tools are the main ways MVX is transmitted.
2. The virus infects cells inside the mushroom tissue, resulting in symptoms like growth retardation and abnormalities.

3. The virus may persist throughout cropping cycles and spread as new fruiting bodies form.

References

Adedokun OM, Odiketa JK, Afieroho OE, Afieroho MC. Importance of mushrooms for food security in Africa. In: Food security for African smallholder farmers. Springer Nature Singapore: Singapore; 2022. p. 343–60.

Baker RL, Dunn DA. The life cycle and control of Phorid flies (Diptera: Phoridae) in food processing facilities. J Econ Entomol. 2011;104(2):594–602. https://doi.org/10.1603/EC10224.

Bell V, Silva CRPG, Guina J, Fernandes TH. Mushrooms as future generation healthy foods. Front Nutr. 2022;9:1050099.

Buxton JH, Mallet JL. The effects of *Lycoriella auripila* (Diptera: Sciaridae) on the mycelial growth of *Agaricus bisporus*. Ann Appl Biol. 1980;95(3):327–32. https://doi.org/10.1111/j.1744-7348.1980.tb03281.x.

Campbell JS, Anselme C. Trichoderma spp. and their role in mushroom cultivation. Mushroom J. 2011;

Carvalho AJS, Zárate-Salazar JR, Jacob MCM, Araújo PL, Gomes SM, Pereira FDO. The role of edible mushrooms in the Brazilian diet: a comprehensive analysis. Nutr Food Sci. 2024;54(4):821–34.

Disney RHL. Natural history of the scuttle fly, *Megaselia scalaris*. Annu Rev Entomol. 2008;53(1):39–60. https://doi.org/10.1146/annurev.ento.53.103106.093415.

Ekbom BS, Olsson K. Bulb mites in agriculture: biology and management. Agric Entomol J. 2021;32(4):467–75.

Fatchurochim S, Geden CJ, Axtell RC. Filth fly (Diptera) oviposition and larval development in poultry manure of various moisture levels. J Entomol Sci. 1989;24(2):224–31.

Ghose A, Mitra S. Spent waste from edible mushrooms offers innovative strategies for the remediation of persistent organic micropollutants: a review. Environ Pollut. 2022;305:119285.

Greenberg B. Flies and disease: ecology, classification, and biotic associations, vol. 1. Princeton University Press; 1973.

Grewal PS, Ehlers RU, Shapiro-Ilan DI, editors. Nematodes as biological control agents. CABI Publishing; 2005.

Harris MA, Oetting RD. Biology of fungus gnats (*Bradysia spp.*) in relation to horticulture. Annu Rev Entomol. 1997;42(1):659–83.

Jeffries P, Hayward C. Wet bubble disease and its impact on mushroom production. Plant Pathol J. 2008;

Jess S, Kilpatrick M. The effect of *Lycoriella mali* (Diptera: Sciaridae) on mushroom yield and quality. Ann Appl Biol. 2000;136(1):47–52. https://doi.org/10.1111/j.1744-7348.2000.tb00008.x.

Kaliyaperumal M, Kezo K, Gunaseelan S. A global overview of edible mushrooms. In: Biology of Macrofungi. Springer; 2018. p. 15–56.

Kosre A, Koreti D, Mahish PK, Chandrawanshi NK. Current perspective of sustainable utilization of agro waste and biotransformation of energy in mushroom. In: Energy: crises, challenges and solutions; 2021. p. 274–302.

Merritt RW, Courtney GW, Keiper JB. Diptera:(flies, mosquitoes, midges, gnats). In: Encyclopedia of insects; 2009. p. 284–97.

Pfeiffer BE, Dove C. Identification and management of Phorid flies in mushroom cultivation. J Econ Entomol. 2009;102(6):2139–45. https://doi.org/10.1603/029.102.0630.

Sánchez JE, Royse DJ, Montes A. Pests and diseases in commercial mushroom production: challenges and solutions. Fungal Biol Rev. 2020;34(2):67–78.

White PF. Pests and diseases of mushrooms and their control. Pest Manag Sci. 1986;17(3):187–97. https://doi.org/10.1002/ps.2780170304.
Zhang Z, Kells SA. Integrated Pest Management of Mites in controlled agricultural environments. J Econ Entomol. 2018;111(6):2485–93.
Zhang J, Yu L. Characterization and control of Pseudomonas gingeri in mushroom cultivation. Plant Dis J. 2010;

Chapter 8
Biotechnological Interventions Revitalizing the Mushroom Industry

Muhmmad Waris, Ahmad Nisar, Nasir Ahmad Khan, and Hayssam Mohamed Ali

Abstract Recent advancements in the genetics of mushrooms have been made possible by technologies such as RNA interference and CRISPR-Cas9 genome editing, which relieve an important amplification in our uncertainty about these complex fungi and allow their applications in agriculture, health, and conservation. Researchers investigate both the regulatory and ethical aspects that are a part of genetic research in mushrooms. CRISPR-Cas9 is an innovative technology because it targets gene knockouts, enables precise genome editing and alters pathways and RNA interference strengthens those endeavors since it reduces genes so crop yield grows and crops reduces pests and diseases impact. As emerging technologies with ethical considerations still influence mushroom research, genetic studies hold the potential for redefining our interaction with these fungi to foster a more sustainable future. Significantly, they also provide opportunities for discovering novel medicinal compounds, expanding possibilities in pharmaceuticals as well as nutraceuticals, aiding in the conservation of endangered species, and contributing to the development of more versatile mushroom varieties for sustainable cultivation.

8.1 Introduction

Worldwide, mushroom production has been on a constant increase in the last four decades. From 1965 to 1997, the total production of mushrooms increased from about 350,000 metric tons to about 6,160,800 metric tons. In the past few years, the bulk of the gradual increase occurred. *Agaricus*, the button mushroom, accounted

M. Waris (✉)
Department of Plant Pathology, Baluchistan Agriculture College Quetta, Quetta, Pakistan

A. Nisar · N. A. Khan
Department of Plant Pathology, University of Agriculture, Faisalabad, Pakistan

H. M. Ali
Department of Botany and Microbiology, College of Science, King Saud University, Riyadh, Saudi Arabia

I. U. Haq et al. (eds.), *Mushrooms*, https://doi.org/10.1007/978-981-95-2539-3_8

for over 70% of the world's mushrooms during the last two decades. However, by the end of twenty-first century, it accounted for only 32% of world production. This gradual decrease of button mushroom was due to the involvement of other mushroom varieties, for example *Volvariella* and *Pleurotus* (Chang 2002). European countries and China both accounted for 96% of the world supply of mushrooms. Canada and the United States, in contrast, provided about 2%. Pakistan, in this regard, was nowhere on the map.

Mushrooms provide significant medicinal potential, indicate diverse ecological purposes and provide essential nutritional advantages. In ecosystems, mushrooms function as symbiotic partners and play a crucial role in the cycling of nutrients. A group of mushrooms called basidiomycetes plays vital roles as mutualistic symbionts and decomposers. These mushrooms are important for their medicinal properties and as a food source (De Mattos-Shipley et al. 2016). Mushrooms are essential to the global carbon cycle because they help recycle nutrients in various ecosystems (Gyulai et al. 2018). Studies show that various health-promoting physiologically active chemicals are present in mushrooms, which include antimicrobial and anti-dementia properties (Kawagishi 2010).

Their relevance in environmental conservation is revealed by the fact that some mushrooms can break down ecological sustainability and are used in bio-remediation to control agricultural waste (Barh et al. 2018). Different developments in genetic research techniques, especially RNA interference (RNAi) and CRISPR-Cas9 genome editing, have remarkably changed the study of mushrooms. These developments have changed the ability to modify the genetic structure of mushrooms and increased their potential for use in agriculture and medicine (Kim and Kim 2016). To repress specific gene functions and improve properties like disease resistance and growth rates, RNA interference technology has been applied to mushrooms (Eş et al. 2019). CRISPR-Cas9 genetic alterations have created opportunities to increase mushroom crop productivity and disease resistance. Moreover, it may be possible to enhance the production of substances with pharmacological significance by altering the genetic pathways in medicinal mushrooms such as *Ganoderma lucidum*. These groundbreaking inventions not only expand our insight into fungal genetics but also explore new possibilities for medicinal and agricultural applications, increasing the utility of mushrooms for both therapeutic and industrial purposes (Wang et al. 2020).

There are certain factors that support the need to perform genetic research on mushrooms: firstly, their complex biology, a wide range of secondary metabolites, and potential medicinal uses. Various types of metabolites that mushrooms produce have important biotechnological uses in fields like agriculture and medicine. The secondary metabolites that have antibacterial, anticancer and antioxidant qualities include polysaccharides and proteoglycans (Ghosh and Acharya 2022). In the research of mushrooms, both traditional and modern medicine have shown interest. To improve our knowledge of Genetic research, its aim is to understand how particular genes regulate the production of bioactive substances (Elkhateeb and Daba 2021). The creation of innovative medications made from mushrooms may result from this research. Anti-inflammatory, anticancer, and neuroprotective therapies are important therapeutic targets (Chang and Wasser 2018). Mushrooms are a desirable topic for genetic research because of their complex genetic makeup and the wide range of uses of their secondary metabolites, to use their qualities for both industrial and health benefits (Zhang et al. 2022).

To increase food security and sustainability, global agriculture depends significantly on farmed mushrooms like *Agaricus bisporus* and *Pleurotus* spp., and it is very essential to improve their genetic makeup. Genetic research also encourages the conservation of biodiversity by identifying and conserving the rare and endangered species of mushrooms (Zhang et al. 2023). The maintenance of ecosystem health and the preservation of genetic variety depend heavily on these efforts. Research has highlighted the ecological processes of mushrooms in bioremediation and rehabilitation (Czaplicki and Gunsch 2016).

As genetic technology advances, techniques like CRISPR-Cas9 provide both ethical issues and potential in the production and genetic enhancement of mushrooms. These include concerns about biodiversity, the ecological implications of genetically modified species, and the social impacts of such technology. These elements need to be carefully considered ethically in order to maintain acceptable scientific operations. For researchers who want to explore the complex field of mushroom genetics, this comprehensive overview is a timely and helpful source. Furthermore, it benefits both policymakers and the ordinary public by educating them about the ethical concerns and the significant environmental aspects of this rapidly evolving subject (Czaplicki and Gunsch 2016).

8.2 Issues and Concerns in the Mushroom Industry

Contamination, spoilage, target audience identification, and expansion were the major issues that producers faced. Production volume, target audience identification, consumer awareness, product development, and entrepreneurial development were the areas of focus for industry concerns at the same time. The term "contamination" describes the unintentional introduction of pathogenic organisms into inoculated spawn, which makes them hazardous for use in production and potentially dangerous for spreading to other spawns. Of the 33 producers, 30 (91%) reported experiencing issues as a result of spawn contamination. The issue was present even for producers who mainly purchased tissue culture and planted spawns. Of the producers, 67% described spawn contamination as occasional, 16% as seldom, and 16% as frequent (Chang 2002).

The term spoilage is described as the severe decline of mushroom fruits caused due to aging, environmental exposure, and improper management. The comparatively short shelf life of mushroom fruits was the primary cause of spoilage. Furthermore, spoilage takes place at the end of process and marketing phase. Almost 73% of the producers reported spoilage. Of the producers, 91% limit their production to easily sellable levels in order to prevent spoilage. Of the producers, 9% solved the spoilage issue by drying fruits or turning them into mushroom dishes like burgers and dumplings. However, none of them turned to canning or bottling fruits due to the lack of resources and expertise. For the producers, spoilage directly resulted in lower sales. Furthermore, the number of impacted producers suggested a significant collective loss, even if spoilage seems to be a rare and manageable occurrence (Soriano 2017). Market identification was seen as a significant issue among manufacturers while not being significantly mentioned in the results. This issue resulted from producers' incompetence to look for new markets for their goods. Transportation expenses, spoilage risk, low production volume, and lack of expertise among most manufacturers appeared to be the primary barriers to market exploration (Soriano 2017).

8.3 Precision Genome Editing with CRISPR-Cas 9

CRISPR-Cas9 has played a major role in mushroom genetics by providing new tools for precision genomic editing. A new technique of high-throughput genome editing in the mushroom-farming basidiomycete was developed using this technology. Genetic research on edible mushrooms can be greatly enhanced by using a strong promoter to express Cas9 and a particular promoter for the guide RNA (Sugano et al. 2017). The first effective application of CRISPR-Cas9 in a commercially and scientifically significant species was in *Lentinula edodes* or shiitake mushrooms. With its ability to precisely target and modify genetic loci, this approach shows promise for widespread use in the genetics of mushroom (Moon et al. 2021). Targeting the pyrG gene with CRISPR/Cas9 was another significant advancement that showed the range of CRISPR technology in *L. edodes* genetic research and resulted in homologous recombination (Kamiya et al. 2023). Researchers have been able to precisely manipulate the DNA of mushrooms by focusing on particular genes of interest, thanks to CRISPR-Cas9 (Singh et al. 2021). Genes can be regenerated or eliminated by cellular regeneration processes, particularly by in fungus and mammalian cells (De Jong et al. 2010). Enhancing the effectiveness of gene repair mechanisms for therapeutic applications is the main focus of the present research (Thorpe et al. 2002). CRISPR-Cas9 precise genome editing in mushrooms has several practical applications, including advancements in the genome editing of the model organism *Coprinopsis cinerea*, which is employed in research on mushrooms. Using CRISPR-Cas9 has enabled us to better understand how genes function in development (Sugano et al. 2017).

Targeted gene knockouts are one of the primary applications of the functional analysis of certain genes in mushrooms. Researchers can, for example, use CRISPR-Cas9 is crucial in the development of mushroom fruiting bodies, and then assess the impact on their growth (Wang et al. 2021). Targeted gene knockouts have enabled the identification of the genetic basis of mushroom traits. CRISPR-Cas9 enables precise and controlled gene editing, such as determining the genes that create bioactive chemicals with medicinal properties or comprehending the mechanisms behind mushrooms' responses to environmental stimuli (Ullah et al. 2020). Similarly, more nutrient-dense mushrooms that create bioactive compounds for medical application can be produced by modifying genes involved in secondary metabolite biosynthesis (Arshadi et al. 2023).

Breeding mushrooms for resistant features without sacrificing agronomic qualities is a breakthrough. This technique shows how individual mushroom genes may be altered by CRISPR/Cas9 to enhance specific characteristics (Liu et al. 2023). This study demonstrates how CRISPR technology can be used for more than gene inactivation; it can also be used to modify fungal metabolic pathways (Shanmugam et al. 2019). Double-gene targeting in *Pleurotus ostreatus* (oyster mushrooms) using a preassembled Cas9 ribonucleoprotein complex induces CRISPR/Cas9-mediated alterations that boost resistance to environmental challenges. This work shows how CRISPR/Cas9 can be applied to enhance a variety of mushroom characteristics (Boontawon et al. 2023). They developed the first CRISPR/Cas9 genome editing system in shiitake mushrooms (*Lentinula edodes*) by targeting the mating-type gene HD1 with gRNAs. This study shows how the genetic circuits of mushrooms can be effectively modified using CRISPR/Cas9 to achieve desired phenotypes (Shanmugam et al. 2019; Moon et al. 2021). Targeted genome engineering can modify the manufacture of secondary metabolites and enhance mushroom resistance (Zhang et al. 2022). Enhancing the nutritional value of mushrooms is a significant use of route modulation. The extensive nutritional profile of mushrooms makes them valuable (Lu et al. 2020; Sambyal and Singh 2021). Scientists can modify the genes using CRISPR-Cas9 that generate nutrients that increase the nutritional value of mushrooms (Kaul et al. 2020; Tiwari et al. 2021). For example, mushrooms with higher vitamin content may result from overexpressing genes linked to vitamin production, such as those for vitamin D or folate (Chakravorty et al. 2022).

Through route modulation, mushrooms can be made more nutrient-dense and more resilient to environmental stressors. Three tropical edible mushrooms were tested by Ayodele et al. (2010) for resistance to abiotic stresses such as dehydration and high solar exposure. The fungus *Flammulina filiformis*, which produces mushrooms, has conserved genes relevant to the formation of fruiting bodies and the response to heat stress (Liu et al. 2020; Wang et al. 2020).

By fungi using CRISPR-Cas9-mediated editing tools, the production of bioactive substances from plants shows how this technology can increase the yield of beneficial pharmaceutical molecules from these sources (Bhagwat et al. 2022). It became the first CRISPR-edited organism to circumvent US regulatory processes, when the US Department of Agriculture chose not to regulate a CRISPR-Cas9-modified mushroom. The safety, moral, and ethical concerns associated with CRISPR-Cas9 technology underscore the need for stringent regulations to ensure ethical application (Shinwari et al. 2018).

8.4 RNA Interference (RNAi)

8.4.1 Gene Silencing

In eukaryotic species, RNA interference (RNAi) is a significant and highly conserved mechanism that is necessary for downregulating target genes and ultimately silencing them. RNAi is the method by which double-stranded RNA (dsRNA) silences genes. The RNA-induced silencing complex (RISC) typically does this by degrading mRNA in a manner analogous to that of siRNA (Xu et al. 2019). Even though effective delivery systems are still problematic, the technology has grown into a vital research tool for studying gene function (Akhtar et al. 2021). RNA interference systems are linked to both gene regulation and antiviral defense. The synthesis of microRNAs, which are structurally and functionally comparable to RNAi, depends on the Dicer and Argonaute proteins, two components of this pathway (Treiber et al. 2019). In many animals, the integrity of the genome preserved by RNA interference, which provides a natural defense mechanism against viruses. Target genes can be deliberately downregulated to study their functions using RNA interference (RNAi), a valuable tool in biotechnology and research (Ahmad et al. 2021). RNA interference (RNAi) provides an effective and ongoing way to protect mushrooms from common enemies and is a promising method for boosting their resistance to pests and diseases. This innovative strategy lessens the need for chemical pesticides and promotes sustainable growing practices by employing RNA interference (RNAi) to confer resistance against diseases and pests of mushrooms (Tafrihi and Hasheminasab 2019).

Through precise regulatory genes, RNA interference, a diverse genetic approach, can boost mushroom harvests. RNA interference targets specific genes involved in the development of mushrooms, stress responses, or fruiting body formation, allowing for the process of mushroom production. A major genetic approach for boosting mushroom yields is RNA interference (RNAi), which precisely regulates genes crucial for fruiting body formation, stress responses, and mushroom development. Although there is limited direct research on mushroom yields, the widespread use of RNA interference in fungal species raises the possibility of improving productivity and stress resistance, which are crucial for commercial mushroom cultivation. Focused adjustments made possible by this genetic approach have the potential to revolutionize mushroom agriculture by boosting its sustainability and efficiency (Thipe et al. 2021).

RNA interference (RNAi) has demonstrated potential in targeted gene control for several uses, such as enhancing agricultural production and preventing disease. When growing mushrooms, RNA interference (RNAi) can control growth processes, increase production, and strengthen infection resistance (Gebremichael et al. 2021). Crop production enhancement is one of the many important uses for RNA interference (RNAi), a potent gene silencing technology (Salame et al. 2011).

This method can increase mushroom harvests by focusing on particular genes related to growth and development. Here, we investigate how RNA interference (RNAi) can increase mushroom productivity.

8.4.2 Conservation and Genetic Diversity

The protection of rare and endangered mushroom species has benefited considerably from recent genetic research. This study provides essential data about the genetic diversity and population dynamics of this fungus, which may be necessary for its maintenance and restoration. For unusual and endangered fungus in culture collections to retain their genetic diversity, ex-situ conservation is essential. Research on commercial species, like *Agaricus bisporus*, has demonstrated unexpected genetic variability within natural populations. Adaptability and resistance to environmental changes depend on this genetic diversity (Xu et al. 2002). The survival of these species depends on maintaining their genetic diversity, and conservation plans can be developed right away using the knowledge gained from genetic research. By integrating molecular data, this tactic increases the effectiveness of measures to save these significant yet endangered species in preservation practices. In wild mushroom populations, the potential for developing more durable and resilient mushroom strains is tremendous when this genetic variation is used (Yadav and Negi 2021). The cultivation of mushrooms and conservation efforts will eventually benefit from this genetic variety.

8.4.3 Regulatory and Ethical Considerations

Like any technology, genetic engineering in mushrooms presents a number of ethical concerns, including potential risks and benefits as well as ecological consequences. To guarantee appropriate scientific methods, these elements are essential. Given that genetic alteration may produce unwanted features, careful observation is required. To suppress any negative effects on natural ecosystems, the consequences of genetically engineered mushroom strains on biodiversity present a significant moral challenge that necessitates actions (Tyagi et al. 2020). In the context of genetic modification, scientific concerns are discussed alongside ethical manipulations, including the presentation and discussion of ethical arguments, which are often influenced by social values and the perceived benefits and risks of genetic engineering (van Tongrren 1991). Given how easy it is to modify genetic material and the potential for their application without sufficient regulatory oversight (Wang et al. 2020).

8.4.4 Gene Flow, Contamination, and Ecological Impacts

The potential effect of introducing genetically engineered mushroom varieties on the environment is significant. The potential for the genetically modified organisms to change the dynamics of the ecosystem among different species is a risk to species dynamics. The resulting changes in fungal communities may affect biodiversity in general.

The distribution of GM mushrooms may result in unexpected changes in ecological stability and biodiversity, as the change variety may suppress local fungal species (Warwick et al. 2009). The gene flow from the genetically engineered mushrooms to the wild varieties raises serious ecological concerns, since it may contaminate local species and impair their ability to survive and adapt (Tsatsakis et al. 2017).

8.5 Future Directions

8.5.1 Emerging Technologies

Research on mushrooms is advancing significantly due to emerging genetic technologies, which are opening up new possibilities for investigation and creativity in various research domains. In understanding the evolution of mushrooms, functional genomics is making significant strides with technologies like genome and transcriptome sequencing. As a result, important transcription factors such as Roc1 in Basidiomycota have been identified. During the development of fruiting bodies, these transcription factors are crucial for comprehending gene function and cellular processes (Nagy 2023). Research on mushrooms is anticipated to grow significantly, as these genetic technologies develop further, revealing new insights about the biology of mushrooms, genetic modification and applicability across a range of industries (Cairns et al. 2019). These advancements have the potential to address global concerns about sustainable agriculture, food security and medicinal discovery, making mushrooms even more valuable to humanity in the upcoming years (Jo et al. 2023).

8.5.2 Novel Compounds for Medicine and Sustainable Cultivation

In 2013, Savoie et al. described that the significance of genetic research in creating eco-friendly mushrooms production technologies is emphasized in recent studies. Genetic research on mushrooms is necessary to comprehend and enhance traits like crop improvement and disease resistance, both of which are critical for sustainable production techniques. Incorporating CO_2 implementation tactics into the mushroom industry, such as using CO_2 from mushroom cultivation to grow other crops like lettuce, demonstrates creative, eco-friendly growth techniques that lessen the environmental impact (Kaur and Kapoor 2023). The research highlights genetics' role in creating environmentally friendly and resource-efficient mushroom farming methods. The research lays the foundations for making economically and environmentally sound methods of growing mushrooms, enhancing ecological sustainability, and ensuring food security.

8.6 Conclusion

Research on the genetic aspects of mushrooms offers a glance into the hidden aspects of nature and holds the potential for significant development. These analyses refine mushrooms' complex biology and genetic composition and provide attainable solutions to crucial global dilemmas. Genetic science is substitute how specific approach mushrooms and their quiescent from the precise genome editing powers of CRISPR-Cas9 to the versatile gene regulation offered by RNA interference. It offers the possibility of more environmentally friendly farming practices, the identification of new pharmaceutical components and the preservation of endangered species through the fusion of research, conservation, medicine, and agriculture. These genetic developments set the stage for a time when mushrooms will be used more and more to address environmental issues, ranging from environmental sustainability to food security, as we traverse the moral and legal environment.

References

Ahmad A, Ghouri MZ, Munawar N, Ismail M, Ashraf S, Aftab SO. Regulatory, ethical, and social aspects of CRISPR crops. In: CRISPR crops: the future of food security; 2021. p. 261–87.

Akhtar A, Andleeb A, Waris TS, Bazzar M, Moradi AR, Awan NR, Yar M. Neurodegenerative diseases and effective drug delivery: a review of challenges and novel therapeutics. J Control Release. 2021;330:1152–67.

Arshadi N, Nouri H, Moghimi H. Increasing the production of the bioactive compounds in medicinal mushrooms: an omics perspective. Microb Cell Factories. 2023;22(1):11.

Ayodele SM, Ohwojero D, Odii IK. Screening for abiotic stress resistance in three tropical edible mushrooms. Agric Biol J North Am. 2010;1(4):711–4.

Barh A, Upadhyay RC, Kamal S, Annepu SK, Sharma VP, Shirur M, Banyal S. Mushroom crop in agricultural waste cleanup. In: Microbial biotechnology in environmental monitoring and cleanup. IGI Global; 2018. p. 252–66.

Bhagwat AC, Patil AM, Saroj SD. CRISPR/Cas 9-based editing in the production of bioactive molecules. Mol Biotechnol. 2022;64:1–7.

Boontawon T, Nakazawa T, Choi YJ, Ro HS, Oh M, Kawauchi M, Honda Y. Double-gene targeting with preassembled Cas9 ribonucleoprotein for safe genome editing in the edible mushroom Pleurotus Ostreatus. FEMS Microbiol Lett. 2023;370:fnad015.

Cairns TC, Zheng X, Zheng P, Sun J, Meyer V. Moulding the mould: understanding and reprogramming filamentous fungal growth and morphogenesis for next generation cell factories. Biotechnol Biofuels. 2019;12(1):77.

Chakravorty M, Nanda M, Arora N, Singh S, Kumar V, Deshwal S. CRISPR-Cas9 mediated genome tailoring to improve nutritional quality and shelf life in crops: a review. Plant Gene. 2022;31:100369.

Chang ST. Past and present trends in the production of Lentinula edodes in Asia. In: Sanchez JE, et al., editors. Mushroom biology and mushroom products, proceedings of the 4th international conference, February 20–23. Cuernavaca, Mexico; 2002. p. 1–8.

Chang ST, Wasser SP. Current and future research trends in agricultural and biomedical applications of medicinal mushrooms and mushroom products. Int J Med Mushrooms. 2018;20(12):1121.

Czaplicki LM, Gunsch CK. Reflection on molecular approaches influencing state-of-the-art bioremediation design: culturing to microbial community fingerprinting to omics. J Environ Eng. 2016;142(10):03116002.

De Jong JF, Ohm RA, De Bekker C, Wösten HA, Lugones LG. Inactivation of ku80 in the mushroom-forming fungus Schizophyllum commune increases the relative incidence of homologous recombination. FEMS Microbiol Lett. 2010;310(1):91–5.

De Mattos-Shipley KM, Ford KL, Alberti F, Banks AM, Bailey AM, Foster GD. The good, the bad and the tasty: the many roles of mushrooms. Stud Mycol. 2016;85(1):125–57.

Elkhateeb WA, Daba GM. Mycotherapy of the good and the tasty medicinal mushrooms Lentinus, Pleurotus, and Tremella. J Pharm Pharmacol Res. 2021;4(3):1–6.

Eş I, Gavahian M, Marti-Quijal FJ, Lorenzo JM, Khaneghah AM, Tsatsanis C, Kampranis SC, Barba FJ. The application of the CRISPR-Cas9 genome editing machinery in food and agricultural science: current status, future perspectives, and associated challenges. Biotechnol Adv. 2019;37(3):410–21.

Gebremichael DE, Haile ZM, Negrini F, Sabbadini S, Capriotti L, Mezzetti B, Baraldi E. RNA interference strategies for future management of plant pathogenic fungi: prospects and challenges. Plants. 2021;10(4):650.

Ghosh S, Acharya K. Milky mushroom: a healthy nutritious diet. Food Res Int. 2022;156:111113.

Gyulai GZ, Malone RP, Czakó M, Murenetz L, Gyulai G. How mushrooms tend to break through the genetic dead end. Ecocycles. 2018;4(2):46–57.

Jo C, Zhang J, Tam JM, Church GM, Khalil AS, Segrè D, Tang TC. Unlocking the magic in mycelium: using synthetic biology to optimize filamentous fungi for biomanufacturing and sustainability. Mater Today Bio. 2023;19:100560.

Kamiya A, Ueshima H, Nishida S, Honda Y, Kamitsuji H, Sato T, Miyamoto H, Sumita T, Izumitsu K, Irie T. Development of a gene-targeting system using CRISPR/Cas9 and utilization of pyrG as a novel selectable marker in Lentinula edodes. FEMS Microbiol Lett. 2023;370:fnad042.

Kaul T, Sony SK, Verma R, Motelb KFA, Prakash AT, Eswaran M, Bharti J, Nehra M, Kaul R. Revisiting CRISPR/Cas-mediated crop improvement: special focus on nutrition. J Biosci. 2020;45:1–37.

Kaur P, Kapoor P. Revolutionizing mushroom cultivation: a comprehensive review of hydroponics in Fungiculture. Curr J Appl Sci Technol. 2023;42(44):19–37.

Kawagishi H. Biologically functional molecules from mushrooms. J Synth Org Chem Jpn. 2010;68:514–21.

Kim J, Kim JS. Bypassing GMO regulations with CRISPR gene editing. Nat Biotechnol. 2016;34(10):1014–5.

Liu XB, Xia EH, Li M, Cui YY, Wang PM, Zhang JX, Xie BG, Xu JP, Yan JJ, Li J, Nagy LG. Transcriptome data reveal conserved patterns of fruiting body development and response to heat stress in the mushroom-forming fungus Flammulina filiformis. PLoS One. 2020;15(10):e0239890.

Liu Q, Meng G, Wang M, Li X, Liu M, Wang F, Yang Y, Dong C. Safe-Harbor-targeted CRISPR/Cas9 system and Cmhyd1 overexpression enhances disease resistance in Cordyceps militaris. J Agric Food Chem. 2023;71(41):15249–60.

Lu H, Lou H, Hu J, Liu Z, Chen Q. Macrofungi: a review of cultivation strategies, bioactivity, and application of mushrooms. Compr Rev Food Sci Food Saf. 2020;19(5):2333–56.
Moon S, An JY, Choi YJ, Oh YL, Ro HS, Ryu H. Construction of a CRISPR/Cas9-mediated genome editing system in Lentinula edodes. Mycobiology. 2021;49(6):599–603.
Nagy LG. Mushroom functional genomics springs up. Innov Life. 2023;1(1):100005–1.
Salame TM, Ziv C, Hadar Y, Yarden O. RNAi as a potential tool for biotechnological applications in fungi. Appl Microbiol Biotechnol. 2011;89:501–12.
Sambyal K, Singh RV. A comprehensive review on Morchella importuna: cultivation aspects, phytochemistry, and other significant applications. Folia Microbiol. 2021;66(2):147–57.
Shanmugam K, Ramalingam S, Venkataraman G, Hariharan GN. The CRISPR/Cas9 system for targeted genome engineering in free-living fungi: advances and opportunities for lichenized fungi. Front Microbiol. 2019;10:62.
Shinwari ZK, Tanveer F, Khalil AT. Ethical issues regarding CRISPR mediated genome editing. Curr Issues Mol Biol. 2018;26(1):103–10.
Singh AK, Ramalingam S, Rao DN, Chandrasegaran S. Genome editing revolution in life sciences. Resonance. 2021;26:971–98.
Soriano EA. The mushroom industry in Central Luzon: Towards Finetuning R&D and extension for industry revitalization. Available at SSRN 4132835. 2017.
Sugano SS, Suzuki H, Shimokita E, Chiba H, Noji S, Osakabe Y, Osakabe K. Genome editing in the mushroom-forming basidiomycete Coprinopsis cinerea, optimized by a high-throughput transformation system. Sci Rep. 2017;7(1):1260.
Tafrihi M, Hasheminasab E. MiRNAs: biology, biogenesis, their web-based tools, and databases. Microrna. 2019;8(1):4–27.
Thipe VC, Maloney V, Klein A, Gokul A, Keyster M, Katti KV. RNA interference and CRISPR/Cas9 techniques for controlling mycotoxins. In: CRISPR and RNAi systems. Elsevier; 2021. p. 463–90.
Thorpe P, Stevenson BJ, Porteous DJ. Optimising gene repair strategies in cell culture. Gene Ther. 2002;9(11):700–2.
Tiwari M, Kumar Trivedi P, Pandey A. Emerging tools and paradigm shift of gene editing in cereals, fruits, and horticultural crops for enhancing nutritional value and food security. Food Energy Secur. 2021;10(1):e258.
Treiber T, Treiber N, Meister G. Regulation of microRNA biogenesis and its crosstalk with other cellular pathways. Nat Rev Mol Cell Biol. 2019;20(1):5–20.
Tsatsakis AM, Nawaz MA, Kouretas D, Balias G, Savolainen K, Tutelyan VA, Golokhvast KS, Lee JD, Yang SH, Chung G. Environmental impacts of genetically modified plants: a review. Environ Res. 2017;156:818–33.
Tyagi S, Kumar R, Das A, Won SY, Shukla P. CRISPR-Cas9 system: a genome-editing tool with endless possibilities. J Biotechnol. 2020;319:36–53.
Ullah M, Xia L, Xie S, Sun S. CRISPR/Cas9-based genome engineering: a new breakthrough in the genetic manipulation of filamentous fungi. Biotechnol Appl Biochem. 2020;67(6):835–51.
van Tongrren PJ. Ethical manipulations: an ethical evaluation of the debate surrounding genetic engineering. Hum Gene Ther. 1991;2(1):71–5.
Wang PA, Xiao H, Zhong JJ. CRISPR-Cas9 assisted functional gene editing in the mushroom Ganoderma lucidum. Appl Microbiol Biotechnol. 2020;104:1661–71.
Wang T, Yue S, Jin Y, Wei H, Lu L. Advances allowing feasible pyrG gene editing by a CRISPR-Cas9 system for the edible mushroom Pleurotus eryngii. Fungal Genet Biol. 2021;147:103509.
Warwick SI, Beckie HJ, Hall LM. Gene flow, invasiveness, and ecological impact of genetically modified crops. Ann N Y Acad Sci. 2009;1168(1):72–99.
Xu J, Desmerger C, Callac P. Fine-scale genetic analyses reveal unexpected spatial-temporal heterogeneity in two natural populations of the commercial mushroom Agaricus bisporus. Microbiology. 2002;148(5):1253–62.
Xu W, Jiang X, Huang L. RNA interference technology. In: Comprehensive biotechnology. Elsevier; 2019. p. 560–75.

Yadav D, Negi PS. Bioactive components of mushrooms: processing effects and health benefits. Food Res Int. 2021;148:110599.

Zhang J, Shen N, Li C, Xiang X, Liu G, Gui Y, Patev S, Hibbett DS, Barry K, Andreopoulos W, Lipzen A. Population genomics provides insights into the genetic basis of adaptive evolution in the mushroom-forming fungus Lentinula edodes. J Adv Res. 2022;38:91–106.

Zhang Y, Chen S, Yang L, Zhang Q. Application progress of CRISPR/Cas9 genome-editing technology in edible fungi. Front Microbiol. 2023;14:1169884.

MIX
Papier aus verantwortungsvollen Quellen
Paper from responsible sources
FSC® C105338

If you have any concerns about our products,
you can contact us on
ProductSafety@springernature.com

In case Publisher is established outside the EU,
the EU authorized representative is:
Springer Nature Customer Service Center GmbH
Europaplatz 3, 69115 Heidelberg, Germany

Printed by Libri Plureos GmbH
in Hamburg, Germany